Engineering Applications of Computational Methods

Volume 21

The book series Engineering Applications of Computational Methods addresses the numerous applications of mathematical theory and latest computational or numerical methods in various fields of engineering. It emphasizes the practical application of these methods, with possible aspects in programming. New and developing computational methods using big data, machine learning and AI are discussed in this book series, and could be applied to engineering fields, such as manufacturing, industrial engineering, control engineering, civil engineering, energy engineering and material engineering.

The book series Engineering Applications of Computational Methods aims to introduce important computational methods adopted in different engineering projects to researchers and engineers. The individual book volumes in the series are thematic. The goal of each volume is to give readers a comprehensive overview of how the computational methods in a certain engineering area can be used. As a collection, the series provides valuable resources to a wide audience in academia, the engineering research community, industry and anyone else who are looking to expand their knowledge of computational methods.

This book series is indexed in both the **Scopus** and **Compendex** databases.

Xiangyu Kong · Dazheng Feng

Efficient Online Learning Algorithms for Total Least Square Problems

Science Press
Beijing

Springer

Xiangyu Kong
Xi'an Institute of Hi-Tech
Xi'an, Shaanxi, China

Dazheng Feng
Xidian University
Xi'an, Shaanxi, China

ISSN 2662-3366 ISSN 2662-3374 (electronic)
Engineering Applications of Computational Methods
ISBN 978-981-97-1764-4 ISBN 978-981-97-1765-1 (eBook)
https://doi.org/10.1007/978-981-97-1765-1

Jointly published with Science Press
The print edition is not for sale in China mainland. Customers from China mainland please order the print book from: Science Press.
ISBN of the Co-Publisher's edition: 978-7-03-062161-0

Translation from the Chinese Simplified language edition: "随机系统总体最小二乘参数估计理论及应用" by Xiangyu Kong and Dazheng Feng, © Science Press, Beijing 2019. Published by Science Press, Beijing. All Rights Reserved.

This Springer imprint is published by the registered company Springer Nature Singapore Pte Ltd.
The registered company address is: 152 Beach Road, #21-01/04 Gateway East, Singapore 189721, Singapore

To all the researchers with original contributions to Efficient Online Learning Algorithms for Total Least Squares Problems

—Xiangyu Kong, Dazheng Feng

Preface

Although its original name is different, the total least squares (TLS) method has a long history. The earliest ideas about TLS can be traced back to the paper of Pearson in 1901 who considered an approximate method for solving the matrix equation $\boldsymbol{A}\boldsymbol{x} = \boldsymbol{b}$ when in both $\boldsymbol{A}$ and $\boldsymbol{b}$ there exist errors. However, as late as 1980, Golub and Van Loan for the first time gave an overall treatment from the point of view of numerical analysis and formally referred to this method as the total least squares. In mathematical statistics, the method is called orthogonal regression or errors-in-variables regression. In system identification, the TLS method is called the characteristic vector method or the Koopmans–Levin method. Now the TLS method is widely used in statistics, physics, economics, biology and medicine, signal processing, automatic control, system science, and many other disciplines and fields.

Over the past few decades, a large number of TLS methods have been developed. Fruitful academic results have been produced and widely used in the actual applications. We have devoted ourselves to the theoretical and applied research work on TLS methods for many years, especially concerned with the flourishing development of online TLS techniques. The objective of this book is to introduce some efficient online learning algorithms for TLS problems. This book focuses on the developments of TLS algorithms for adaptive filtering and parameter estimation. Specifically, this book introduces the authors' latest achievements on online TLS algorithms in the past 20 years, including recursive TLS algorithm, inverse power iteration TLS algorithm, neural-based minor component analysis (MCA) algorithm, neural-based SVD algorithm, neural-based TLS algorithm, TLS algorithms under non-Gaussian noises, performance analysis methods of TLS algorithms, etc.

The plan of the book is as follows:

Chapter 1: An overview of total least square problems and methods is given, and the aim and main feature of this book are illustrated.

Chapter 2: TLS problems are shown, and a sketch of their history is given. Most formulas are presented because they are used in later chapters. This is mainly an introduction to problems addressed in other chapters.

Chapter 3: Conventional TLS methods are briefly reviewed, including partial TLS algorithm, iterative computation methods, Rayleigh quotient minimization non-neural TLS method, neural-based TLS methods, and other extended TLS methods.

Chapter 4: This chapter presents two fast recursive TLS algorithms for adaptive filter, the one is a fast recursive TLS algorithm for adaptive FIR filtering and the other is a fast recursive TLS algorithm for adaptive IIR filtering.

Chapter 5: The approximate inverse power (AIP) iterative algorithm for TLS problem is studied, a fast AIP iterative algorithm for adaptive TLS FIR filtering is developed, and an AIP algorithm for adaptive extraction of minor subspace is proposed.

Chapter 6: This chapter is devoted to neural-based MCA algorithm for adaptive TLS, and several novel neural-based MCA algorithms, including a novel parallel multiple minor components (MCs) extraction algorithm by diagonal matrix, a weighted information criterion, and multiple MCs algorithm and a coupled MCA algorithm, are proposed.

Chapter 7: The neural-based SVD algorithms for adaptive TLS are reviewed, and a neural network algorithm for SVD of non-squared data matrix, a fast and effective neural network algorithm to perform SVD and a coupled neural network algorithm for principal singular triplet extraction are introduced.

Chapter 8: This chapter is devoted to the neural-based TLS algorithms. Several known TLS algorithms are reviewed, and a self-stabilizing neural-based TLS algorithm is proposed.

Chapter 9: TLS algorithms under non-Gaussian noises are researched. The LS method under non-Gaussian noises and TLS method with input and output non-Gaussian noises is reviewed. Several mixed LS-TLS algorithms under non-Gaussian noises are proposed.

Chapter 10: This chapter is focused on different performance analysis methods of TLS algorithms, especially the convergence/stability analysis of iterative and recursive TLS/MCA algorithms.

To Whom This Book Is Addressed

The high potentiality of the TLS approach and the possibility of using very reliable, accurate, and fast neural algorithms can be very attractive for researchers in many fields. In particular, the book we propose can be useful to:

- Statisticians, because of the new tools in data analysis.
- Applied mathematics experts, because of the novel methods and techniques that we propose.
- Engineers, above all for the applications in automatic control, system identification, computer vision, and signal processing.

Prerequisites

Readers should be familiar with basic linear algebra, matrix computations, and numerical analysis as well as elementary statistics, such as the basics of least squares, and preferably, but not necessarily, stochastics algorithms. Although the book is focused on neural-based MCA/SVD/TLS algorithms, these are presented only by their learning law, which is simply an iterative algorithm: therefore, no a priori knowledge of neural networks is required.

Xi'an, China

Xiangyu Kong
Dazheng Feng

Acknowledgments

The material presented in this book has been the outcome of over twenty years of research efforts by the authors. Many chapters of the book have gone through several rounds of revisions by discussions with a number of colleagues, collaborators, graduate and Ph.D. students. We would like specifically to thank our colleagues and collaborators, Prof. Hongguang Ma, Prof. Qiusheng An, Vice-Prof. Hongzeng Li, Vice-Prof. Qi Zhang, who have inspired many discussions in this or related topics over the past years; Former and current graduate and Ph.D. students, Dr. Zhongyin Xu, Dr. Yingbin Gao, Dr. Boyang Du, Dr. Jianqiang Qin, Dr. Jiayu Luo, Dr. Xiaowei Feng, Master Ben Wang, for their participations in the discussions; and other supporting staffs of the Department of Control Science and Engineering at Xi'an Institute of Hi-Tech. We also express our gratitude to the financial support from the National Natural Science Foundation of China (NSFC) (62273354, 61673387, 61903375, 62106242).

Finally, we wish to Springer Nature Publishers, in particular the editing staff, for their assistance in the publication of this book.

Contents

Notations

Abbreviations

AIP	Approximate inverse power
ALE	Adaptive line enhancement
AR	Autoregressive
AWGN	Additive white Gaussian noise
BFGS	Broyden–Fletcher–Goldfarb–Shanno procedure
CNN	Cross-correlation neural network
c-RQ	Constrained Rayleigh quotient
CSVD	Coupled SVD
CTLS	Constrained TLS
DC	Direction cosine
DCT	Deterministic continuous-time system
DDT	Deterministic discrete-time system
DGHA	Double generalized Hebbian algorithm
DLS	Data LS
EEC	Error entropy criterion
EIV	Errors-in-variables regression
FGV	Fast gain vector
FIR	Finite impulse response filtering
GTLS	Generalized TLS
ICI	Inverse Chebyshev iteration
II	Inverse iteration
IIR	Infinite impulse response filtering
ILZ	Inverse Lanczos method
IP	Inverse power
LNN	Linear neural networks
LS	Least squares approach
LZ	Lanczos methods
MA	Moving average
MC	Minor component

MCA	Minor component analysis
MSE	Mean squared error
NG-TLS	Non-Gaussian noise TLS
NIC	Novel information criterion
NOja	Normalized Oja
NOOja	Normalized orthogonal Oja
NQC	Non-quadratic criterion
OCI	Ordinary Chebyshev iteration
ODE	Ordinary differential equation
OLS	Ordinary LS
OOja	Orthogonal Oja
PDF	Probability distribution function
PSS	Principal singular subspace
PST	Principal singular triplet
PSVD	Partial SVD
PTLS	Partial TLS
RLS	Recursive least squares
RQ	Rayleigh quotient
RQI	Rayleigh quotient iteration
RTLS	Recursive TLS
SCG	Scaled conjugate gradient method
SDT	Stochastic discrete-time system
SVD	Singular value decomposition
TLMS	Total least mean squares
TLS NN	TLS neural networks
TLS	Total least squares
WTLS	Weighted TLS

Mathematical Notations

$\forall$	For all
$\in$	Belongs to
$\approx$	Approximately equal to
x	A scalar
$\boldsymbol{x}$	A vector
$\hat{\boldsymbol{x}}$	Estimate of x
$\boldsymbol{X}$	Data matrix
$\boldsymbol{X}^{\mathrm{T}}$	Transpose of $\boldsymbol{X}$
$\boldsymbol{X}^{-1}$	Inverse of $\boldsymbol{X}$
$\mathrm{diag}(x_1, \ldots, x_n)$	A diagonal matrix formed with $x_1, \ldots, x_n$
$\otimes$	The Kronecker product
$\mathbf{I}_m$	m by m identity matrix

R	The set of real numbers
R^n	The set of n-dimensional real vectors
$R^{n \times m}$	The set of $n \times m$ real matrices
$\delta_{i,j}$	The Kronecker delta function

List of Figures

List of Tables

Chapter 1
Introduction

1.1 An Overview of Total Least Square

1.1.1 Total Least Square Problems

The problem of linear parameter estimation gives rise to an overdetermined set of linear equations $\boldsymbol{Ax} \approx b$, where $\boldsymbol{A}$ is the data matrix and $\boldsymbol{b}$ is the observation vector. In the ***classical least squares*** (LS) approach, there is the underlying assumption that all errors are confined to the observation vector. This assumption is often unrealistic: The data matrix is not error-free because of sampling errors, human errors, modeling errors, and instrument errors [1]. Thus, the estimation of LS methods may include extra errors. The ***total least squares*** (TLS) method is a technique devised to compensate for data errors [1]. The TLS method has a long history though its original name is different. The earliest ideas about TLS can be traced back to the paper of Pearson in 1901 [2], who considered an approximate method for solving the matrix equation $\boldsymbol{Ax} = \boldsymbol{b}$ when in both $\boldsymbol{A}$ and $\boldsymbol{b}$ there exist errors. However, as late as 1980, Golub and Van Loan [3] for the first time gave an overall treatment from the point of view of numerical analysis and formally referred to this method as the total least squares. In mathematical statistics, the method is called ***orthogonal regression*** or ***errors-in-variables*** (EIV) regression [4]. In system identification, the TLS method is called the ***characteristic vector method*** or the ***Koopmans–Levin method*** [5]. Later, the technique was extended to multivariate problems and to multidimensional problems [6].

In most problems, the TLS problem has a unique solution which can be obtained from a simple scaling of the right singular vector of $[\boldsymbol{A}; \boldsymbol{b}]$ corresponding to its smallest singular value. Extensions of this basic TLS problem to multidimensional TLS problems $\boldsymbol{AX} \approx \boldsymbol{B}$ having more than one right-hand side vector, to problems in which the TLS solution is no longer unique or fails to have a solution altogether, and to mixed LS-TLS problems that assume some of the columns of $\boldsymbol{A}$ to be error-free,

X. Kong and D. Feng, *Efficient Online Learning Algorithms for Total Least Square Problems*, Engineering Applications of Computational Methods 21,
https://doi.org/10.1007/978-981-97-1765-1_1

are also considered in [6]. In 1988, Van Huffel and Vandewalle [7] generalized the algorithm of Golub and Van Loan [3] to all cases in which their algorithm failed to produce a solution and described the properties of these so-called ***non-generic TLS problems*** [8].

It is known that the statistical model of basic TLS is the EIV model with the restrictive condition that the measurement errors are zero-mean independent and identically distributed. In order to relax these restrictions, several extensions of the TLS problem have been proposed. In particular, the ***mixed LS-TLS problem*** formulation allows to extend consistency of the TLS estimator in EIV models, where some of the variables $\boldsymbol{a}_i$ are measured without error. The ***data least squares problem*** [9] refers to the special case in which the $\boldsymbol{A}$ matrix is noisy and the $\boldsymbol{B}$ matrix is exact. Whenever the errors are independent but unequally sized, ***weighted TLS problems*** [10] should be considered using appropriate diagonal scaling matrices in order to maintain consistency. If additionally, the errors are also correlated then the ***generalized TLS problem*** [11] formulation allows to extend consistency of the TLS estimator in EIV models, provided the corresponding error covariance matrix is known up to a factor of proportionality. More general problem formulations, such as ***restricted TLS problem*** [12], which also allow the incorporation of equality constraints, have been proposed, as well as equivalent problem formulations using other L_p norms and resulting in the so-called ***total L_p approximations***. The latter problems proved to be useful in the presence of outliers. Robustness of the TLS solution is also improved by adding regularization, resulting in the ***regularized TLS methods*** [13, 14], or by exploiting the matrix structure (e.g., Toeplitz, Hankel) in the problem formulation, leading to the so-called ***structured TLS problems***.

Furthermore, the ***constrained TLS problems*** have been formulated [15, 16]. In the so-called ***structured TLS problems*** [17], the data matrix $[\boldsymbol{A}; \boldsymbol{B}]$ is structured. In order to preserve the maximum likelihood properties of the solution, the TLS problem formulation is extended [18] with the additional constraint that the structure of the data matrix $[\boldsymbol{A}; \boldsymbol{B}]$ is preserved in the correction matrix $[\Delta\boldsymbol{A}; \Delta\boldsymbol{B}]$. Later, the important extension to ***nonlinear ElV models*** was studied, including various types of nonlinear EIV models such as bilinear, polynomial, nonlinear functional, semi-linear, Volterra series, and Cox's proportional Hazards models, and consistent estimators are derived. In these models, the relationship between the variables is assumed to be nonlinear. The above-mentioned TLS problems are mainly directed against the impact of Gaussian noises in linear/nonlinear systems, and the used are the mean–variance criterion. However, the pure white noises only exist in the theoretical study. Recently, the ***TLS problem under non-Gaussian noises*** has been researched and some efficient methods have been developed.

Now the TLS method is widely used in statistics, physics, economics, biology and medicine, signal processing, automatic control, system science, and many other disciplines and fields [19].

1.1.2 Total Least Square Methods

The TLS problem can be solved by using direct and iterative methods. ***The direct methods*** compute directly the singular value decomposition (SVD). In the computation, only a partial SVD is needed because the TLS solution is given by only one right singular vector. The partial TLS [6] exploits this fact to save a lot of computations. Among ***the iterative methods***, which are good for slowly varying set of equations, the most important use is the inverse iteration, the ordinary and inverse Chebyshev iteration, the Rayleigh quotient (RQ) iteration, and the Lanczos methods [6].

To adaptively compute the generalized eigenvector associated with the smallest eigenvalue of an autocorrelation matrix, a number of algorithms have been proposed, which can be used to solve the TLS problem. These algorithms fall into two broad categories. The first category involves ***stochastic-type adaptive algorithms***. The neural approaches can be considered as this type of method. There are two neural ways of solving the TLS problem, the one is a ***linear neuron for the minor component analysis (MCA)*** which finds the eigenvector that corresponds to the smallest eigenvalues of the autocorrelation matrix of the data vectors with a gradient learning; a subsequent normalization (the last component of neural network weight $\boldsymbol{x}_{n+1} = -1$) is needed. Many neural MCA learning algorithms [20–31] can be used to solve the TLS problems. In neural network algorithms for MCA, the only nonlinear network is the Hopfield network by Mathew and Reddy [31, 32], and all other existing neural networks are made up of one simple linear neuron. This type of neural MCA algorithms can be divided into four types: (1) Oja's MCA algorithms [20–22]; (2) self-stabilizing MCA algorithms [23–25]; (3) orthogonal Oja's MCA algorithms [26–28]; (4) coupled MCA algorithms [29, 30]. The other neural ways of solving the TLS problem are a linear neural network acting directly on the $\boldsymbol{x}_{n+1} = -1$ hyperplane (***TLS neural networks***, called as TLS NN). The existing TLS NNs have Luo's Hopfield-like neural network [31, 32], Gao's linear neuron [33, 34], Cichochi's linear neurons [35, 36], Bruce's LS-TLS neural algorithm [37, 38], and the TLS EXIN linear neuron [1].

A large variety of the second-category algorithms are called ***the recursive TLS (RTLS) algorithms***, which usually have $O(N^2)$ computational complexity per iteration. Other algorithms such as the inverse power method [39], the conjugate-gradient method [40], and the least squares-like method [41] also require $O(N^2)$ multiplication operations per eigenvector update. In particular, for online solution of the TLS problems in adaptive FIR filtering, Davila [42] proposed a fast RTLS algorithm that can fast track the eigenvector associated with the smallest eigenvalue of the augmented correlation matrix, which has $O(N)$ computational complexity per iteration. Later, two new RTLS (N-RTLS) algorithm for adaptive FIR/IIR filtering was proposed in [43, 44] by using the fast computation of the fast gain vector and the adaptation minimization of the constrained Rayleigh quotient (c-RQ), and a fast recursive algorithm by approximating the well-known inverse power iteration along the input data vector was developed in [45].

Most of the algorithms mentioned above are used to solve the basic TLS problem. Here, the problem is defined as basic if (1) it is unidimensional, (2) it is solvable, and (3) it has a unique solution. As for the various extensions of basic TLS problems, many new TLS algorithms have been developed, such as regularized TLS algorithms and new algorithms for solving TLS problems with elementwise weighting. Similarly to the classical TLS estimator, ***the generalized TLS estimator*** is computed reliably using the SVD [8]. An iterative procedure for its computation of ***the weighted TLS estimator*** is proposed in [10]. Similarly to the weighted TLS problem, ***the structured TLS solution***, in general, has no closed-form expression in terms of the SVD. An important exception is the circulant structured TLS, which can be solved using the fast Fourier transform, see [46]. In the general case, a structured TLS solution is searched via numerical optimization methods. However, fast and robust algorithms have been presented. In particular for Toeplitz or Hankel structured problems, a significant reduction in computation time is achieved by exploiting the displacement structure. Regularized structured TLS solution methods are proposed in [47, 48]. In addition, solution methods for nonlinearly structured TLS methods are developed in [49, 50]. New properties of ***EIV models in the presence of colored and non-Gaussian noise distributions***, outliers, and latency errors have been investigated, various types of ***nonlinear EIV models***, including bilinear, polynomial, nonlinear functional, semi-linear, and Cox's proportional Hazards models, are considered and consistent estimators are derived [51]. In addition, various types of bounded uncertainties have been proposed in order to improve robustness of the estimators under various noise conditions and algorithms are outlined [51]. In system identification, the well-known ***Kalman filtering is extended to the EIV context*** in which noise on the inputs as well as on the outputs is taken into account thereby improving the filtering performance. In recent years, ***the TLS methods under non-Gaussian noises*** are researched and a learning theory approach to minimum error entropy criterion (EEC) is proposed in [52]. Based on the EEC criterion, a minimum total error entropy method for parameter estimation is presented, which is called ***non-Gaussian noise TLS (NG-TLS) method*** [53]. In this algorithm, the parameters are estimated by minimizing the error of quadratic Renyi's entropy, the iterative algorithms based on gradient were derived, and the batch processing mode and online mode application methods were obtained.

1.1.3 Related Monographs for Total Least Square

For decades, several books about the TLS problems or related methods have been published. Some are dedicated to the TLS problems such as references [6, 51, 54]. Van Huffel and Vandewalle presented a comprehensive description of the state of the art on TLS from its conception up to the summer of 1990 and its use in parameter estimation in [6]. Extensions of this basic TLS problem to multidimensional TLS problems having more than one right-hand side vector, to problems in which the TLS solution is no longer unique or fails to have a solution altogether and to mixed

LS- TLS problems that assume some of the columns of augmented data matrix to be error-free, are also considered in [6]. Reference [54] presented the progress in TLS and in the broader field of EIV modeling from 1990 till 1996, including efficient algorithms of the TLS computations directly by computing the SVD only partially or iteratively, recursive TLS algorithms, rank-revealing TLS algorithms, regularized TLS algorithms, TLS algorithms for large scale problems, etc. Reference [51] provided a complete overview of the main achievements since 1996 till 2002 in TLS and EIV modeling, including rank-revealing, neural-based and regularized TLS algorithms, fast algorithms for solving TLS problems with elementwise weighting, fast, and robust algorithms of structured TLS problems, algorithms for EIV models in the presence of colored and non-Gaussian noise distributions, outliers and latency errors, algorithms for extensions to nonlinear and polynomial EIV models and to Cox's proportional hazards model, algorithms for TLS problem with various types of bounded uncertainties, faster algorithms for computing orthogonal curve fitting, etc.

Luo and Unbehauen [32] presented a Hopfield-like neural network to deal with the TLS problem, the principal limit of which is the fact that it is linked to the dimensions of the data matrix and cannot be used without structural changes for other TLS problems. Cichocki and Unbehauen [35] presented a linear neuron with different learning laws to deal with ordinary LS (OLS), TLS, and data LS (DLS) problems. For each algorithm, an appropriate cost energy to be minimized is devised, which contains the classical minimization [e.g., the minimization of $E_{\mathrm{TLS}}(\boldsymbol{x})$ for the TLS] plus regularization terms for ill-conditioned problems and nonlinear functions for the robustness of the method. Then the system of differential equations describing the gradient flow of this energy is implemented in an analog network for the continuous-time learning law and in a digital network for the discrete-time learning law. The neurons for the TLS do not work on the exact gradient flow, but on its linearization.

Later, from the perspective of the neural-based orthogonal data fitting, Cirrincione and Cirrincione [55] presented their EXIN neural networks, which is actually still a monograph about the TLS problems. This book is not oriented toward all TLS methods, only to some iterative ones which in the literature are defined as neural. Among them, the EXIN neurons are the main topic. Here they are introduced, explained, analyzed (convergence, stability, transient, dynamics, numerical considerations), and compared to the other neural approaches. The work is mainly theoretical; that is, it deals with the mathematical and numerical aspects of the EXIN neurons.

Recently, several books about matrix analysis and its applications have been published, in which the special chapters about the TLS problems or methods are analyzed or discussed. In [19], Sect. 6.3 is devoted to the TLS methods and Sect. 6. 4 deals with the constrained TLS. In the two sections, the TLS problems, the TLS solution, the performances of TLS solution, the generalized TLS, the TLS fitting, the total maximum likelihood method and the constrained TLS method have been discussed in details. In [56], Sect. 4.7 is also devoted to the TLS method, in which the TLS solution, the performances of TLS solution, and the generalized TLS are analyzed in details.

Reference [57] gave a comprehensive overview of EIV problems in system identification. The problem is about modeling of dynamic systems when all measured variables and signals are noise-corrupted, and a number of different approaches are described and analyzed. This book starts with giving a background for the EIV problem. First static systems are treated in some detail. The dominating part of this book copes with dynamic systems. The EIV problem as such is carefully analyzed, and it is demonstrated that some additional assumption(s) must be imposed if a unique solution is to be found. Several approaches and EIV methods are presented and analyzed.

In addition, for decades, several books [58–62] about matrix computation or eigenvalues of matrices provided lots of methods for TLS computation.

1.2 Aim and Main Features of This Book

1.2.1 Aim of This Book

This book is not oriented toward all TLS methods, only to some iterative ones which in the literature are defined as neural, and to some recursive ones. The aim of this book is to introduce some efficient online learning algorithms for TLS problem. This book focuses on the developments of the online TLS algorithms for adaptive filtering and parameter estimation. Specifically, this book introduces the authors' latest achievements in the past 20 years, including fast recursive TLS algorithms, approximate inverse power (AIP) iteration TLS algorithm, neural-based MCA algorithms, neural-based SVD algorithms, neural-based TLS algorithms, TLS algorithm under non-Gaussian noise, performance analysis methods of TLS algorithms, etc. However, due to the limited space, only some methods have been selected for detailed description in this book.

1.2.2 Main Features of This Book

This book can be divided into five parts: Part I is the review of TLS problems, TLS methods, and related monographs, which includes in this chapter, Chaps. 2 and 3. Part II presents several efficient recursive TLS algorithms, which includes Chaps. 4 and 5. Part III introduces many neural-based algorithms for adaptive TLS, which includes Chaps. 6–8. Part IV is focused on the TLS algorithms under non-Gausses noises, which includes Chap. 9. Part V reviews and studies performance analysis methods of neural-based TLS algorithms, which includes Chap. 10. Wherein, Part I is the foundations for subsequent chapters of this book, Parts II and III are the core content of this book, and Parts IV and V are supplementary and expanded content of this book.

The key features of this book are given as follows:

1. The TLS problems, methods, and the related monographs are reviewed and discussed comprehensively and systematically in Chapter 1. Especially lots of formulas of different TLS problems are presented, including ordinary LS problem, basic TLS problem, multidimensional TLS problem, non-generic unidimensional TLS problem, mixed OLS-TLS problem, basic data LS problem, generalized TLS problem, weighted TLS problem, constrained TLS problem, structured TLS problem, TLS in nonlinear EIV models, TLS under non-Gaussian noises, etc., in Chap. 2. Also especially, a lots of steps and characteristics of different types of TLS methods are presented, including partial TLS algorithm, iterative computation methods, RQ minimization non-neural, neural-based methods for TLS, and other methods of extended TLS, etc., in Chap. 3. These constitute the basic TLS theory, which is the foundation of later chapters.
2. The recursive TLS algorithms are reviewed, and several fast recursive TLS algorithms are presented. Firstly, a new fast recursive TLS algorithm to recursively compute the TLS solution for adaptive FIR filtering is proposed. The algorithm is based on the minimization of the c-RQ in which the last entry of the parameter vector is constrained to the negative one and has high computational efficiency. The global convergence of the algorithm is studied by LaSalle's invariance principle. Secondly, a new fast recursive TLS algorithm is developed to recursively compute the TLS solution for adaptive IIR filtering. The algorithm is also based on the minimization of the c-RQ in which the first entry of the parameter vector is fixed to the negative one. Using the shift structure of the input data vectors, a fast algorithm for computing the gain vector is established. Thirdly, an AIP iteration algorithm for adaptive TLS FIR filtering is developed, which is introduced in combination with Galerkin method so that the TLS solution can be updated adaptively at a lower computational cost. The regular form of the TLS solution is used to constrain the last element of the filter parameter vector to the negative one. The computational complexity of the algorithm is reduced by making efficient computation of the fast gain vector and using rank-one update of the augmented autocorrelation matrix.
3. Three types neural-based methods for adaptive TLS are introduced: (1) One is a linear neuron for the MCA, which finds the eigenvector associated with the smallest eigenvalue of the correlation matrix of augmented input data, and a subsequent normalization ($\boldsymbol{x}_{n+1} = -1$) is needed; (2) another is a linear neural for the SVD, which finds the right singular vector associated with the smallest singular value of the augmented input data matrix, and a subsequent normalization ($\boldsymbol{x}_{n+1} = -1$) is needed; (3) the third is linear neural network acting directly on the $\boldsymbol{x}_{n+1} = -1$ hyperplane. Firstly, three neural-based MCA algorithms for adaptive TLS are presented, including a novel multiple minor components (MCs) extraction algorithm by adding a diagonal matrix into the Douglas minor subspace tracking, a weighted information criterion for multiple MC and adaptive algorithms, and a novel method to derive neural network algorithms based on a special information criterion. Secondly, three neural-based SVD algorithms for adaptive

TLS are presented, including a neural network for SVD of a non-squared data matrix, a fast, and effective neural network algorithm to perform SVD of a non-squared matrix, and a novel coupled neural network algorithm to extract the principal singular triplet (PST) of a non-square data matrix. Thirdly, a linear neuron with self-stabilizing TLS algorithm for FIR filtering is presented.

4. The TLS problem under non-Gaussian noises is researched. Firstly, the LS method under non-Gaussian noises and TLS method with input and output of the same noise intensity are reviewed and presented. Secondly, on the basis of TLS method under non-Gaussian noises, several mixed LS-TLS algorithms under non-Gaussian noises, including several individual scenarios such as partial input and output contaminated with non-Gaussian noises, which means some column vectors contain non-Gaussian noises and another column vectors do not contain non-Gaussian noises, disparate input and output non-Gaussian noises, and input non-Gaussian noises only, are proposed, some simulations experiments verified the efficiency of algorithms.
5. The performance analysis methods of iterative MCA or TLS algorithms are reviewed, and typical formulas or steps of analyses are presented, including deterministic continuous-time (DCT) system method, stochastic discrete-time (SDT) system method, Lyapunov function approach, fixed pointed method, and deterministic discrete-time system (DDT) method. As an example of the performance analysis of algorithms, a novel self-stabilizing algorithm for generalized minor eigenpairs extraction is presented and the self-stability property and the dynamic characteristic are analyzed in details by the DDT system method.

1.3 Organization of This Book

The chapter organization of this book is detailed as follows.

In this chapter gives an overview of TLS problems and methods and illustrates the aim and main features of this book.

Chapter 2 introduces the TLS problems and a sketch of their history. Most formulas are presented because they are used in later chapters. This is mainly an introduction to problems addressed in other chapters.

Chapter 3 briefly reviews the conventional TLS methods including partial TLS algorithm, iterative computation methods, RQ minimization non-neural method, and neural networks methods, etc.

Chapter 4 introduces two recursive TLS algorithms for adaptive filter, the one is a fast recursive TLS algorithm for adaptive FIR filtering and the other is a fast recursive TLS algorithm for adaptive IIR filtering.

Chapter 5 deals with the AIP iterative algorithms of TLS problem, a fast AIP iterative algorithm for adaptive TLS FIR filtering is developed, and an AIP algorithm for adaptive extraction of minor subspace is proposed.

Chapter 6 is devoted to the neural-based MCA algorithm for adaptive TLS. In this chapter, several novel neural-based MCA algorithms are introduced, including a novel parallel multiple MC extraction algorithm by diagonal matrix, a weighted information criterion and multiple MC algorithm, and a coupled MCA algorithm.

Chapter 7 is devoted to the neural-based SVD algorithms for adaptive TLS. Several neural-based SVD algorithms are proposed, including a neural network for SVD of non-squared data matrix, a fast and effective neural network algorithm to perform SVD, and a coupled neural network algorithm for principal singular triplet extraction.

Chapter 8 is focused on the neural-based TLS algorithms. Self-stabilizing neural-based TLS algorithms is proposed.

Chapter 9 is devoted to TLS algorithms under non-Gaussian noises. The LS method under non-Gaussian noise and TLS algorithm with input and output non-Gaussian noises is reviewed. Several mixed LS-TLS algorithms under non-Gaussian noise are proposed.

Chapter 10 is focused on different performance analysis methods of TLS or MCA algorithms, especially the DCT, SDT, Lyapunov function, fixed analysis method, and the DDT method.

References

1. G. Cirrincione, M. Cirrincione, Linear system identification using the TLS EXIN neuron. Neurocomputing **28**, 53–74 (1999)
2. K. Pearson, On lines and planes of closest fit to points in space. Philosophic. Mag. 559–572 (1901)
3. G.H. Golub, C.F. Van Loan, An analysis of the total least squares problem. SIAM J. Numeric. Anal. **17**, 883–893 (1980)
4. L.J. Gleser, Estimation in a multivariate "errors in variables" regression model: large sample results. Ann. Statistic. **9**, 24–44 (1981)
5. S. Van Huffel, J. Vandewalle, Analysis and properties of the generalized total least squares problem $Ax = b$ when some or all columns in A are subject to error. SIAM J. Matrix Anal. Appl. **10**, 294–315 (1989)
6. S. Van Huffel, J. Vandewalle, in *The Total Least Squares Problems: Computational Aspects and Analysis. Frontiers in Applied Mathematics* (SIAM, Philadelphia, 1991)
7. S. Van Huffel, J. Vandewalle, Analysis and solution of the nongeneric total least squares problem. SIAM J. Matrix Anal. Appl. **9**, 360–372 (1988)
8. I. Markovsky, S. Van Huffel, Overview of total least-squares methods. Sign. Process. **87**, 2283–2302 (2007)
9. R. Degroat, E. Dowling, The data least squares problem and channel equalization. IEEE Trans. Sign. Process. **41**, 407–411 (1993)
10. A. Kukush, S. Van Huffel, Consistency of element wise weighted total least squares estimator in a multivariate errors-in-variables model AX=B. Metrika **59**(1), 75–97 (2004)
11. S. Van Huffel, J. Vandewalle, Analysis and properties of the generalized total least squares problem $AX \approx B$ when some or all columns in A are subject to error. SIAM J. Matrix Anal. Appl. **10**(3), 294–315 (1989)
12. S. Van Huffel, H. Zha, The restricted total least squares problem: formulation, algorithm and properties. SIAM J. Matrix Anal. Appl. **12**(2), 292–309 (1991)
13. R. Fierro, G. Golub, P. Hansen, D. O'Leary, Regularization by truncated total least squares. SIAM J. Sci. Comput. **18**(1), 1223–1241 (1997)
14. G. Golub, P. Hansen, D. OLeary, Tikhonov regularization and total least squares. SIAM J. Matrix Anal. Appl. 21(1), 185–194 (1999)

15. K. Arun, A unitarily constrained total least-squares problem in signal-processing. SIAM J. Matrix Anal. Appl. **13**, 729–745 (1992)
16. T. Abatzoglou, J. Mendel, G. Harada, The constrained total least squares technique and its application to harmonic superresolution. IEEE Trans. Sign. Process. **39**, 1070–1087 (1991)
17. B. De Moor, Structured total least squares and L2 approximation problems. Linear Algebra Appl. **188–189**, 163–207 (1993)
18. A. Kukush, I. Markovsky, S. Van Huffel, Consistency of the structured total least squares estimator in a multivariate errors-in-variables model. J. Statistic. Plann. Inferen. **133**(2), 315–358 (2005)
19. X.D. Zhang, in *Matrix Analysis and Applications* (Cambridge University Press, 2017)
20. L. Xu, E. Oja, C. Suen, Modified Hebbian learning for curve and surface fitting. Neural Netw. **5**(3), 441–457 (1992)
21. F. Luo, R. Unbehauen, A. Cichocki, A minor component analysis algorithm. Neural Netw. **10**(2), 291–297 (1997)
22. G. Cirrincione, M. Cirrincione, The MCA EXIN neuron for the minor component analysis. IEEE Trans. Neural Netw. **13**(1), 160–187 (2002)
23. R. Möller, A self-stabilizing learning rule for minor component analysis. Int. J. Neural Syst. **14**(1), 1–8 (2004)
24. S.C. Douglas, S.Y. Kung, S. Amari, A self-stabilized minor subspace rule. IEEE Sign. Process. Lett. **5**(12), 328–330 (1998)
25. D.Z. Feng, W.X. Zheng, Y. Jia, Neural network learning algorithms for tracking minor subspace in high-dimensional data stream. IEEE Trans. Neural Netw. **16**(3), 513–521 (2005)
26. K. Abed-Meraim, S. Attallah, A. Chkeif, Y. Hua, Orthogonal Oja algorithm. IEEE Sign. Process. Lett. **7**(5), 116–119 (2000)
27. S. Attallah, K. Abed-Meraim, Fast algorithms for subspace tracking. IEEE Sign. Process. Lett. **8**(7), 3–206 (2001)
28. S. Attallah, The generalized Rayleigh's quotient adaptive noise subspace algorithm: a Householder transformation-based implementation. IEEE Trans. Circ. Syst. II **53**(1), 3–7 (2006)
29. R. Moller, A. Konies, Coupled principal component analysis. IEEE Trans. Neural Netw. **15**(1), 214–222 (2004)
30. X.W. Feng, X.Y. Kong, H.G. Ma, H.M. Liu, Unified and coupled self-stabilizing algorithm for minor and principal eigen-pair extraction. Neural Process. Lett. (2016). https://doi.org/10.1007/s11063-016-9520-3
31. F. Luo, Y. Li, C. He, Neural network approach to the TLS linear prediction frequency estimation problem. Neurocomputing **11**, 31–42 (1996)
32. F. Luo, R. Unbehauen, *Applied neural networks for signal processing* (Cambridge University Press, Cambridge, 1997)
33. K. Gao, M.O. Ahmad, M.N. Swamy, Learning algorithm for total least-squares adaptive signal processing. Electron. Lett. **28**(4), 430–432 (1992)
34. K. Gao, M.O. Ahmad, M.N. Swamy, A constrained anti-hebbian learning algorithm for total least-squares estimation with applications to adaptive FIR and IIR Filtering. IEEE Trans. Circ. Syst. II: Analog Digit. Process. **41**(11), 718–729 (1994)
35. A. Cichocki, R. Unbehauen, *Neural networks for optimization and signal processing* (Wiley, New York, 1993)
36. A. Cichocki, R. Unbehauen, Simplified neural networks for solving linear least squares and total least squares problems in real time. IEEE Trans. Neural Netw. **5**(6), 910–923 (1994)
37. B.E. Dunne, G.A. Williamson, Analysis of gradient algorithms for TLS-based adaptive IIR filter. IEEE Trans. Sign. Process. **52**(12), 3345–3356 (2004)
38. B.E. Dunne, G.A. Williamson, QR-based TLS and mixed LS-TLS algorithms with applications to adaptive IIR filtering. IEEE Trans. Sign. Process. **51**(2), 386–394 (2003)
39. R. Vaccaro, On adaptive implementations of Pisarenko's harmonic retrieval method, in *Procceding ICASSP* (1984), pp. 224–227

40. H. Chen, T. Sarkar, S. Dianat, J. Brule, Adaptive spectral estimation by the conjugate gradient method. IEEE Trans. Acoustic Speech Sign. Process. **34**, 272–284 (1986)
41. V.U. Reddy, B. Edgardt, T. Kailath, Least squares-type algorithm for adaptive implementations of Pisarenko's harmonic retrieval method. IEEE Trans. Acoustic Speech Sign. Process. **30**, 399–405 (1982)
42. C.E. Davila, An efficient recursive total least squares algorithm for FIR adaptive filtering. IEEE Trans. Sign. Process. **42**(2), 268–280 (1994)
43. D.Z. Feng, X.D. Zhang, D.X. Chang, W.X. Zheng, A fast recursive total least squares algorithm for adaptive FIR filtering. IEEE Trans. Sign. Process. **52**(10), 2729–2737 (2004)
44. D.X. Chang, D.Z. Feng, W.X. Zheng, L. Li, A fast recursive total least squares algorithm for adaptive IIR filtering. IEEE Trans. Sign. Process. **53**(3), 957–965 (2005)
45. D.Z. Feng, W.X. Zheng, Fast approximate inverse power iteration algorithm for adaptive total least-squares fir filteing. IEEE Trans. Sign. Process. **54**(10), 4032–4039 (2006)
46. A. Beck, A. Ben-Tal, A global solution for the structured total least squares problem with block circulant matrices. SIAM J. Matrix Anal. Appl. **27**(1), 238–255 (2006)
47. N. Younan, X. Fan, Signal restoration via the regularized constrained total least squares. Sign. Process. **71**, 85–93 (1998)
48. N. Mastronardi, P. Lemmerling, S. Van Huffel, Fast regularized structured total least squares algorithm for solving the basic deconvolution problem. Numeric. Linear Algebra Appl. **12**(2–3), 201–209 (2005)
49. J. Rosen, H. Park, J. Glick, Structured total least norm for nonlinear problems. SIAM J. Matrix Anal. Appl. **20**(1), 14–30 (1998)
50. P. Lemmerling, S. Van Huffel, B. De Moor, The structured total least squares approach for nonlinearly structured matrices. Numeric. Linear Algebra Appl. **9**(1–4), 321–332 (2002)
51. S. Van Huffel, P. Lemmerling (eds.), *Total Least Squares and Errors-in-variables Modeling: Analysis* (Algorithms and Applications. Kluwer Academic Publishers, Dordrecht, 2002)
52. T. Hu, J. Fan, Q. Wu, D.X. Zhou, Learning theory approach to minimum error entropy criterion. J. Mach. Learn. Res. **14**(1), 377–397 (2013)
53. P. Shen, C. Li, Minimum total error entropy method for parameter estimation. IEEE Trans. Sign. Process. **63**(15), 4079–4090 (2015)
54. S. Van Huffel (ed.), *Recent Advances in Total Least Squares Techniques and Errors in Variables Modeling* (SIAM, Philadelphia, 1997)
55. G. Cirrincione, M. Cirrincione, *Neural-Based Orthogonal Data Fitting-the EXIN Neural Networks* (John Wiley & Sons Inc., Publication, 2010)
56. X. D. Zhang, *A Matrix Algebra Approach to Artificial Intelligence* (Springer, 2020)
57. T. Söderström, *Errors-in-variables methods in system identification* (Springer, 2018)
58. J.H. Wilkinson, *The Algebraic Eigenvalue Problem* (Clarendon Press, Oxford, UK, 1965)
59. B.N. Parlett. *The Symmetric Eigenvalue Problem* (Englewood Cliffs, NJ, Prentice Hall, 1980)
60. F. Chatelin, *Eigenvalues of Matrices* (Wiley, New York, 1993)
61. G.H. Golub, C.F. Van Loan, *Matrix Computations*, 2nd edn. (Johns Hopkins University Press, Baltimore, 1989)
62. G.W. Stewart. *Matrix Algorithms*, vol. II, (Eigensystems, SIAM, Philadelphia, 2001)

Chapter 2
Total Least Square Problems

The earliest ideas about the total least squares (TLS) can be traced back to the paper of Pearson in 1901 [1], who considered an approximate method for solving the matrix equation $\boldsymbol{A}\boldsymbol{x} = \boldsymbol{b}$ when in both $\boldsymbol{A}$ and $\boldsymbol{b}$ there exist errors. However, as late as 1980, an overall treatment from the point of view of numerical analysis was given by Golub and Van Loan [2] for the first time, and this method was formally referred to as the total least squares. For decades, TLS has achieved significant development, and lots of TLS problems and their solution methods have been proposed and have been applied in many engineering applications.

In this chapter, TLS problems and most formulas are presented, including the ordinary LS problems, basic TLS problem, multidimensional TLS problem, non-generic unidimensional TLS problem, mixed OLS-TLS problem, data LS problem, generalized TLS problem, weighted TLS problem, constrained TLS problem, structured TLS problem, TLS in nonlinear EIV models, TLS problem under non-Gaussian noises, etc., most of which is based on [1, 3–5]. This is mainly an introduction to problems addressed in other chapters.

2.1 Preliminaries

2.1.1 Overdetermined Linear Equations and SVD of Matrix

Suppose that a set of m linear equations are represented in matrix form by

$$\boldsymbol{A}\boldsymbol{X} \approx \boldsymbol{B} \tag{2.1}$$

X. Kong and D. Feng, *Efficient Online Learning Algorithms for Total Least Square Problems*, Engineering Applications of Computational Methods 21,
https://doi.org/10.1007/978-981-97-1765-1_2

where data matrix $\boldsymbol{A} \in \Re^{m\times n}$, observation matrix $\boldsymbol{B} \in \Re^{m\times d}$ and the unknown variables $\boldsymbol{X} \in \Re^{n\times d}$. If $m > n$, the system is overdetermined and does not have an exact solution. If $d > 1$, the problem is multidimensional. If $n > 1$, the problem is multivariate.

It is not difficult to find an error correction matrix $\boldsymbol{E}$ so that the following equation holds true:

$$\boldsymbol{A}\boldsymbol{X} = \boldsymbol{B} + \Delta\boldsymbol{B}.$$

Thus, such an error correction matrix $\Delta\boldsymbol{B}$ is not unique. Therefore, different error correction matrices correspond to different solutions.

As is well known, since Gaussian, there has been the following **minimum error correction rule, principle, or axiom:** ***an error correction matrix with the minimum Frobenius norm is the best.*** With the minimum error correction rule, principle, or axiom, ones can find the unique error correction matrix $\Delta\boldsymbol{B}$.

Let $\boldsymbol{X}_{\mathrm{LS}} \in R^{n\times d}$ represent the least squares (LS) solution of Eq. (2.1). We assume that $n + d \le m$ and the augmented matrix $\tilde{\boldsymbol{A}}=[\boldsymbol{A}, \boldsymbol{B}]$ has row-full rank $n + d$. The LS tries to correct the errors that existed only in $\boldsymbol{B}$ so that there are the following relations

$$\boldsymbol{A}\boldsymbol{X}_{\mathrm{LS}} = \boldsymbol{B} + \Delta\boldsymbol{B} \text{ or } \Delta\boldsymbol{B}=\boldsymbol{A}\boldsymbol{X}_{\mathrm{LS}}-\boldsymbol{B} \tag{2.2}$$

$$\min_{\boldsymbol{X}_{LS},\Delta\boldsymbol{B}} \|\Delta\boldsymbol{B}\|_F^2, \tag{2.3}$$

where $\Delta\boldsymbol{B} \in R^{m\times d}$ denotes the perturbation matrix, and $\|\bullet\|_F^2$ indicates Frobenius norm that represents the sum of the squares of all entries in a matrix. The elimination of perturbation matrix variables $\Delta\boldsymbol{B}$ in (2.2) is derived as follows:

$$\min_{\boldsymbol{X}_{LS}} \|\boldsymbol{A}\boldsymbol{X}_{\mathrm{LS}}-\boldsymbol{B}\|_F^2 \tag{2.4}$$

Hence, The LS techniques can be called the single perturbation methods for correcting the unilateral errors. It is well known that the solution to (2.2) and (2.3) is given by

$$\boldsymbol{X}_{\mathrm{LS}}=\left(\boldsymbol{A}^T\boldsymbol{A}\right)^{-1}\boldsymbol{A}^T\boldsymbol{B} \tag{2.5}$$

Hence, ones usually consider that such a LS correction $\Delta\boldsymbol{B}=\boldsymbol{A}\boldsymbol{X}_{\mathrm{LS}}-\boldsymbol{B}$ preserves the minimum correction rule, principle, or axiom. However, Feng has proved that the above fact is false, which will be showed in Theorem 5A and its proof.

Singular value decomposition (SVD) of the matrix $\boldsymbol{A}$ $(m > n)$ in Eq. (2.1) is denoted by

$$\boldsymbol{A} = \boldsymbol{U}'\boldsymbol{\Sigma}'\boldsymbol{V}'^T, \tag{2.6}$$

where

$$\begin{gathered}\boldsymbol{U}'=[\boldsymbol{U}'_1;\boldsymbol{U}'_2],\ \boldsymbol{U}'_1=[\boldsymbol{u}'_1,...,\boldsymbol{u}'_n],\ \boldsymbol{U}'_2=[\boldsymbol{u}'_{n+1},...,\boldsymbol{u}'_m],\\ \boldsymbol{u}'_i\in\Re^m,\ \ \boldsymbol{U}'^T\boldsymbol{U}'=\boldsymbol{U}'\boldsymbol{U}'^T=\boldsymbol{I}_m\end{gathered}$$

$$\boldsymbol{V}'=[\boldsymbol{v}'_1,...,\boldsymbol{v}'_n],\ \boldsymbol{v}'_i\in\Re^n,\ \boldsymbol{V}'^T\boldsymbol{V}'=\boldsymbol{V}'\boldsymbol{V}'^T=\boldsymbol{I}_n$$

$$\boldsymbol{\Sigma}'=\ \text{diag}\,(\sigma'_1,...,\sigma'_n)\in\Re^{m\times n},\ \sigma'_1\ge\cdots\ge\sigma'_n\ge 0\,.$$

Moreover, the SVD of the $m\times(n+d)$ augmented matrix $[\boldsymbol{A};\boldsymbol{B}]$ $(m>n)$ in Eq. (2.1) is denoted by

$$[\boldsymbol{A};\boldsymbol{B}]=\boldsymbol{U}\boldsymbol{\Sigma}\boldsymbol{V}^T \tag{2.7}$$

Here, if $m\ge n+d$, Eq. (2.1) indicates an overdetermined system, and there are

$$\boldsymbol{U}=[\boldsymbol{U}_1;\boldsymbol{U}_2],\ \boldsymbol{U}_1=[\boldsymbol{u}_1,...,\boldsymbol{u}_n],\ \boldsymbol{U}_2=[\boldsymbol{u}_{n+1},...,\boldsymbol{u}_m],\ \boldsymbol{u}_i\in\Re^m,\ \ \boldsymbol{U}^T\boldsymbol{U}=\boldsymbol{U}\boldsymbol{U}^T=\boldsymbol{I}_m$$

$$\boldsymbol{V}=[\boldsymbol{v}_1,...,\boldsymbol{v}_{n+d}],\ \boldsymbol{v}_i\in\Re^{n+d},\ \boldsymbol{V}^T\boldsymbol{V}=\boldsymbol{V}\boldsymbol{V}^T=\boldsymbol{I}_{n+d}\,,\boldsymbol{\Sigma}\quad=\quad\begin{bmatrix}\boldsymbol{\Sigma}_1\\ \boldsymbol{0}\end{bmatrix}\quad\in$$
$\Re^{m\times(n+d)}$, $\boldsymbol{\Sigma}_1=$ diag $(\sigma_1,...,\sigma_{n+d})\in\Re^{(n+d)\times(n+d)}$, $\sigma_1\ge\cdots\sigma_{n+d}\ge 0$.

When $m<n+d$, Eq. (2.1) denotes an underdetermined system, and we have

$$\boldsymbol{U}=[\boldsymbol{u}_1,...,\boldsymbol{u}_m],\ \boldsymbol{u}_i\in\Re^m,\ \ \boldsymbol{U}^T\boldsymbol{U}=\boldsymbol{U}\boldsymbol{U}^T=\boldsymbol{I}_m$$

$$\begin{aligned}&\boldsymbol{V}=[\boldsymbol{V}_1;\boldsymbol{V}_2],\boldsymbol{V}_1=[\boldsymbol{v}_1,...,\boldsymbol{v}_m],\boldsymbol{V}_2=[\boldsymbol{v}_{m+1},...,\boldsymbol{v}_{n+d}],\boldsymbol{v}_i\in\Re^{m+d},\\ &\boldsymbol{V}^T\boldsymbol{V}=\boldsymbol{V}\boldsymbol{V}^T=\boldsymbol{I}_{n+d}\\ &\boldsymbol{\Sigma}=\begin{bmatrix}\boldsymbol{\Sigma}_1 & \boldsymbol{0}\end{bmatrix}\in\Re^{m\times(n+d)},\quad\boldsymbol{\Sigma}_1=\ \text{diag}\,(\sigma_1,...,\sigma_m)\in\Re^{m\times m},\ \sigma_1\ge\cdots\sigma_m\ge 0\end{aligned}$$

It is seen that $(\boldsymbol{u}'_i,\sigma'_i,\boldsymbol{v}'_i)$ and $(\boldsymbol{u}_i,\sigma_i,\boldsymbol{v}_i)$ are, respectively, the singular triplet of $\boldsymbol{A}$ and $[\boldsymbol{A};\boldsymbol{B}]$.

For the unidimensional $(d=1)$ case, Eq. (2.1) is written as

$$\boldsymbol{A}\boldsymbol{x}\approx\boldsymbol{b}, \tag{2.8}$$

where $\boldsymbol{A}\in\Re^{m\times n}$, $\boldsymbol{b}\in\Re^m$ and $\boldsymbol{x}\in\Re^n$. In Sect. 2.2.3, the multidimensional case is analyzed. Here, the problem is defined as basic if (1) it is unidimensional, (2) it is solvable, and (3) it has a unique solution [1, 3].

2.1.2 Ordinary Least Squares (OLS) Problems

Definition 2.1 (*OLS Problem*) Given the overdetermined system (2.8), the LS problem searches for

$$\min_{b' \in \Re^m} \left\| b - b' \right\|_2 \text{ subject to } b' \in R(A), \tag{2.9}$$

where $R(A)$ is the column space of **A**. Once a minimizing b' is found, then any x' satisfying

$$Ax' = b' \tag{2.10}$$

is called a LS solution, and the corresponding LS correction $\Delta b' = b - b'$ is called the minimum error.

Equations (2.9) and (2.10) are satisfied if b' is the orthogonal projection of b into $R(A)$.

Theorem 2.1 (Closed-Form OLS Solution): *If rank(A) = n, Eqs.(2.9) and (2.10) are satisfied for the unique LS solution given by*

$$x' = (A^T A)^{-1} A^T b = A^+ b. \tag{2.11}$$

For an underdetermined system, this is also the minimal L_2 norm solution.

A^+ is called the Moore–Penrose pseudoinverse of matrix A. The underlying assumption is that errors occur only in the observation vector and that the data matrix is known exactly.

The OLS solution can be computed as the minimizer of the following error function:

$$E_{\text{OLS}} = \frac{1}{2}(Ax - b)^T (Ax - b). \tag{2.12}$$

2.2 Classical TLS Problem

In the OLS problem, the underlying assumption that all errors are confined to the observation vector is often unrealistic: The data matrix is not error-free because of sampling errors, human errors, modeling errors, and instrument errors. The classical TLS problem looks for the minimal (in the Frobenius norm sense) corrections ΔA and ΔB on the given data A and B that make the corrected system of equations $\hat{A}X = \hat{B}$, $\hat{A} = A + \Delta A$, $\hat{B} = B + \Delta B$ solvable [6]. In the generic case, the classical TLS problem has a unique solution, which is given in analytic form in terms of the SVD of the data matrix [6]. The extension of the TLS problem, i.e., the weighted TLS and structured TLS problems, has no such analytic solution and is currently solved numerically by local optimization methods [6].

2.2.1 Basic TLS Problems

Definition 2.2 (*Basic TLS Problem*) Given the overdetermined set (2.8), the TLS problem search for

$$\min_{[\hat{A};\hat{b}]\in\Re^{m\times(n+1)}} \left\| [A; b] - [\hat{A}; \hat{b}] \right\|_F \text{ subject to } \hat{b} \in R(\hat{A}) \tag{2.13}$$

where $\|*\|_F$ is the Frobenius norm. Once a minimizing $[\hat{A}; \hat{b}]$ is found, then any $\hat{x}$ Satisfying

$$\hat{A}\hat{x} = \hat{b} \tag{2.14}$$

is called an TLS solution, and the corresponding TLS correction $[\Delta\hat{A}, \Delta\hat{b}] = [A; b] - [\hat{A}; \hat{b}]$ has the minimum Frobenius norm.

Theorem 2.2 (Solution of Basic TLS problem) *Given Eq.* (2.6) *[respectively,* (2.7)*] as the SVD of* A *(respectively,* $[A;b]$*), if* $\sigma_n' > \sigma_{n+1}$, *then solving the TLS problem* (2.13) *follows:*

$$[\hat{A}; \hat{b}] = U\hat{\Sigma}V^T \text{ where } \hat{\Sigma} = diag(\sigma_1, ..., \sigma_n, 0), \tag{2.15}$$

the corresponding TLS correction matrix

$$[\Delta\hat{A}; \Delta\hat{b}] = \sigma_{n+1} u_{n+1} v_{n+1}^T, \tag{2.16}$$

and

$$\hat{x} = -\frac{1}{v_{n+1,n+1}} [v_{1,n+1}, ..., v_{n,n+1}]^T, \tag{2.17}$$

and the solution to $\hat{A}\hat{x} = \hat{b}$ *is unique.*

Proof See [3].

The interlacing theorem for singular values [7] implies that

$$\sigma_1 \geq \sigma_1' \geq \cdots \geq \sigma_n \geq \sigma_n' \geq \sigma_{n+1} \tag{2.18}$$

In [3], it is proved that $\sigma_n' > \sigma_{n+1} \Leftrightarrow \sigma_n > \sigma_{n+1}$ and $v_{n+1,n+1} \neq 0$.

If $\sigma_{n+1} = 0$, rank $[\boldsymbol{A};\boldsymbol{b}] = n$, then the following equation recasted by (2.8)

$$[\boldsymbol{A};\boldsymbol{b}][\boldsymbol{x}^T; -1]^T \approx 0 \tag{2.19}$$

is compatible and no approximation is needed to obtain the exact solution (2.17).

The TLS solution is obtained by finding the closest subspace $\boldsymbol{R}([\hat{\boldsymbol{A}};\hat{\boldsymbol{b}}])$ to the $n + 1$ column of $[\boldsymbol{A};\boldsymbol{b}]$ such that the sum of the squared perpendicular distances from each column of $[\boldsymbol{A};\boldsymbol{b}]$ to $\boldsymbol{R}([\hat{\boldsymbol{A}};\hat{\boldsymbol{b}}])$ is minimized, and each column is approximated by its orthogonal projection onto that subspace.

Theorem 2.3 (Closed-Form Basic TLS Solution) *Given Eq.* (2.6) *[respectively,* (2.7)*] as the SVD of* $\boldsymbol{A}$ *(respectively,* $[\boldsymbol{A};\boldsymbol{b}]$*), if* $\sigma'_n > \sigma_{n+1}$*, then*

$$\hat{\boldsymbol{x}} = (\boldsymbol{A}^T\boldsymbol{A} - \sigma_{n+1}^2\boldsymbol{I})^{-1}\boldsymbol{A}^T\boldsymbol{b}. \tag{2.20}$$

Proof The condition $\sigma'_n > \sigma_{n+1}$ assures the existence and the uniqueness of the solution. Since the singular vectors $\boldsymbol{v}_i$ are eigenvectors of $[\boldsymbol{A};\boldsymbol{b}]^T[\boldsymbol{A};\boldsymbol{b}]$, $\hat{\boldsymbol{x}}$ satisfies the following set:

$$[\boldsymbol{A};\boldsymbol{b}]^T[\boldsymbol{A};\boldsymbol{b}]\begin{bmatrix}\hat{\boldsymbol{x}}\\-1\end{bmatrix} = \begin{bmatrix}\boldsymbol{A}^T\boldsymbol{A} & \boldsymbol{A}^T\boldsymbol{b}\\ \boldsymbol{b}^T\boldsymbol{A} & \boldsymbol{b}^T\boldsymbol{b}\end{bmatrix}\begin{bmatrix}\hat{\boldsymbol{x}}\\-1\end{bmatrix} = \sigma_{n+1}^2\begin{bmatrix}\hat{\boldsymbol{x}}\\-1\end{bmatrix} \tag{2.21}$$

Formula (2.16) is directly derived from the top part of (2.21).

If $v_{n+1,n+1} \neq 0$, the TLS problem is solvable and is then called generic.

If $\sigma_p > \sigma_{p+1} = \cdots = \sigma_{n+1}$, any vector in the subspace created by the right singular vectors associated with the smallest singular vector is a solution of the TLS problem (2.9); the same happens in the case $m < n$ (underdetermined system), since then the conditions $\sigma_{m+1} = \cdot = \sigma_{n+1} = 0$ hold.

2.2.2 OLS and TLS Geometric Considerations

In the subsection, the difference between TLS and OLS is considered from a geometric point of view in the row space $\boldsymbol{R}_r([\boldsymbol{A};\boldsymbol{b}])$(see Fig. 2.1 for the case $n = 2$).

If no errors are present in data (EIV model), the set $\boldsymbol{Ax} \approx \boldsymbol{b}$ is compatible, rank$[\boldsymbol{A};\boldsymbol{b}] = n$, and the space $\boldsymbol{R}_r([\boldsymbol{A};\boldsymbol{b}])$ is n-dimensional (hyperplane). If errors occur, the set is no longer compatible and the rows $r_1, r_2,\ldots, r_m$ are scattered around the hyperplane. The normal of the hyperplane, corresponding to the minor component of $[\boldsymbol{A};\boldsymbol{b}]$, gives the corresponding solution as its intersection with the hyperplane $\boldsymbol{x}_{n+1} = -1$.

Definition 2.3 The TLS hyperplane is the hyperplane $\mathbf{x}_{n+l} = -1$.

The LS approach [for $n = 2$, see Fig. 2.1a] looks for the best approximation $\boldsymbol{b}'$ to $\boldsymbol{b}$ satisfying (2.9) such that the space $\boldsymbol{R}_r([\boldsymbol{A};\boldsymbol{b}'])$ generated by the LS approximation

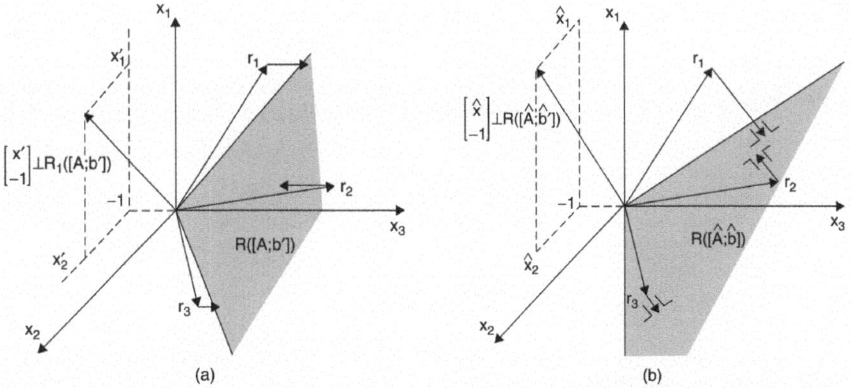

Fig. 2.1 Geometry of the LS solution $\boldsymbol{x}'(a)$ and of the TLS solution $\hat{\boldsymbol{x}}(b)$ for $n = 2$. Part (b) shows the TLS hyperplane

is a hyperplane. Only the last components of r_1, r_2,…, r_{m} can vary. This approach assumes random errors along one coordinate axis only. The TLS approach [for $n = 2$, see Fig. 2.1b] looks for a hyperplane $\boldsymbol{R}_r([\hat{\boldsymbol{A}}; \hat{\boldsymbol{b}}])$ such that (2.13) will be satisfied. The data changes are not restricted to being along one coordinate axis $\boldsymbol{x}_{\text{n+1}}$. All correction vectors $\Delta \boldsymbol{r}_i$ given by the rows of $[\Delta \hat{\boldsymbol{A}}; \Delta \hat{\boldsymbol{b}}]$ are parallel with the solution vector $[\hat{\boldsymbol{x}}^T; -1]^T$[it follows from (2.16) and (2.17)]. The TLS solution is parallel to the right singular vector corresponding to the minimum singular value of $[\boldsymbol{A}; \boldsymbol{b}]$, which can be expressed as a Rayleigh quotient:

$$\sigma_{n+1}^2 = \frac{\left\| [\boldsymbol{A}; \boldsymbol{b}][\hat{\boldsymbol{x}}^T; -1]^T \right\|_2^2}{\left\| [\hat{\boldsymbol{x}}^T; -1]^T \right\|_2^2} = \frac{\sum_{i=1}^m \left| \boldsymbol{a}_i^T \hat{\boldsymbol{x}} - \boldsymbol{b}_i \right|^2}{1 + \hat{\boldsymbol{x}}^T \hat{\boldsymbol{x}}} = \sum_{i=1}^m \| \Delta \boldsymbol{r}_i \|_2^2 \tag{2.22}$$

with $\boldsymbol{a}_i^T$ the ith row of $\boldsymbol{A}$ and $\| \Delta \boldsymbol{r}_i \|_2^2$ the square of the distance from $[\boldsymbol{a}_i^T; \boldsymbol{b}_i]^T \in \Re^{n+1}$ to the nearest point in the subspace:

$$\boldsymbol{R}_r\big([\hat{\boldsymbol{A}}; \hat{\boldsymbol{b}}]\big) = \left\{ \begin{bmatrix} \tilde{\boldsymbol{a}} \\ \tilde{b} \end{bmatrix} \middle| \tilde{\boldsymbol{a}} \in \Re^n, \tilde{b} \in \Re, \tilde{b} = \hat{\boldsymbol{x}}^T \tilde{\boldsymbol{a}} \right\} \tag{2.23}$$

Thus, the TLS solution $\hat{\boldsymbol{x}}$ minimizes the sum of the squares of the orthogonal distances (weighted squared residuals):

$$E_{\text{TLS}(x)} = \frac{\sum_{i=1}^m \left| \boldsymbol{a}_i^T \boldsymbol{x} - \boldsymbol{b}_i \right|^2}{1 + \boldsymbol{x}^T \boldsymbol{x}} \tag{2.24}$$

which is the Rayleigh quotient of $[\boldsymbol{A}; \boldsymbol{b}]^T [\boldsymbol{A}; \boldsymbol{b}]$ constrained to $\boldsymbol{x}_{n+1} = -1$. It can also be rewritten as

$$E_{\mathrm{TLS}(x)} = \frac{(\boldsymbol{Ax}-\boldsymbol{b})^T(\boldsymbol{Ax}-\boldsymbol{b})}{1+\boldsymbol{x}^T\boldsymbol{x}} \tag{2.25}$$

This formulation is very important from the neural point of view because it can be considered as the energy function to be minimized for the training of a neural network whose final weights represent the TLS solution.

2.2.3 *Multidimensional TLS Problem*

2.2.3.1 Unique Solution

Definition 2.4 (Multidimensional TLS Problem) *Given the overdetermined set* (2.1), *the TLS problem searches for*

$$\min_{[\hat{\boldsymbol{A}};\hat{\boldsymbol{B}}]\in\Re^{m\times(n+d)}} \left\|[\boldsymbol{A};\boldsymbol{B}]-[\hat{\boldsymbol{A}};\hat{\boldsymbol{B}}]\right\|_F \text{ subject to } \boldsymbol{R}(\hat{\boldsymbol{B}}) \in \boldsymbol{R}(\hat{\boldsymbol{A}}) \tag{2.26}$$

Once a minimizing $[\hat{\boldsymbol{A}};\hat{\boldsymbol{B}}]$ *is found, then any* $\hat{\boldsymbol{X}}$ *satisfying*

$$\hat{\boldsymbol{A}}\hat{\boldsymbol{X}} = \hat{\boldsymbol{B}} \tag{2.27}$$

is called a TLS solution, and the corresponding TLS correction is $[\Delta\hat{\boldsymbol{A}};\Delta\hat{\boldsymbol{B}}] = [\boldsymbol{A};\boldsymbol{B}]-[\hat{\boldsymbol{A}};\hat{\boldsymbol{B}}]$.

Theorem 2.4 (Solution of the Multidimensional TLS Problem) *Given* (2.7) *as the SVD of* $[\boldsymbol{A};\boldsymbol{B}]$, *if* $m \geq n+d$ *and* $\sigma_n > \sigma_{n+1}$, *then solving* (2.26) *maintains*

$$[\hat{\boldsymbol{A}};\hat{\boldsymbol{B}}] = [\boldsymbol{u}_1,\ldots,\boldsymbol{u}_n]\,\mathrm{diag}\,(\sigma_1,\ldots,\sigma_n)[\boldsymbol{v}_1,\ldots,\boldsymbol{v}_n]^T \tag{2.28}$$

and the corresponding TLS correction matrix

$$\begin{aligned}\left[\Delta\hat{\boldsymbol{A}};\Delta\hat{\boldsymbol{B}}\right] &= \left[\boldsymbol{u}_{n+1},\ldots,\boldsymbol{u}_{n+d}\right]\mathrm{diag}\,(\sigma_{n+1},\ldots,\sigma_{n+d})\left[\boldsymbol{v}_{n+1},\ldots,\boldsymbol{v}_{n+d}\right]^T\\ &= \boldsymbol{U}_2\boldsymbol{\Sigma}_2\boldsymbol{V}_2^T = \boldsymbol{U}_2\boldsymbol{\Sigma}_2\begin{bmatrix}\boldsymbol{V}_{21}\\ \boldsymbol{V}_{22}\end{bmatrix}^T\end{aligned} \tag{2.29}$$

TLS solution

$$\hat{\boldsymbol{X}}_{\mathrm{TLS}} = -\boldsymbol{V}_{21}\boldsymbol{V}_{22}^{-1} \tag{2.30}$$

is the unique solution to $\hat{\boldsymbol{A}}\hat{\boldsymbol{X}}_{\mathrm{TLS}} = \hat{\boldsymbol{B}}$.

Proof See [3].

It is deduced from (2.28) and (2.29) that

$$[\boldsymbol{A};\,\boldsymbol{B}]=[\hat{\boldsymbol{A}};\,\hat{\boldsymbol{B}}]+[\Delta\hat{\boldsymbol{A}};\,\Delta\hat{\boldsymbol{B}}] \tag{2.31}$$

The fitting error of TLS solution is followed by

$$\begin{aligned}\boldsymbol{E}_{\text{TLS}}&=\boldsymbol{A}\boldsymbol{X}_{\text{TLS}}-\boldsymbol{B}=[\boldsymbol{A};\,\boldsymbol{B}]\begin{bmatrix}\boldsymbol{X}_{\text{TLS}}\\-\boldsymbol{I}_d\end{bmatrix}\\&=\hat{\boldsymbol{A}}\boldsymbol{X}_{\text{TLS}}-\hat{\boldsymbol{B}}+\Delta\hat{\boldsymbol{A}}\boldsymbol{X}_{\text{TLS}}-\Delta\hat{\boldsymbol{B}}=[\Delta\hat{\boldsymbol{A}};\,\Delta\hat{\boldsymbol{B}}]\begin{bmatrix}\boldsymbol{X}_{\text{TLS}}\\-\boldsymbol{I}_d\end{bmatrix}.\end{aligned} \tag{2.32}$$

Similarly, the fitting error of LS solution is given by

$$\boldsymbol{E}_{\text{LS}}=\boldsymbol{A}\boldsymbol{X}_{\text{LS}}-\boldsymbol{B}=[\boldsymbol{A};\,\boldsymbol{B}]\begin{bmatrix}\boldsymbol{X}_{\text{LS}}\\-\boldsymbol{I}_d\end{bmatrix}. \tag{2.33}$$

To eliminate the additional scaling effects caused by non-orthogonal matrix $\begin{bmatrix}\boldsymbol{X}_{\text{TLS}}\\-\boldsymbol{I}_d\end{bmatrix}$, the renormalized error of TLS solution should be followed by

$$\tilde{\boldsymbol{E}}_{\text{TLS}}=\boldsymbol{E}_{\text{TLS}}(-\boldsymbol{V}_{22})=[\Delta\hat{\boldsymbol{A}};\,\Delta\hat{\boldsymbol{B}}]\boldsymbol{V}_2 \tag{2.34}$$

Similarly, the renormalized error of LS solution should be given by

$$\tilde{\boldsymbol{E}}_{\text{LS}}=\boldsymbol{E}_{\text{LS}}\boldsymbol{G}=[\boldsymbol{A};\,\boldsymbol{B}]\tilde{\boldsymbol{V}}_2 \tag{2.35}$$

where $\tilde{\boldsymbol{V}}_2=\begin{bmatrix}\boldsymbol{X}_{\text{LS}}\\-\boldsymbol{I}_d\end{bmatrix}\boldsymbol{G}$ is column-orthonormal matrix and $\boldsymbol{G}$ is the column-orthonormalized matrix of $\begin{bmatrix}\boldsymbol{X}_{\text{LS}}\\-\boldsymbol{I}_d\end{bmatrix}$.

Theorem 2.5 (Feng inequality) *Assume* $\sigma_M>\sigma_{M+1}$ and $\text{span}(\boldsymbol{X}_{TLS})\neq\text{span}(\boldsymbol{X}_{LS})$, then the *two errors* (2.34) *and* (2.35) *must satisfy the following inequality*

$$\|\tilde{\boldsymbol{E}}_{\text{TLS}}\|_F^2<\|\tilde{\boldsymbol{E}}_{\text{LS}}\|_F^2. \tag{2.36}$$

Proof First, Condition $\text{span}(\boldsymbol{X}_{TLS})\neq\text{span}(\boldsymbol{X}_{LS})$ deduces $\tilde{\boldsymbol{V}}_2\neq\boldsymbol{V}_2$. Second, we have

$$\begin{aligned}\|\tilde{\boldsymbol{E}}_{\text{TLS}}\|_F^2&=\text{trace}\left\{([\Delta\hat{\boldsymbol{A}};\,\Delta\hat{\boldsymbol{B}}]\boldsymbol{V}_2)([\Delta\hat{\boldsymbol{A}};\,\Delta\hat{\boldsymbol{B}}]\boldsymbol{V}_2)^T\right\}\\&=\text{trace}\left\{(\boldsymbol{U}_2\boldsymbol{\Sigma}_2\boldsymbol{V}_2^T\boldsymbol{V}_2)(\boldsymbol{U}_2\boldsymbol{\Sigma}_2\boldsymbol{V}_2^T\boldsymbol{V}_2)^T\right\}\end{aligned}$$

$$= \text{trace}\{U_2\boldsymbol{\Sigma}_2^2 U_2^T\} = \text{trace}\{\boldsymbol{\Sigma}_2^2 U_2^T U_2\}$$
$$= \text{trace}\{\boldsymbol{\Sigma}_2^2\} = \sum_{i=n+1}^{n+d} \sigma_i^2 \tag{2.37}$$

Third, we follow

$$\begin{aligned}\|\tilde{E}_{\text{LS}}\|_F^2 &= \text{trace}\left\{([A; B]\tilde{V}_2)([A; B]\tilde{V}_2)^T\right\}\\ &= \text{trace}\left\{(U\Sigma V^T\tilde{V}_2)(U\Sigma V^T\tilde{V}_2)^T\right\}\\ &= \text{trace}\left\{(V^T\tilde{V}_2)^T\Sigma^T U^T U\Sigma(V^T\tilde{V}_2)\right\}\\ &= \text{trace}\left\{(V^T\tilde{V}_2)^T\Sigma^T\Sigma(V^T\tilde{V}_2)\right\} = \text{trace}\left\{\overline{V}_2^T\overline{\Sigma}\overline{V}_2\right\}\end{aligned} \tag{2.38}$$

where $\overline{V}_2 = V^T\tilde{V}_2$ and $\overline{\Sigma} = \Sigma^T\Sigma = \text{diag}(\sigma_1^2, \ldots, \sigma_{n+d}^2)$. Since V is orthonormal and $\tilde{V}_2$ is column-orthonormal, $\overline{V}_2$ must also be column-orthonormal. This yields

$$\|\overline{V}_2\|_F^2 = \sum_{p=1}^{n+d}\sum_{q=1}^{d}\overline{v}_{p,q}^2 = \text{trace}\left\{\overline{V}_2^T\overline{V}_2\right\} = \text{trace}\{I_d\} = d. \tag{2.39}$$

Let $\beta_p = \sum_{q=1}^{d}\overline{v}_{p,q}^2$, then (2.39) and column-orthonormal matrix $\overline{V}$ can deduce relations

$$0 \le \beta_q \le 1 \tag{2.40}$$

$$\sum_{p=1}^{n+d}\beta_p = d. \tag{2.41}$$

The expression (2.39) can be rewritten into

$$\begin{aligned}\|\tilde{E}_{\text{LS}}\|_F^2 &= \text{trace}\left\{\overline{V}_2^T\overline{\Sigma}\overline{V}_2\right\} = \sum_{p=1}^{n+d}\sum_{q=1}^{d}\sigma_p^2\overline{v}_{p,q}^2\\ &= \sum_{p=1}^{n+d}\sigma_p^2\sum_{q=1}^{d}\overline{v}_{p,q}^2 = \sum_{p=1}^{n+d}\sigma_p^2\beta_p = \sum_{p=1}^{n}\sigma_p^2\beta_p + \sum_{p=n+1}^{n+d}\sigma_p^2\beta_p\end{aligned} \tag{2.42}$$

Consider (2.40) and let $\beta_p + \overline{\beta}_p = 1$ $(0 \le \overline{\beta}_p \le 1)$ for $p = m+1, \ldots, m+d$, then we have

$$\sum_{p=n+1}^{n+d}\beta_p+\sum_{p=n+1}^{m+d}\overline{\beta}_p=d=\sum_{p=1}^{n}\beta_p+\sum_{p=n+1}^{n+d}\beta_p \tag{2.43}$$

$$\sum_{p=n+1}^{m+d}\overline{\beta}_p=\sum_{p=1}^{n}\beta_p. \tag{2.44}$$

Since $\tilde{\boldsymbol{V}}_2 \neq \boldsymbol{V}_2$ implies that $\overline{v}_{p,q}$ for $q = 1,\ldots,d$ and $p = 1,\ldots,n$ are not all equal to zero, and condition $\sum_{q=n+1}^{n+d}\overline{\beta}_q=\sum_{q=1}^{n}\beta_q>0$ is preserved. Thus, consider $\sigma_n^2-\sigma_q^2>0$ for $q=n+1,\ldots,n+d$, then expression (2.42) can be changed into

$$\begin{aligned}\|\tilde{\boldsymbol{E}}_{\text{LS}}\|_F^2&=\sum_{q=1}^{n}\sigma_q^2\beta_q+\sum_{p=n+1}^{n+d}\sigma_q^2\beta_q=\sum_{q=1}^{n}\sigma_q^2\beta_q-\sum_{q=n+1}^{n+d}\sigma_q^2\overline{\beta}_q+\sum_{q=n+1}^{n+d}\sigma_q^2\\&\geq\sigma_M^2\sum_{q=1}^{M}\beta_q-\sum_{q=n+1}^{n+1}\sigma_q^2\overline{\beta}_q+\sum_{q=n+1}^{n+d}\sigma_q^2\\&=\sigma_M^2\sum_{q=n+1}^{n+d}\overline{\beta}_q-\sum_{q=n+1}^{n+d}\sigma_q^2\overline{\beta}_q+\sum_{q=n+1}^{n+d}\sigma_q^2\\&=\sum_{q=n+1}^{n+d}(\sigma_M^2-\sigma_q^2)\overline{\beta}_q+\sum_{q=n+1}^{n+d}\sigma_q^2>\sum_{q=n+1}^{n+d}\sigma_q^2=\|\tilde{\boldsymbol{E}}_{\text{TLS}}\|_F^2\end{aligned} \tag{2.45}$$

This completes the proof of Theorem 2.5.

Relation (2.36) shows that such a LS correction $\Delta\boldsymbol{B}=\boldsymbol{A}\boldsymbol{X}_{\text{LS}}-\boldsymbol{B}$ does not preserve the minimum correction rule, principle, or axiom. In contrast, expression (2.37) denotes that the TLS error correction follows the minimum correction rule, principle, or axiom.

2.2.4 Non-generic Unidimensional TLS Problem

If $\sigma_p > \sigma_{p+1} = \cdots = \sigma_{n+1}$, $p \leq n$ and all $v_{n+1,i} = 0$, $i = p+1,\ldots,n+1$, the TLS problem (2.9) fails to have a solution, this is so-called non-generic TLS problem. In this case, the existence conditions of the preceding theorems are not satisfied. These problems occur whenever $\boldsymbol{A}$ is rank-deficient ($\sigma_p' \approx 0$) or when the set of equations is highly conflicting ($\sigma_p' \approx \sigma_{p+1}$ large). The latter situation is detected by inspecting the size of the smallest σ_i for which $v_{n+1,i} \neq 0$.

Alternatively, the problem can be made generic by adding more equations. This is the case if the model is EIV and the observation errors are statistically independent and equally sized (same variance). For $\sigma_p' \approx 0$, it is possible to remove the dependency among the columns of $\boldsymbol{A}$ by removing appropriate columns in $\boldsymbol{A}$ such that the remaining submatrix has full rank and then apply TLS to the reduced problem (subset selection [8, 9]).

Although exact non-generic TLS problems seldom occur, close-to-non-generic TLS problems are not uncommon. The generic TLS solution can still be computed, but it is unstable and becomes even very sensitive to data errors when $\sigma'_p - \sigma_{p+1}$ is very close to zero [10].

Theorem 2.6 (Properties of Non-generic Unidimensional TLS) *(See* [3]*) Let (2.6) [respectively,* (2.7)] *be the SVD of* $\mathbf{A}$ *(respectively,* $[\mathbf{A}; \boldsymbol{b}]$*); let* $\boldsymbol{b}'$ *be the orthogonal projection of* $\boldsymbol{b}$ *onto* $\boldsymbol{R}(\boldsymbol{A})$ *and* $[\hat{\mathbf{A}}; \hat{\boldsymbol{b}}]$ *the rank n approximation of* $[\mathbf{A}; \boldsymbol{b}]$*, as given by* (2.15)*. If* $\mathbf{V}'(\sigma_j)$*[respectively,* $\mathbf{U}'(\sigma_j)$*] is the right (respectively, left) singular subspace of* $\mathbf{A}$ *associated with* σ_j*, then the following relations can be proven:*

$$v_{n+1,j} = 0 \Leftrightarrow \boldsymbol{v}_j = \begin{bmatrix} \boldsymbol{v}' \\ 0 \end{bmatrix} \text{ with } \boldsymbol{v}' \in \mathbf{V}'(\sigma_j) \tag{2.46}$$

$$v_{n+1,j} = 0 \Rightarrow \sigma_j = \sigma'_k \text{ with } k = j - 1 \text{ or } k = j \text{ and } 1 \leq k \leq n \tag{2.47}$$

$$v_{n+1,j} = 0 \Rightarrow \boldsymbol{b} \bot u' \text{ with } \boldsymbol{u}' \in \mathbf{U}'(\sigma_j) \tag{2.48}$$

$$v_{n+1,j} = 0 \Leftrightarrow \boldsymbol{u}_j = \boldsymbol{u}' \text{ with } \boldsymbol{u}' \in \mathbf{U}'(\sigma_j) \tag{2.49}$$

$$v_{n+1,j} = 0 \Rightarrow \boldsymbol{b}' \bot \boldsymbol{u}' \text{ with } \boldsymbol{u}' \in \mathbf{U}'(\sigma_j) \tag{2.50}$$

$$v_{n+1,j} = 0 \Rightarrow \hat{\boldsymbol{b}} \bot \boldsymbol{u}' \text{ with } \boldsymbol{u}' \in \mathbf{U}'(\sigma_j) \tag{2.51}$$

If σ_j *is an isolated singular value, the converse of relations* (2.48) and (2.50) *also holds.*

Proof See [11, Th. 1–3].

Definition 2.5 (*Non-generic Unidimensional TLS Problem*) Given the set (2.8), let (2.7) be the SVD of $[\mathbf{A}; \mathbf{b}]$; the non-generic TLS problem searches for

$$\begin{cases} \min\limits_{[\hat{\boldsymbol{A}};\hat{\boldsymbol{b}}] \in \Re^{m \times (n+1)}} \left\| [\boldsymbol{A}; \boldsymbol{b}] - [\hat{\boldsymbol{A}}; \hat{\boldsymbol{b}}] \right\|_F \text{ s.t. } \boldsymbol{b}' \in \boldsymbol{R}(\hat{\boldsymbol{A}}) \\ \begin{bmatrix} \hat{\boldsymbol{x}} \\ -1 \end{bmatrix} \bot \boldsymbol{v}_j, j = p + 1, \ldots, n + 1 \text{ (provided} \textit{that } u_{n+1,p} \neq 0) \end{cases} \tag{2.52}$$

Once a minimizing $[\hat{\boldsymbol{A}}; \hat{\boldsymbol{b}}]$ is found, then any $\hat{\boldsymbol{x}}$ satisfying

$$\hat{\boldsymbol{A}}\hat{\boldsymbol{x}} = \hat{\boldsymbol{b}} \tag{2.53}$$

is called a non-generic TLS solution, and the corresponding non-generic TLS correction is $[\Delta\hat{A};\ \Delta\hat{b}] = [A; b] - [\hat{A}; \hat{b}]$.

Theorem 2.7 (Non-generic Unidimensional TLS Solution) *Let (2.7) be the SVD of $[A;b]$, and assume that $v_{n+1,j} = 0$ for $j = p+1,\dots,n+1$, $p \leq n$. If $\sigma_{p-1} > \sigma_p$ and $v_{n+1,p} \neq 0$, then*

$$[\hat{A}; \hat{b}] = U\hat{\Sigma}V^T \text{ where } \hat{\Sigma} = \operatorname{diag}(\sigma_1, \dots, \sigma_{p-1}, 0, \sigma_{p+1}, \dots, \sigma_{n+1}) \tag{2.54}$$

with corresponding non-generic TLS correction matrix

$$[\Delta\hat{A};\ \Delta\hat{b}] = \sigma_p u_p v_p^T \tag{2.55}$$

solves the non-generic TLS problem (2.52) and

$$\hat{x} = -\frac{1}{v_{n+1,p}}[v_{1,p}, \dots, v_{n,p}]^T \tag{2.56}$$

exists and is the unique solution to $\hat{A}\hat{x} = \hat{b}$.

Proof See [3, p.72].

If $v_{n+1,n+1} = 0 \wedge v_{n+1,n} \neq 0 \wedge \sigma_{n-1} > \sigma_n$, then (2.56) becomes

$$\begin{bmatrix} \hat{x} \\ -1 \end{bmatrix} = -\frac{v_n}{v_{n+1,n}}. \tag{2.57}$$

Theorem 2.8 (Closed-Form Non-generic TLS Solution) *Let* (2.6) [*respectively,* (2.7)] *be the SVD of A (respectively, $[A;\ b]$), and assume that $v_{n+1,j} = 0$ for $j = p+1,\dots,n+1$, $p \leq n$. If $\sigma_{p-1} > \sigma_p$ and $v_{n+1,p} \neq 0$, the non-generic TLS solution is*

$$\hat{x} = \left(A^T A - \sigma_p^2 I_n\right)^{-1} A^T b. \tag{2.58}$$

Proof See [3, p. 74].

2.3 Extension of TLS Problem

2.3.1 Mixed OLS-TLS Problem

The statistical model that corresponds to the basic TLS approach is the EIV model with the restrictive condition that the measurement errors are zero-mean independent and identically distributed. In order to relax these restrictions, several extensions of the TLS problem have been investigated. For example, the mixed LS-TLS problem formulation allows to extend consistency of the TLS estimator in EIV models, where some of the variables are measured without error [6].

If n_1 columns of the $m \times n$ data matrix $\boldsymbol{A}$ are known exactly, the problem is called mixed OLS-TLS [3]. It is natural to require that the TLS solution not perturb the exact columns. After some column permutations in $\boldsymbol{A}$ such that $\boldsymbol{A} = [\boldsymbol{A}_1;\boldsymbol{A}_2]$, where $\boldsymbol{A}_1 \in \Re^{m\times n_1}$ is made of the exact n_1 columns and $\boldsymbol{A}_2 \in \Re^{m\times n_2}$, perform n_1 Householder transformations $\boldsymbol{Q}$ on the matrix $[\boldsymbol{A}; \boldsymbol{B}]$ (QR factorization) so that

$$[\boldsymbol{A}_1; \boldsymbol{A}_2; \boldsymbol{B}] = \boldsymbol{Q}\underset{n_1 \;\; n-n_1 \;\; d}{\begin{bmatrix} \boldsymbol{R}_{11} & \boldsymbol{R}_{12} & \boldsymbol{R}_{1b} \\ 0 & \boldsymbol{R}_{22} & \boldsymbol{R}_{2b} \end{bmatrix}} \begin{matrix} n_1 \\ m-n_1 \end{matrix} \tag{2.59}$$

where $\boldsymbol{R}_{11}$ is a $n_1 \times n_1$ upper triangular matrix. Then compute the TLS solution $\hat{\boldsymbol{X}}_2$ of $\boldsymbol{R}_{22}\boldsymbol{X} \approx \boldsymbol{R}_{2b}$. $\hat{\boldsymbol{X}}_2$ yields the last $n - n_1$ components of each solution vector $\hat{\boldsymbol{X}}_i$. To find the first n_1 rows $\hat{\boldsymbol{X}}_1$ of the solution matrix $\hat{\boldsymbol{X}} = [\hat{\boldsymbol{X}}_1^T; \hat{\boldsymbol{X}}_2^T]^T$, solves (OLS) $\boldsymbol{R}_{11}\hat{\boldsymbol{X}}_1 \approx \boldsymbol{R}_{1b} - \boldsymbol{R}_{12}\hat{\boldsymbol{X}}_2$. Thus, the entire method amounts to a preprocessing step, a TLS problem, an OLS problem, and a post-processing step (inverse row permutations) [12].

Theorem 2.9 (Closed-Form Mixed OLS-TLS Solution) *Let rank* $\mathrm{A}_1 = n_1$; *denote by* σ'*(respectively,* σ*) the smallest [respectively,* $(n_2 + 1)$*th] singular value of* $\boldsymbol{R}_{22}$ *(respectively,* $[\boldsymbol{R}_{22}; \boldsymbol{R}_{2b}]$*); assume that the smallest singular values* $\sigma = \sigma_{n2+1} = \cdot = \sigma_{n2+d}$ *coincide. If* $\sigma' > \sigma$*, then the mixed OLS-TLS solution is*

$$\hat{\boldsymbol{X}} = \left(\boldsymbol{A}^T\boldsymbol{A} - \sigma^2\begin{bmatrix} 0 & 0 \\ 0 & \boldsymbol{I}_{n_2} \end{bmatrix}\right)^{-1} \boldsymbol{A}^T\boldsymbol{B}. \tag{2.60}$$

Proof It is a special case of [13, Th. 4].

2.3.2 *Statistical Properties and Validity*

Under the assumption that *all* errors in the augmented matrix $[\boldsymbol{A}; \boldsymbol{B}]$ are row-wise independently and identically distributed (iid) with zero-mean and common covariance matrix of the form $\sigma_v^2\boldsymbol{\Xi}_0$, $\boldsymbol{\Xi}_0$ known and positive definite (e.g., the identity matrix), the TLS method offers the *best* estimate and is more accurate than the LS solution in estimating the parameters of a model. The most suitable model for the TLS concept is the errors-in-variables (EIV) model, which assumes an unknown but exact linear relation (zero residual problems) among the true variables that can only be observed with errors.

Definition 2.6 (Multivariate Linear EIV Model)

$$
\begin{aligned}
&\boldsymbol{B}_0 = \boldsymbol{l}_m\boldsymbol{\alpha}^T + \boldsymbol{A}_0\boldsymbol{X}_0,\ \boldsymbol{B}_0 \in \Re^{m\times d},\ \boldsymbol{A}_0 \in \Re^{m\times n},\ \boldsymbol{\alpha} \in \Re^d \\
&\boldsymbol{A} = \boldsymbol{A}_0 + \Delta\boldsymbol{A} \\
&\boldsymbol{B} = \boldsymbol{B}_0 + \Delta\boldsymbol{B},
\end{aligned}
\tag{2.61}
$$

where $\mathbf{1}_m = [1,\ldots,1]^T$. $\mathbf{X}_0$ is the n × d matrix of the true but unknown parameters to be estimated. The intercept vector $\boldsymbol{\alpha}$ is either zero (no-intercept model) or unknown (intercept model) and must be estimated.

The following remarks can be given about the multivariate linear EIV model:

Strong Consistency: In the EIV model, if it is assumed that the rows of $[\boldsymbol{A}; \boldsymbol{B}]$ are iid with common zero-mean vector and common covariance matrix of the form $\Xi = \sigma_v^2 \boldsymbol{I}_{n+d}$, where $\sigma_v^2 > 0$ is unknown, then the TLS method is able to compute strongly consistent estimates of the unknown parameters $\mathbf{X}_0$, $\mathbf{A}_0$, $\boldsymbol{\alpha}$, and σ_v.

Scaling: The assumption about the errors seems somewhat restrictive: It requires that all measurements in $\boldsymbol{A}$ and $\boldsymbol{B}$ be affected by errors and, moreover, that these errors must be uncorrelated and equally sized. If these conditions are not satisfied, the classical TLS solution is no longer a consistent estimate of the model parameters. Provided that the error covariance matrix Ξ is known up to a factor of proportionality, the data $[\boldsymbol{A}; \boldsymbol{B}]$ can be transformed to the new data $[\boldsymbol{A}^*; \boldsymbol{B}^*] = [\boldsymbol{A};\boldsymbol{B}]\boldsymbol{C}^{-1}$, where $\boldsymbol{C}$ is a square root (Cholesky factor) of $\Xi (= \boldsymbol{C}^T\boldsymbol{C})$, such that the error covariance matrix of the transformed data is now diagonal with equal error variances. Then the TLS algorithm can be used on the new set, and finally, its solution must be converted to a solution of the original set of equations [14, 15].

Covariance Knowledge: The knowledge of the form of the error covariance matrix up to a constant scalar multiple may still appear too restrictive, as this type of information is not always available to an experimenter. Assuming that independent repeated observations are available for each variable observed with error, this type of replication provides enough information about the error covariance matrix to derive consistent unbiased estimates of Ξ [16, 17]. Using these consistent estimates instead of Ξ does not change the consistency properties of the parameter estimators [16].

2.3.3 Basic Data Least Squares Problem

The TLS problem can be viewed as an unconstrained perturbation problem because all columns of $[\boldsymbol{A}; \boldsymbol{b}]$ can have error perturbations. The OLS problem constrains the columns of $\boldsymbol{A}$ to be errorless; the opposite case is the data least squares (DLS) problem, because the error is assumed to lie only in the *data* matrix $\boldsymbol{A}$.

Definition 2.7 (*Basic DLS Problem*) Given the overdetermined set (2.8), the data LS (DLS) problem searches for.

$$
\min_{\boldsymbol{A}'' \in \Re^{m\times n}} \left\| \boldsymbol{A} - \boldsymbol{A}'' \right\|_F \ \text{subject to}\ \hat{\boldsymbol{b}} \in \boldsymbol{R}(\boldsymbol{A}''). \tag{2.62}
$$

Once a minimizing $\boldsymbol{A}''$ is found, then any $\boldsymbol{x}''$ satisfying

$$\boldsymbol{A}''\boldsymbol{x}'' = \boldsymbol{b} \tag{2.63}$$

is called a DLS solution, and the corresponding DLS correction is $\Delta\boldsymbol{A}'' = \boldsymbol{A} - \boldsymbol{A}''$.

The DLS case is particularly appropriate for certain deconvolution problems, such as those that may arise in system identification or channel equalization [18].

Theorem 2.10 *The DLS problem (2.62) is solved by*

$$\boldsymbol{x}'' = \frac{\boldsymbol{b}^T\boldsymbol{b}}{\boldsymbol{b}^T\boldsymbol{A}\boldsymbol{v}_{\min}}\boldsymbol{v}_{\min}\ (\boldsymbol{b}^T\boldsymbol{A}\boldsymbol{v}_{\min} \neq 0)\,, \tag{2.64}$$

where $\boldsymbol{v}_{\min}$ is the right singular vector corresponding to the smallest singular value of the matrix $\boldsymbol{P}_b^{\perp}\boldsymbol{A}$, where $\boldsymbol{P}_b^{\perp} = (\boldsymbol{I} - \boldsymbol{b}(\boldsymbol{b}^T\boldsymbol{b})^{-1}\boldsymbol{b}^T)$ *is a projection matrix that projects the column space of* **A** *into the orthogonal complement of* **b**. *If the smallest singular vector is repeated, then the solution is not unique. The minimum norm solution is given by*

$$\boldsymbol{x}'' = \frac{\boldsymbol{b}^T\boldsymbol{b}}{(\boldsymbol{b}^T\boldsymbol{A}\boldsymbol{V}_{\min})(\boldsymbol{V}_{\min}^T\boldsymbol{A}^T\boldsymbol{b})}\boldsymbol{V}_{\min}(\boldsymbol{V}_{\min}^T\boldsymbol{A}^T\boldsymbol{b})\ ((\boldsymbol{b}^T\boldsymbol{A}\boldsymbol{V}_{\min})(\boldsymbol{V}_{\min}^T\boldsymbol{A}^T\boldsymbol{b}) \neq 0)\,, \tag{2.65}$$

where $\boldsymbol{V}_{\min}$ is the right singular space corresponding to the repeated smallest singular value of $\boldsymbol{P}_b^{\perp}\boldsymbol{A}$.

Proof See [18], which derives its results from the constrained TLS (CTLS) [19, 20].

2.3.4 *Generalized TLS Problem*

The OLS, the DLS, Tikhonov regularization, and the TLS method can be derived and explained by a unified theoretical framework.

Definition 2.8 (*Generalized TLS Problem*) Consider the minimization problem:

$$\min_{\Delta\boldsymbol{A},\Delta\boldsymbol{b},\boldsymbol{x}} \{\|\Delta\boldsymbol{A};\ \Delta\boldsymbol{b}\|_F^2 + \lambda\|\boldsymbol{x}\|_2^2\} \text{ subject to } (\boldsymbol{A} + \alpha\Delta\boldsymbol{A})\boldsymbol{x} = \boldsymbol{b} + \beta\Delta\boldsymbol{b}, \tag{2.66}$$

where α and β are the weighting coefficients of respectively the perturbation $\Delta\mathbf{A}$ of the data matrix **A** and the perturbation $\Delta\mathbf{b}$ of the data vector **b**, and λ is the Tikhonov regularization parameter. The above minimization problem is called the generalized TLS (GTLS) problem.

Comparison of Optimization Problem [4]:

(1) OLS: $\alpha = 0$, $\beta = 1$, and $\lambda = 0$, which gives

$$\min_{\Delta A, \Delta b, x} \|\Delta b\|_2^2 \text{ subject to } Ax = b + \Delta b. \tag{2.67}$$

(2) DLS: $\alpha = 1$, $\beta = 0$, and $\lambda = 0$, which gives

$$\min_{\Delta A, \Delta b, x} \|\Delta b\|_F^2 \text{ subject to } (A + \Delta A)x = b. \tag{2.68}$$

(3) Tikhonov regularization: $\alpha = 0$, $\beta = 1$, and $\lambda > 0$, which gives

$$\min_{\Delta A, \Delta b, x} \{\|\Delta b\|_2^2 + \lambda\|x\|_2^2\} \text{ subject to } Ax = b + \Delta b. \tag{2.69}$$

(4) TLS: $\alpha = \beta = 1$ and $\lambda = 0$, which gives

$$\min_{\Delta A, \Delta b, x} \|\Delta A; \Delta b\|_2^2 \text{ subject to } (A + \Delta A)x = b + \Delta b. \tag{2.70}$$

Comparison of Solution Vectors [4]:

The constraint condition $(A + \alpha\Delta A)x = (b + \beta\Delta b)$ can be represented as

$$([\alpha^{-1}A; \beta^{-1}b] + [\Delta A; \Delta b])\begin{bmatrix} \alpha x \\ -\beta \end{bmatrix} = 0.$$

Letting $D = [\Delta A; \Delta b]$ and $z = \begin{bmatrix} \alpha x \\ -\beta \end{bmatrix}$, the above equation becomes

$$Dz = -[\alpha^{-1}A; \beta^{-1}b]z. \tag{2.71}$$

Under the assumption $z^H z = 1$, we have

$$\min\|D\|_F^2 = \min\|Dz\|_2^2 = \min\left\|[\alpha^{-1}A; \beta^{-1}b]z\right\|_2^2 \tag{2.72}$$

and thus the solution to the GTLS problem (2.49) can be rewritten as

$$\hat{x}_{\text{GTLS}} = \arg\min_z \left\{ \frac{\left\|[\alpha^{-1}A; \beta^{-1}b]z\right\|_2^2}{z^H z} + \lambda\|x\|_2^2 \right\}. \tag{2.73}$$

Noting that $z^H z = \alpha^2 x^H x + \beta^2$ and $[\alpha^{-1}A; \beta^{-1}b]z = [\alpha^{-1}A; \beta^{-1}b]\begin{bmatrix} \alpha x \\ -\beta \end{bmatrix} = Ax - b$ the solution to the GTLS problem (2.73) is given by

$$\hat{x}_{\mathrm{GTLS}} = \arg\min_{x}\left\{\frac{\|Ax-b\|_2^2}{\alpha^2\|x\|_2^2+\beta^2} + \lambda\|x\|_2^2\right\}. \tag{2.74}$$

When the weighting coefficients α, β and the Tikhonov regularization parameter λ take appropriate values, the GTLS solution (2.74) gives the following results:

$$\hat{x}_{LS} = \arg\min_{x}\|Ax-b\|_2^2 \tag{2.75}$$

$$\hat{x}_{\mathrm{DLS}} = \arg\min_{x}\frac{\|Ax-b\|_2^2}{\|x\|_2^2} \tag{2.76}$$

$$\hat{x}_{\mathrm{Tik}} = \arg\min_{x}\left\{\|Ax-b\|_2^2 + \lambda\|x\|_2^2\right\} \tag{2.77}$$

$$\hat{x}_{TLS} = \arg\min_{x}\frac{\|Ax-b\|_2^2}{\|x\|_2^2+1}. \tag{2.78}$$

Comparison of Perturbation Methods [4]:

(1) *OLS method:* This uses the possible small correction term Δb to perturb the data vector b in such a way that $b - \Delta b \approx b_0$, and thus compensates for the observed noise e in b. The correction vector is selected as $\Delta b = Ax - b$, and the analytical solution is

$$\hat{x}_{\mathrm{LS}} = (A^H A)^{-1} A^H b.$$

(2) *DLS method:* The correction term $\Delta A = (Ax - b)x^H/(x^H x)$ compensates for the observed error matrix E in the data matrix A. The data LS solution is

$$\hat{x}_{\mathrm{DLS}} = \arg\min_{x}\frac{(Ax-b)^H(Ax-b)}{x^H x}.$$

(3) *Tikhonov regularization method:* This adds the same perturbation term $\lambda > 0$ to every diagonal entry of the matrix $A^H A$ to avoid the numerical instability of the LS solution $(A^H A)^{-1} A^H b$. The analytical solution is

$$\hat{x}_{\mathrm{Tik}} = (A^H A - \lambda I)^{-1} A^H b.$$

(4) *TLS method:* By subtracting the perturbation matrix λI, the noise or perturbation in the covariance matrix of the original data matrix is restrained. There are three kinds of TLS solution: the minimum norm solution, the anti-regularization solution with n components

$$\hat{x}_{\mathrm{TLS}} = (A^H A - \lambda I)^{-1} A^H b$$

and the SVD-TLS solution with only $p = \mathrm{rank}([A, b])$ principal parameters.

Comparison of Application Ranges [4]:

(1) The OLS method is applicable for a data matrix $\boldsymbol{A}$ with full column rank and a data vector $\boldsymbol{b}$ containing iid Gaussian errors.
(2) The DLS method is applicable for a data vector $\boldsymbol{b}$ without error and a data matrix $\boldsymbol{A}$ that has full column rank and iid Gaussian error column vectors.
(3) The Tikhonov regularization method is applicable for a data matrix $\boldsymbol{A}$ with deficient column rank.
(4) The TLS method is applicable for a data matrix $\boldsymbol{A}$ with full column rank, where both $\boldsymbol{A}$ and the data vector $\boldsymbol{b}$ contain iid Gaussian error.

2.3.5 Weighted TLS Problem

The weighted TLS problems are developed for the case where the measurement errors are differently sized. Consistency of the weighted TLS estimator is proven and an iterative procedure for its computation is proposed in [21].

Definition 2.9 *Weighted TLS Problem*) [6] For a given positive definite weight matrix $\boldsymbol{W} \in \Re^{m(n+d)\times m(n+d)}$, define the weighted matrix norm

$$\|\boldsymbol{C}\|_W = \sqrt{\mathrm{vec}^T(\boldsymbol{C}^T)\boldsymbol{W}\mathrm{vec}(\boldsymbol{C}^T)} \tag{2.79}$$

and the Weighted TLS Misfit Function

$$M_{\mathrm{WTLS}}(\boldsymbol{C}, \boldsymbol{\Omega}) = \min_{\hat{\boldsymbol{C}}\in\boldsymbol{\Omega}}\left\|\boldsymbol{C} - \hat{\boldsymbol{C}}\right\|_W, \tag{2.80}$$

where $\boldsymbol{C} \in \Re^{m(n+d)}$ is the data matrix $\mathbf{C} = [\mathbf{A}, \mathbf{B}]$, $\boldsymbol{\Omega} \subseteq \Re^{n+d}$ is referred to as a linear static model. The approximation problem with weighted TLS misfit function is called the weighted TLS problem.

Definition 2.10 (Weighted total least squares) [6]. *Given a data matrix* $\boldsymbol{C} \in \Re^{m(n+d)}$, *a positive definite weight matrix* $\mathbf{W}$, *and a complexity specification n, solve the optimization problem.*

$$\{\hat{\boldsymbol{\Omega}}_{\mathrm{WTLS}}, \hat{\boldsymbol{C}}_{\mathrm{WTLS}}\} = \arg\min_{\boldsymbol{\Omega}\in\varphi_n} M_{WTLS}(\boldsymbol{C}, \boldsymbol{\Omega}) \tag{2.81}$$

The motivation for considering the weighted TLS problem is that it defines the maximum likelihood estimator for the EIV model when the measurement noise $\tilde{\boldsymbol{C}} = [\tilde{\boldsymbol{A}}, \tilde{\boldsymbol{B}}]$ is zero mean, normally distributed, with a covariance matrix

$$\mathrm{cov}(\mathrm{vec}(\tilde{\boldsymbol{C}}^T)) = \sigma^2\mathbf{W}^{-1}, \tag{2.82}$$

i.e., the weight matrix $\boldsymbol{W}$ is up to a scaling factor σ^2 the inverse of the measurement noise covariance matrix.

Element-wise weighted TLS [6]: The special case when the weight matrix $\boldsymbol{W}$ is diagonal is called element-wise weighted TLS. It corresponds to an EIV problem with uncorrelated measurement errors. Let $\boldsymbol{W} = \text{diag}\,(\boldsymbol{w}_1, ..., \boldsymbol{w}_{m(n+d)})$ and define the $m \times (n+d)$ weight matrix $\boldsymbol{\Sigma}$ by $\boldsymbol{\Sigma}_{ij} = \boldsymbol{w}_{(i-1)(n+d)+j}$. Denote by $\boldsymbol{A} \odot \boldsymbol{B} = [a_{ij}b_{ij}]$ the element-wise product. Then $\|\Delta \boldsymbol{C}\|_W = \|\boldsymbol{\Sigma} \odot \Delta \boldsymbol{C}\|_F$.

TLS as an unweighted weighted TLS [6]: The extreme special case when $\boldsymbol{W} = \boldsymbol{I}$ is called unweighted. Then the weighted TLS problem reduces to the TLS problem. The TLS misfit M_{tls} weights equally all elements of the correction matrix $\Delta \boldsymbol{C}$. It is a natural choice when there is no prior knowledge about the data. In addition, the unweighted case is computationally easier to solve than the general weighted case.

The weighted TLS problem has no closed-form solution in terms of the SVD. As an optimization problem, it is non-convex, so that the currently available solution methods do not guarantee convergence to a global optimum solution. Similarly to the TLS and generalized TLS problems, the weighted TLS problem is a double minimization problem. The inner minimization is the search for the best approximation of the data in a given model and an outer minimization is the search for the model [6].

2.3.6 Constrained TLS Problem

The DLS method and the TLS method for solving the matrix equation $\boldsymbol{Ax} = \boldsymbol{b}$ consider the case where the data matrix includes observed errors or noise, but the two methods assume that the errors are iid random variables and have the same variances. However, in some important applications, the noise coefficients of the data matrix $\boldsymbol{A}$ may be statistically correlated or, although statistically independent, they have different variances. The constrained TLS problem considers the former case, and the weighted TLS problem considers the latter. The constrained TLS problem is also called as the structured TLS problem.

The matrix equation $\boldsymbol{A}_{m\times n}\boldsymbol{x}_n = \boldsymbol{b}_m$ can be rewritten as

$$[\boldsymbol{A}; \boldsymbol{b}]\begin{bmatrix} \boldsymbol{x} \\ -1 \end{bmatrix} = 0 \text{ or } \boldsymbol{C}\begin{bmatrix} \boldsymbol{x} \\ -1 \end{bmatrix} = 0, \tag{2.83}$$

where $\boldsymbol{C} = [\boldsymbol{A}; \boldsymbol{b}] \in \Re^{m\times(n+1)}$ is the augmented data matrix.

Consider the noise matrix $\boldsymbol{D} = [\boldsymbol{E},\boldsymbol{e}]$ of $[\boldsymbol{A},\mathbf{b}]$. If the column vectors of the noise matrix are statistically correlated then the column vectors of the correction matrix should also be statistically correlated in order to use an augmented correction matrix $\Delta \boldsymbol{C} = [\Delta \boldsymbol{A},\ \Delta \boldsymbol{b}]$ to suppress the effects of the noise matrix $\boldsymbol{D} = [\boldsymbol{E},\boldsymbol{e}]$.

A simple way to make the column vectors of the correction matrix $\Delta \boldsymbol{C}$ statistically correlated is to let every column vector be linearly dependent on the same vector (e.g., $\boldsymbol{u}$):

$$\Delta C = [G_1 u, ..., G_{n+1} u] \in \Re^{m \times (n+1)}, \tag{2.84}$$

where $G_i \in \Re^{m \times m}$, $i = 1, 2, ..., n + 1$, are known matrices, while u is to be determined.

Definition 2.11 (Constrained TLS problem) [22] Determine a solution vector x and a minimum norm perturbation vector u such that

$$(C + [G_1 u, ..., G_{n+1} u]) \begin{bmatrix} x \\ -1 \end{bmatrix} = \mathbf{0} \tag{2.85}$$

which can be equivalently expressed as the constrained optimization problem

$$\min_{u,x} u^T W u \quad \text{subject to } (C + [G_1 u, ..., G_{n+1} u]) \begin{bmatrix} x \\ -1 \end{bmatrix} = \mathbf{0}, \tag{2.86}$$

where $\mathbf{W}$ is a weighting matrix and is usually diagonal or the identity matrix.

The correction matrix ΔA is constrained as $\Delta A = [G_1 u, \ldots, G_n u]$, while the correction vector Δb is constrained as $\Delta b = G_{n+1} u$. In constrained TLS problems, the linear correlation structure between column vectors of the augmented correction matrix $[\Delta A,\ \Delta b]$ is kept by selecting appropriate matrices $G_i (i = 1,\ldots,n + 1)$. The key to the constrained TLS method is how to choose appropriate matrices $G_1,\ldots,G_{n+1}$, depending on the application.

Theorem 2.11 [22] *Let*

$$\mathbf{W}_x = \sum_{i=1}^{n} x_i G_i - G_{n+1}. \tag{2.87}$$

then the constrained TLS solution is given by

$$\min_x \left\{ F(x) = \begin{bmatrix} x \\ -1 \end{bmatrix}^H C^H (\mathbf{W}_x \mathbf{W}_x^H)^+ C \begin{bmatrix} x \\ -1 \end{bmatrix} \right\}, \tag{2.88}$$

where $\mathbf{W}_x^+$ is the Moore–Penrose inverse matrix of $\mathbf{W}_x$.

2.3.7 Structured TLS Problem

Abatzoglou et al. [20] are considered to be the first who formulated a structured total least squares (STLS) problem. They called their approach constrained TLS and motivated the problem as an extension of the TLS method to matrices with structure.

In the STLS problems [23], the data matrix $[\boldsymbol{A}; \boldsymbol{B}]$ is structured. In order to preserve the maximum likelihood properties of the solution, the TLS problem formulation is extended [24] with the additional constraint that the structure of the data matrix $[\boldsymbol{A}; \boldsymbol{B}]$ is preserved in the correction matrix $[\Delta \boldsymbol{A}; \Delta \boldsymbol{B}]$.

Let $\psi : \Re^{n_p} \to \Re^{m\times(n+d)}$ be an injective function. The vector $\boldsymbol{p}$ for which $\boldsymbol{C} = \psi(\boldsymbol{p})$ is called the parameter vector of the structured matrix $\boldsymbol{C}$. Respectively, $\Re^{n_p}$ is called the parameter space of the structure ψ.

The aim of the STLS problem is to perturb as little as possible a given parameter vector $\boldsymbol{p}$ by a vector $\Delta \boldsymbol{p}$, so that the perturbed structured matrix $\psi(\boldsymbol{p} + \Delta \boldsymbol{p})$ becomes rank deficient with rank at most n.

Definition 2.12 (*Structured total least squares*). Given a data vector $\boldsymbol{p} \in \Re^{n_p}$, a structure specification $\psi : \Re^{n_p} \to \Re^{m\times(n+d)}$, and a rank specification n, solve the optimization problem

$$\Delta \boldsymbol{p}_{stls} = \arg\min_{\Delta \boldsymbol{p}} \|\Delta \boldsymbol{p}\| \text{ subject to rank } (\psi(\boldsymbol{p} - \Delta \boldsymbol{p})) \leq n. \tag{2.89}$$

The STLS problem is said to be affine structured if the function ψ is affine, i.e.,

$$\psi(\boldsymbol{p}) = \boldsymbol{S}_0 + \sum_{i=1}^{n_p} \boldsymbol{S}_i \boldsymbol{p}_i \text{ for all } \boldsymbol{p} \in \Re^{n_p}. \tag{2.90}$$

and for some $\boldsymbol{S}_i$, $i = 1, 2, \ldots, n_p$.

The structured TLS solution, in general, has no closed-form expression in terms of the SVD. In the general case, a structured TLS solution is searched via numerical optimization methods.

2.3.8 TLS in Nonlinear ElV Models

The extension to *nonlinear ElV* models has been studied in depth, various types of nonlinear EIV models, including bilinear, polynomial, nonlinear functional, semi-linear, Volterra series, and Cox's proportional Hazards models, etc., are considered and consistent estimators are derived. In these models, the relationship between the variables is assumed to be nonlinear. It is important to notice here that the close relationship between nonlinear TLS and EIV stops to exist [25].

For example, consider the bilinear EIV model $\boldsymbol{AXB} \approx \boldsymbol{C}$, in which $\boldsymbol{A}$, $\boldsymbol{B}$, and $\mathbf{C}$ are noisy. Applying the TLS principle to this model leads to the following bilinear TLS problem [25]:

$$\begin{aligned} &\min_{\Delta \boldsymbol{A}, \Delta \boldsymbol{B}, \Delta \boldsymbol{C}, \boldsymbol{X}} \|\Delta \boldsymbol{A}\|_F + \|\Delta \boldsymbol{B}\|_F + \|\Delta \boldsymbol{C}\|_F \\ &\text{s.t. } (\boldsymbol{A} - \Delta \boldsymbol{A})\boldsymbol{X}(\boldsymbol{B} - \Delta \boldsymbol{B}) = \boldsymbol{C} - \Delta \boldsymbol{C}. \end{aligned} \tag{2.91}$$

However, solving this problem yields inconsistent estimates of $\boldsymbol{X}$. Later, Kukush has proved that a consistent estimate can be obtained using the adjusted LS (ALS) estimator (the full rank case is considered here for reasons of simplicity):

$$\boldsymbol{X}_{ALS} = (\boldsymbol{A}^T\boldsymbol{A} - \boldsymbol{V}_A)^{-1}(\boldsymbol{A}^T\boldsymbol{C}\boldsymbol{B}^T)(\boldsymbol{B}\boldsymbol{B}^T - \boldsymbol{V}_B)^{-1}. \tag{2.92}$$

With $\boldsymbol{V}_A = \boldsymbol{E}[\tilde{\boldsymbol{A}}^T\tilde{\boldsymbol{A}}]$ and $\boldsymbol{V}_B = \boldsymbol{E}[\tilde{\boldsymbol{B}}^T\tilde{\boldsymbol{B}}]$ where $\boldsymbol{E}$ denotes the expected value and $\tilde{\boldsymbol{A}}$ and $\tilde{\boldsymbol{B}}$ represent the errors on $\boldsymbol{A}$ and $\boldsymbol{B}$ respectively.

2.3.9 TLS Under Non-gaussian Noises

A linear system can be described as

$$y_i = \boldsymbol{x}_i^T\boldsymbol{w}^0, \tag{2.93}$$

where $\boldsymbol{w}^0 \in \Re^{D\times 1}$ is a parameter vector of the system, the input $\boldsymbol{x}_i \in \Re^{D\times 1}$, output $y_i \in \Re$, and they are the non-Gaussian noise signals of set $\{\boldsymbol{x}_\mathrm{i}, y_\mathrm{i}\}$ in ith moment.

The input signal and output signal usually both contain error. In this case, the EIV model is adopted, and the mathematical model is described as

$$\begin{cases} \hat{y}_i = y_i + v_i \\ \hat{\boldsymbol{x}}_i = \boldsymbol{x}_i + \boldsymbol{u}_i \end{cases}, \tag{2.94}$$

where $v_i \in \Re$ is the output noise signal, which is unrelated with the input signal $\boldsymbol{x}_\mathrm{i}$, and $\boldsymbol{u}_i \in \Re^{D\times 1}$ is the non-Gaussian input noise signal and it has the stability and ergodicity..

Under the non-Gaussian noises, based on TLS principal and minimum error entropy (EEC) criterion, the TLS parameters estimation problem is researched [26]. This is called as NG-TLS problem, which can deal with the parameter estimation problem when the input and output both contain the non-Gaussian noises.

Definition 2.13 (*TLS Problem under Non-Gaussian Noises* [26]) The NG-TLS method is designed to minimize Renyi's entropy of the output error, that is

$$\min_{\boldsymbol{w}} H_\alpha(\boldsymbol{w}) = H_\alpha\left(e^{tot}\right), \tag{2.95}$$

where α is the order of Renyi's entropy, e^{tot} is the total errors of the input and output, and $\{e_i^{tot} = (\underset{i}{\breve{y}} - \breve{\boldsymbol{x}}_i\boldsymbol{w})/\sqrt{\|\boldsymbol{w}\|^2 + \varepsilon_0^{-2}}, i = 1, 2, ..., N\}$. The cost function is given as

$$\boldsymbol{J}(\boldsymbol{w}) = H_{\alpha=2}(\boldsymbol{w}) = -\log\int p^2\left(e^{tot}\right)de^{\mathrm{tot}}, \tag{2.96}$$

where $p(e^{tot})$ *is probability distribution function (PDF) of the total errors* e^{tot},$\varepsilon_0 = \sigma_{in}/\sigma_{out}$, *and* σ_{in}^2 *and* σ_{out}^2 *are the input noise intensity and output noise intensity, respectively.*

2.4 Summary

In this chapter, TLS problems including OLS problem, basic TLS problem, multidimensional TLS problem, non-generic unidimensional TLS problem, mixed OLS-TLS problem, basic DLS problem, generalized TLS problem, weighted TLS problem, constrained TLS problem, structured TLS problem, TLS in nonlinear EIV models, TLS under non-Gaussian noises, etc. are discussed and most formulas are presented. This chapter is an introduction to problems addressed in other chapters.

References

1. G. Cirrincione, M. Cirrincione, *Neural-Based Orthogonal Data Fitting-the EXIN Neural Networks* (John Wiley & Sons Inc., Publication, 2010)
2. J. Lampe, *Solving Regularized Total Least Squares Problems Based on Eigenproblems.* Zugl.: Hamburg-Harburg, Techn. Univ., Diss., 2010, ISBN 978-3-86624-504-4
3. S. Van Huffel, J. Vandewalle, *The Total Least Squares Problems: Computational Aspects and Analysis* (Frontiers in Applied Mathematics. SIAM, Philadelphia, 1991)
4. X. D. Zhang, *Matrix Analysis and Applications* (Springer, 2017)
5. P.C. Shen, C.G. Li, Minimum total error entropy method for parameter estimation. IEEE Trans. Signal Process. **63**(15), 4079–4090 (2025)
6. I. Markovsky, Overview of total least-squares methods. Signal Process. **87**, 2283–2302 (2007)
7. A. Taleb, G. Cirrincione, Against the convergence of the minor component analysis neurons. IEEE Trans. Neural Netw. **10**(1), 207–210 (1999)
8. R.R. Hocking, The analysis and selection of variables in linear regression. Biometrics **32**, 1–49 (1976)
9. S. Van Huffel, J. Vandewalle, Subset selection using the total least squares approach in collinearity problems with errors in the variables. Linear Algebra Appl. **88–89**, 695–714 (1987)
10. G.H. Golub, C.F. Van Loan, *Matrix Computations*, 2nd edn. (Johns Hopkins University Press, Baltimore, 1989)
11. S. Van Huffel. *Analysis of the Total Least Squares Problem and Its Use in Parameter Estimation.* Ph.D. dissertation. Department of Electrical Engineering, Katholieke Universiteit Leuven, Leuven, Belgium (1987)
12. G.H. Golub, A. Hoffman, G.W. Stewart, A generalization of the Eckart–Young–Mirsky matrix approximation theorem. Linear Algebra Appl. **88–89**, 322–327 (1987)
13. S. Van Huffel, J. Vandewalle, Analysis and properties of the generalized total least squares problem *ax=b* when some or all columns of *a* are subject to errors. SIAM J. Matrix Anal. Appl. **10**, 294–315 (1989)
14. L.J. Gleser, Estimation in a multivariate "errors in variables" regression model: large sample results. Ann. Stat. **9**, 24–44 (1981)
15. L.J. Gleser. *Calculation and Simulation in Errors-in-Variables Regression Problems.* Mimeograph Ser. 78–5. Statistics Department, Purdue University, West Lafayette, IN, 1978.

16. W.A. Fuller, *Error Measurement Models* (Wiley, New York, 1987)
17. P.P. Gallo, Consistency of regression estimates when some variables are subject to error. Commun. Statistical-Theory and Methods **11**, 973–983 (1982)
18. R.D. Degroat, E. Dowling, The data least squares problem and channel equalization. IEEE Trans. Signal Process. **41**(1), 407–411 (1993)
19. T.J. Abatzoglou, J.M. Mendel, Constrained total least squares. Proc. ICASSP (1987), pp. 1485–1488
20. T.J. Abatzoglou, J.M. Mendel, G.A. Harada, The constrained total least squares technique and its application to harmonic super resolution. IEEE Trans. Signal Process. **39**(5), 1070–1087 (1991)
21. A. Kukush, S. Van Huffel, Consistency of element wise weighted total least squares estimator in a multivariate errors-in-variables model AX=B. Metrika **59**(1), 75–97 (2004)
22. T.J. Abatzoglou, J.M. Mendel, G.A. Harada, The constrained total least squares technique and its applications to harmonic super resolution. IEEE Trans. Signal Process. **39**, 1070–1087 (1991)
23. B. De Moor, Structured total least squares and L_2 approximation problems. Linear Algebra Appl. **188–189**, 163–207 (1993)
24. A. Kukush, I. Markovsky, S. Van Huffel, Consistency of the structured total least squares estimator in a multivariate errors-in-variables model. J. Statist. Plann. Inference **133**(2), 315–358 (2005)
25. S. Van Huffel, P. Lemmerling (eds.), *Total Least Squares and Errors-in-variables Modeling: Analysis Algorithms and Applications* (Kluwer Academic Publishers, Dordrecht, 2002)
26. P. Shen, C. Li, Minimum total error entropy method for parameter estimation. IEEE Trans. Signal Process. **63**(15), 4079–4090 (2015)

Chapter 3
Total Least Square Methods

In this chapter, different typical TLS methods are briefly introduced and important formulas and steps are presented, most of which is based on [1–3]. This is mainly an introduction to methods addressed in other chapters.

3.1 TLS Solution

For the unidimensional case, the over-determined linear equation can be written as follows:

$$\boldsymbol{Ax} \approx \boldsymbol{b}, \tag{3.1}$$

where $\boldsymbol{A} \in \Re^{m\times n}$, $\boldsymbol{b} \in \Re^{m}$, and $\boldsymbol{x} \in \Re^{n}$. The TLS problem can be expressed as the constrained optimization problem

$$\min_{\Delta \boldsymbol{A},\, \Delta \boldsymbol{b}, \boldsymbol{x}} \|\Delta \boldsymbol{A},\ \Delta \boldsymbol{b}\|_F^2 \quad s.t. \quad (\boldsymbol{A} + \Delta \boldsymbol{A})\boldsymbol{x} = \boldsymbol{b} + \Delta \boldsymbol{b} \tag{3.2}$$

or

$$\min_{z} \|\boldsymbol{D}\|_F^2 \quad s.t. \quad \boldsymbol{D}z = -\boldsymbol{B}z, \tag{3.3}$$

where $\boldsymbol{D} = [\Delta \boldsymbol{A},\ \Delta \boldsymbol{b}]$, $\boldsymbol{B} = [\boldsymbol{A},\ \boldsymbol{b}]$ and $z = \begin{bmatrix} \boldsymbol{x} \\ -1 \end{bmatrix}$ is an $(n + 1) \times 1$ vector.

Under the assumption $\|z\|_2 = 1$, from $\|\boldsymbol{D}\|_2 \leq \|\boldsymbol{D}\|_F$ and $\|\boldsymbol{D}\|_2 = \sup_{\|z\|_2=1} \|\boldsymbol{D}z\|_2$, it holds that $\min_{z} \|\boldsymbol{D}\|_F^2 = \min_{z} \|\boldsymbol{D}z\|_2^2$ and thus Eq. (3.3) can be written as follows:

X. Kong and D. Feng, *Efficient Online Learning Algorithms for Total Least Square Problems*, Engineering Applications of Computational Methods 21,
https://doi.org/10.1007/978-981-97-1765-1_3

$$\min_{z} \|Dz\|_2^2 \quad s.t. \quad \|z\| = 1 \tag{3.4}$$

or

$$\min_{z} \|Bz\|_2^2 \quad s.t. \quad \|z\| = 1 \tag{3.5}$$

There are two possible cases in the TLS solution of over-determined equations.

Case 1 Single Smallest Singular Value

$\sigma_n > \sigma_{n+1}$, there is only one smallest singular value.

The TLS problem (3.5) is easily solved via the Lagrange multiplier method. To this end, define the objective function

$$J(z) = \|Bz\|_2^2 + \lambda(1 - z^H z), \tag{3.6}$$

where λ is the Lagrange multiplier. Noticing that $\|Bz\|_2^2 = z^H B^H Bz$, from $\partial J(z)/\partial z^* = 0$, it follows that

$$B^H Bz = \lambda z \tag{3.7}$$

This shows that the Lagrange multiplier should be selected as the smallest eigenvalue $\lambda_{\min}$ of the matrix $B^H B = [A, b]^H [A, b]$ (i.e., the square of the smallest singular value of B), while the TLS solution vector z is the eigenvector corresponding to $\lambda_{\min}$. In other words, the TLS solution vector $[x^T, -1]^T$ is the solution of the Rayleigh quotient minimization problem

$$\min_{x} J(x) = \frac{\begin{bmatrix} x \\ -1 \end{bmatrix}^H [A, b]^H [A, b] \begin{bmatrix} x \\ -1 \end{bmatrix}}{\begin{bmatrix} x \\ -1 \end{bmatrix}^H \begin{bmatrix} x \\ -1 \end{bmatrix}} = \frac{\|Ax - b\|_2^2}{\|x\|_2^2 + 1} \tag{3.8}$$

Let the SVD of the $m \times (n + 1)$ augmented matrix B be $B = U\Sigma V^H$; let its singular values be arranged in the order $\sigma_1 \geq \cdot \geq \sigma_{n+1}$ and let their corresponding right singular vectors be $v_1, \ldots, v_{n+1}$. Then, according to the above analysis, the TLS solution is $z = v_{n+1}$, namely

$$x_{TLS} = -\frac{1}{v_{n+1,n+1}} \begin{bmatrix} v_{1,n+1} \\ \vdots \\ v_{n,n+1} \end{bmatrix}, \tag{3.9}$$

where $v_{i,n+1}$ is the ith entry of $(n + 1)$th column of V.

Case 2 Multiple Smallest Singular Values

Let

$$\sigma_1 \geq \sigma_2 \geq \cdots \geq \sigma_p > \sigma_{p+1} \approx \cdots \approx \sigma_{n+1}, \tag{3.10}$$

and $\boldsymbol{v}_i$ be any column in the subspace $\boldsymbol{S} = \text{Span}\{\boldsymbol{v}_{p+1}, \boldsymbol{v}_{p+2}, \ldots, \boldsymbol{v}_{n+1}\}$. Then above any right singular vector $\boldsymbol{v}_{p+1} = \begin{bmatrix} \boldsymbol{y}_{p+1} \\ \alpha_{p+1} \end{bmatrix}$ gives a TLS solution

$$\boldsymbol{x}_i = -\boldsymbol{y}_{p+i}/\alpha_{p+i}, \quad i = 1, 2, \ldots, n+1-p. \tag{3.11}$$

Hence, there are $n + 1 - p$ TLS solutions. The interesting solution is the TLS solution that is unique in some sense. The possible unique TLS solutions are of two kinds: (1) the minimum norm solution, composed of n parameters; (2) the optimal LS approximate solution, containing only p parameters.

(1) *The Minimum Norm Solution Algorithm:*

The TLS algorithm for finding the minimum norm solution was proposed by Golub and Van Loan [4]; it is given as Algorithm 3.1.

Algorithm 3.1 TLS algorithm for minimum norm solution Input: A: $\boldsymbol{A} \in \mathbb{C}^{m\times n}, \boldsymbol{b} \in \mathbb{C}^m, \alpha > 0$.

repeat

1. Compute $\boldsymbol{B} = [\boldsymbol{A}, \boldsymbol{b}] = \boldsymbol{U\Sigma V}^H$, and save $\boldsymbol{V}$ and all singular values.
2. Determine the number p of principal singular values.
3. Put $\boldsymbol{V}_1 = [\boldsymbol{v}_{p+1}, \ldots, \boldsymbol{v}_{n+1}]$ and compute the Householder transformation

$$\boldsymbol{V}_1\boldsymbol{Q} = \begin{bmatrix} \boldsymbol{y} & \times \\ \alpha & 0 \cdots 0 \end{bmatrix},$$

where α is a scalar, and × denote the irrelevant block.

4. **exit if** $\alpha \neq 0$.

return $p \leftarrow p - 1$.
Output: $\boldsymbol{x}_{\text{TLS}} = -\boldsymbol{y}/\alpha$.

It is should be noticed that the minimum norm solution of the TLS problem contains some redundant parameters that is linearly dependent on other parameters.

(2) *The Optimal LS Approximate Solution*

Because the number of parameters in this solution and the effective rank are the same, it is also called a *low-rank TLS solution*, the basic idea of which was proposed by Cadzow in 1982 [5]. It is given in Algorithm 3.2.

Algorithm 3.2 SVD-TLS algorithm Input: A: $\boldsymbol{A} \in \mathbb{C}^{m\times n}, \boldsymbol{b} \in \mathbb{C}^n$.

1. Compute the SVD $\boldsymbol{B} = [\boldsymbol{A}, \boldsymbol{b}] = \boldsymbol{U\Sigma V}^H$, and save $\boldsymbol{V}$.
2. Determine the effective rank p of $\boldsymbol{B}$.

3. Calculate the $(p+1) \times (p+1)$ matrix $\boldsymbol{S}^{(p)}$ using the following equations.

$$\boldsymbol{v}_k^i = [v(i,k), v(i+1,k), \ldots, v(i+p,k)]^T \tag{3.12}$$

$$\boldsymbol{S}^{(p)} = \sum_{j=1}^{p} \sum_{i=1}^{n+1-p} \sigma_j^2 \boldsymbol{v}_j^i (\boldsymbol{v}_j^i)^H \tag{3.13}$$

4. Compute $\boldsymbol{S}^{-(p)}$ and $\boldsymbol{x}_{\mathrm{TLS}}(i) = -\boldsymbol{S}^{-(p)}(i,1)/\boldsymbol{S}^{-(p)}(p+1,1), i = 1, 2, \ldots, p$.

Output: $\boldsymbol{x}_{\mathrm{TLS}}$.

3.2 Partial TLS Algorithm

From Sect. 3.1, we can see that the TLS solution of $\boldsymbol{AX} \approx \boldsymbol{B}$ is deduced from a basis of the right singular subspace associated with the smallest singular values of [$\boldsymbol{A}$; $\boldsymbol{B}$]. If only the basis vectors desired is calculated, it is possible to save considerable computation time. This can be done in a direct way by modifying the SVD algorithm or in an iterative way by making use of start vectors [2].

An improved algorithm PSVD (partial SVD) is presented in [1], which computes this singular subspace. There are three reasons for its higher efficiency vs. the classical SVD [6]:

1. The Householder transformations of the bidiagonalization are applied only to the basis vectors of the desired singular subspace.
2. The bidiagonal is only partially diagonalized.
3. An appropriate choice is made between QR and QL iteration steps [7].

The computing efficiency of PSVD is related to the motor gap, which is the gap between the singular values associated with the desired and undesired vectors, the desired numerical accuracy and the dimension of the desired subspace. Depending on the motor gap (which must be large), the desired numerical accuracy, and the dimension of the desired subspace (it must be small), PSVD can be three times as fast as the classical SVD, while the same accuracy can be maintained. Incorporating the PSVD algorithm into the TLS computations results in an improved partial TLS (PTLS) [8]. Typically, PTLS can reduce the computation time by a factor of 2.

3.3 Iterative Computation Methods

The iterative method is usually used in the estimation of parameters of non-stationary systems that vary slowly with time, space, or frequency. If the changes in these systems are of small norm but of full rank (e.g., when all elements of the data matrix change slowly from step to step), the computation time can be better reduced by using an iterative method [2]. There are also other advantages in using this type of algorithm [2]:

1. Each step supplies a new and better estimate of the solution, permitting us to control the level of convergence depending on the perturbations of the data.
2. It is easy to code.
3. Some iterative routines use the given matrix over and over again without modification, exploiting its particular structure or sparsity.

3.3.1 Direct Versus Iterative Computation Methods

The direct computation methods are the classical TLS and the PTLS, and their efficiency is determined essentially by the dimensions of the data matrix, the desired accuracy, the dimension p of the desired singular subspace, and the motor gap [2]. In contrast to the iterative methods, the direct methods always converge to the desired solution.

In solving TLS problems, the iterative methods are efficient if:

1. The start matrix is good and the problem is generic.
2. The desired accuracy is low.
3. The dimension p of the desired singular subspace is known.
4. The dimension d of the problem is low.
5. The data matrix dimensions are moderate.
6. The gap is sufficiently large.

3.3.2 Inverse Iteration

The inverse iteration (II) method in (non)symmetric eigenvalue problems have been studied by many researchers, see [6, 9]. According to Wilkinson [10], it is the most powerful and accurate method for computing eigenvectors.

Define $\boldsymbol{S} = \boldsymbol{C}^T\boldsymbol{C}$, where $\boldsymbol{C} = [\boldsymbol{A}; \boldsymbol{B}]$, the iteration matrix $\boldsymbol{Q}_k$ is given by the following:

$$\boldsymbol{Q}_k = (\boldsymbol{S} - \lambda_0\boldsymbol{I})^{-k}\boldsymbol{Q}_0 \tag{3.14}$$

with $\boldsymbol{Q}_0$ is a start matrix and λ_0 is a chosen shift. Given the dimension p of the TLS eigensubspace of $\boldsymbol{S}$, taking λ_0 zero or such that the ratio $\left|\sigma_{n-p}^2 - \lambda_0\right| / \left|\sigma_{n-p+1}^2 - \lambda_0\right|$ is high enough, matrix $\boldsymbol{Q}_k$ converges to the desired minor eigensubspace. The iteration destroys the structure of the matrix $\boldsymbol{S}$ (for details, see [1]). If the motor gap (the gap between σ_{n-p}^2 and σ_{n-p+1}^2) is large, the inverse iteration method has the fast convergence [2].

Table 3.1 Link between the singular value spectrum and the best convergence

	II	OCI	ICI	RQI	LZ	ILZ
Motor gap	Large	Small	Small	Small	Small	Small
Undesired SV spread	Independent	Small	Independent	Independent	Small	Independent
Bounds	No	Optimal	Optimal	No	No	No

3.3.3 *Chebyshev Iteration*

When the motor gap is small, the convergence can be accelerated by applying the Chebyshev polynomials (see [9–12]) to the matrix $\boldsymbol{C}^T\boldsymbol{C}$ for the ordinary Chebyshev iteration (OCI) and to the matrix $(\boldsymbol{C}^T\boldsymbol{C})^{-1}$ for the inverse Chebyshev iteration (ICI).

By choosing the interval $[y, z]$ as small as possible so that it contains all the undesired eigenvalues of a matrix $\boldsymbol{S}$, the ordinary Chebyshev iteration method will converge to a basis of the eigensubspace associated with the remaining eigenvalues outside $[y, z]$, in particular, in multidimensional TLS problems.

The inverse Chebyshev iteration is applied to the matrix $\widehat{\boldsymbol{S}} = (\boldsymbol{C}^T\boldsymbol{C})^{-1}$ and then the interval $[\hat{z}, \hat{y}]$ must be chosen as small as possible such that it contains all undesired eigenvalues of $\widehat{\boldsymbol{S}}$ (i.e., the inverses of the squares of the singular values of $[\boldsymbol{A}; \boldsymbol{B}]$). Only OCI does not alter the matrix $\boldsymbol{C}$.

Usually, for small motor gaps, ICI is recommended. Indeed, this method is proven [1, p. 160] always to converge faster than OCI. ICI is proved [1, pp. 152–157] always to converge faster than II, provided that an optimal estimation of the bound $\hat{y}$ is given. The smaller the motor gap, the larger the gain in speed. Table 3.1 summarizes these comparisons[2].

In Table 3.1, II is referred to the inverse iteration, OCI to the ordinary Chebyshev iteration, ICI to the inverse Chebyshev iteration, RQI to Rayleigh quotient iteration, LZ to the Lanczos methods, and ILZ to the inverse Lanczos method.

3.3.4 *Lanczos Methods*

Lanczos methods are iterative routines that bidiagonalize an arbitrary matrix $\boldsymbol{C}$ or tridiagonalize a symmetric matrix $\boldsymbol{S}$ directly without any orthogonal updates as in the Householder approach to the classical SVD. The original matrix structure is not destroyed and needs little storage; thus, these methods work well for large, sparse, or structured matrices [1]. They are used for TLS problems with a small motor gap; if the Lanczos process is applied to $\boldsymbol{S}$(LZ), it has a minimally faster convergence than OCI with optimal bounds. Like OCI, the convergence rate depends not only on the motor gap but also on the relative spread of the undesired singular value spectrum. If it is applied to $\boldsymbol{S}^{-1}$ [the inverse Lanczos method(ILZ)], it has the same convergence properties as ICI with optimal bounds. However, round-off errors make all Lanczos methods difficult to use in practice [6]. For a summary, see Table 3.1.

3.3.5 Rayleigh Quotient Iteration

The Rayleigh quotient iteration (RQI) can be used to accelerate the convergence rate of the inverse iteration process when the motor gap is small (see Table 3.1), no adequate shift λ_0 can be computed, and convergence to only one singular vector of the desired singular subspace is required.

RQI is a variant of II by applying a variable shift $\lambda_0(k)$, which is the RQ $r(\boldsymbol{q}_k)$ of the iteration vector $\boldsymbol{q}_k$, defined by

$$r(\boldsymbol{q}_k) = \frac{\boldsymbol{q}_k^T \boldsymbol{S} \boldsymbol{q}_k}{\boldsymbol{q}_k^T \boldsymbol{q}_k} \quad \text{minimizing} \quad f(\lambda) = \left\| (\boldsymbol{S} - \lambda \boldsymbol{I}) \boldsymbol{q}_k \right\|_2 \tag{3.15}$$

Then, the RQI is given by the following:

$$(\boldsymbol{S} - \lambda_0(k)\boldsymbol{I})\boldsymbol{q}_k = \boldsymbol{q}_{k-1} \tag{3.16}$$

With regard to II, RQI need not estimate the fixed shift λ_0, because it generates the best shift in each step automatically and therefore converges faster. Parlett [13] has shown that the RQI convergence is ultimately cubic. However, RQI requires more computations per iteration step than II. Furthermore, RQI applies only to square-symmetric matrices $\boldsymbol{S}$ and it is impossible to avoid the explicit formation of $\boldsymbol{S} = \boldsymbol{C}^T \boldsymbol{C}$, which may affect the numerical accuracy of the solution. A good start vector $\boldsymbol{q}_0$ should be available to allow the RQI to converge to the desired solution. If convergence to several basis vectors of the desired singular subspace is required, RQI extensions must be used (e.g., inverse subspace iteration with Ritz acceleration) [14].

3.4 Rayleigh Quotient Minimization Non-neural Methods

The TLS solution is parallel to the right singular vector corresponding to the minimum singular value of $[\boldsymbol{A}; \boldsymbol{b}]$, which can be expressed as a Rayleigh quotient:

$$\sigma_{n+1}^2 = \frac{\left\| [\boldsymbol{A}; \boldsymbol{b}][\hat{\boldsymbol{x}}^T; -1]^T \right\|_2^2}{\left\| [\hat{\boldsymbol{x}}^T; -1]^T \right\|_2^2} = \frac{\sum_{i=1}^m \left| \boldsymbol{a}_i^T \hat{\boldsymbol{x}} - \boldsymbol{b}_i \right|^2}{1 + \hat{\boldsymbol{x}}^T \hat{\boldsymbol{x}}} = \sum_{i=1}^m \|\Delta \boldsymbol{r}_i\|_2^2 \tag{3.17}$$

with $\boldsymbol{a}_i^T$ the ith row of $\boldsymbol{A}$ and $\|\Delta \boldsymbol{r}_i\|_2^2$ the square of the distance from $[\boldsymbol{a}_i^T; \boldsymbol{b}_i]^T \in \Re^{n+1}$ to the nearest point in the subspace:

$$R_r\left([\hat{A};\hat{b}]\right) = \left\{\begin{bmatrix}\tilde{a}\\ \tilde{b}\end{bmatrix}\middle|\tilde{a} \in \Re^n, \tilde{b} \in \Re, \tilde{b} = \hat{x}^T\tilde{a}\right\}$$

Thus, the TLS solution $\hat{x}$ minimizes the sum of the squares of the orthogonal distances (weighted squared residuals):

$$E_{TLS(x)} = \frac{\sum_{i=1}^{m} \left|a_i^T x - b_i\right|^2}{1 + x^T x} \tag{3.18}$$

which is the Rayleigh quotient of $[A; b]^T[A; b]$ constrained to $x_{n+1} = -1$.

From Eqs.(3.18) and (3.17), it is evident that $E_{\text{TLS}}(x)$ corresponds to the Rayleigh quotient (RQ) of $[A;b]$, and therefore, the TLS solution corresponds to its minimization. This minimization is equal to the minor components analysis (MCA), because it corresponds to the search of the eigenvector associated with the minimum eigenvalue of $[A; b]^T[A; b]$, followed by a scaling of the solution into the TLS hyperplane.

3.4.1 Davila's Recursive TLS Algorithm

Among the non-neural methods, the first to use such an idea in a recursive TLS algorithm was Davila [15]; here, an algorithm for recursively computing the TLS solution to the adaptive filtering problem is described. The desired N-dimensional eigenvector associated with the minimum eigenvalue of an augmented sample covariance matrix was updated with a correction vector chosen as a Kalman filter gain vector, and the scalar step size was determined by minimizing RQ. Bose et al. [16] applied recursive TLS to reconstruct high-resolution images from under sampled low resolution noisy multiframes. In [17], the minimization is performed by using the conjugate gradient method, which requires more operations but is very fast and is particularly suitable for large matrices.

3.4.2 Feng's Fast Recursive TLS Algorithm

A new fast recursive total least-squares (N-RTLS) algorithm is proposed in [18] to recursively compute the TLS solution for adaptive FIR filtering. The N-RTLS algorithm is based on the minimization of the constrained Rayleigh quotient (c-RQ) in which the last entry of the parameter vector is constrained to the negative one. The high computational efficiency of the new algorithm depends on the efficient computation of the fast gain vector (FGV) and the adaptation of the c-RQ. Since the last entry of the parameter vector in the c-RQ has been fixed as the negative one, a minimum point of the c-RQ is searched only along the input data vector, and a more efficient N-RTLS algorithm is obtained by using the FGV. In addition, a N-RTLS algorithm to recursively compute the TLS solution for adaptive IIR filtering is proposed in [19].

3.4.3 *Feng's Fast Approximate Inverse Power Iterative TLS Algorithm*

A fast recursive algorithm with a view to finding the TLS solution for adaptive FIR filtering is proposed in [20]. Given the fact that the TLS solution is obtainable via inverse power iteration, a novel but approximate inverse power iteration is introduced in combination with Galerkin method so that the TLS solution can be updated adaptively at a lower computational cost. The TLS solution is constrained the last element of the filter parameter vector to the negative one. By making efficient computation of the fast gain vector and using rank-one update of the augmented autocorrelation matrix, the computational complexity of the developed algorithm is reduced. The proposed algorithm has the simpler structure and may be more easily implemented.

3.5 Neural Networks Methods for TLS

The neural networks applied to TLS problems can be divided into three categories: the neural networks for the MCA, which are described in Chap. 6, the neural networks for the SVD, which are described in Chap. 7, and the neural networks that iterate only in the TLS hyperplane and therefore give the TLS solution directly, which are described in Chap. 8. In the following, the above-mentioned methods can be briefly discussed.

3.5.1 *Neural Networks Methods for MCA*

- *Oja's MCA Algorithms* [21–24].

The well-known algorithms for extracting the first minor component include the anti-Hebbian learning rule (OJA) [21], the normalized anti-Hebbian learning rule (OJAn) [21], OJA + algorithm [22], Luo algorithm [23], and the MCA EXIN algorithm [24], etc. The OJA algorithm tends rapidly to infinite magnitudes of the weights, and the OJAn algorithm leads to better convergence, but it may also lead to infinite magnitudes of weights before the algorithm converges [24].

Consider a linear unit with input $\boldsymbol{x}(t) = [x_1(t), x_2(t), \ldots, x_n(t)]^T$ and output $y(t) = \sum_{i=1}^{n} w_i(t)x_i(t) = \boldsymbol{w}^T(t)\boldsymbol{x}(t)$, where $\boldsymbol{w}(t) = [w_1(t), w_2(t), \ldots, w_n(t)]^T$ is the weight vector. In the MCA analysis, $\boldsymbol{x}(t)$ is a bounded continuous-valued stationary ergodic data vector with finite second moments. The existing well-known learning laws for the MCA EXIN of the autocorrelation matrix $\boldsymbol{R} = E[\boldsymbol{x}(t)\boldsymbol{x}^T(t)]$ of the input vector $\boldsymbol{x}(t)$ are listed below.

$$\boldsymbol{w}(t+1) = \boldsymbol{w}(t) - \frac{\alpha(t)y(t)}{\|\boldsymbol{w}(t)\|_2^2}\left[x(t) - \frac{y(t)\boldsymbol{w}(t)}{\|\boldsymbol{w}(t)\|_2^2}\right] \tag{3.19}$$

In [24], the analysis of the temporal behavior of all the above MCA neurons is analyzed by using not only the ODE approximation, but especially the stochastic discrete laws. Using only the ODE approximation, it does not reveal some of the most important features of these algorithms. For instance, it will be shown that the constancy of the weight modulus for OJAn and Luo, which is the consequence of the use of the ODE, is not valid, except, as a very first approximation, in approaching the minor component. This study has led to the very important problem of sudden divergence [24].

- *Self-stabilizing MCA Algorithms* [25–28].

The concept of self-stability was presented in [25]. If the norm of the weight vector in the algorithm converges toward a fixed value independent of the presented input vector, then this algorithm is called self-stabilizing. Since all algorithms lacking self-stability are prone to fluctuations and divergence in the weight vector norm, the self-stability is an indispensable property for adaptive algorithms.

A self-stabilizing MCA algorithm was given in [26] as follows:

$$\boldsymbol{W}(t+1) = \boldsymbol{W}(t) - \eta[\boldsymbol{x}(t)\boldsymbol{y}^T(t)\boldsymbol{W}^T(t)\boldsymbol{W}(t)\boldsymbol{W}^T(t)\boldsymbol{W}(t) - \boldsymbol{W}(t)\boldsymbol{y}(t)\boldsymbol{y}^T(t)] \tag{3.20}$$

Algorithm (3.20) is self-stabilizing, such that none of $\|\boldsymbol{w}_i(t)\| = 1$ deviates significantly from unity. Other self-stabilizing MCA algorithms see reference [27, 28].

- *Orthogonal OJA's MCA Algorithms* [29–31].

The orthogonal Oja (OOja) algorithm consists of Oja's MSA plus an orthogonalization of $\boldsymbol{W}(t)$ at each iteration [29] $\boldsymbol{W}^{\mathrm{T}}(t)\boldsymbol{W}(t) = \boldsymbol{I}$. A Householder transform-based implementation of the MCA algorithm was given as [31]

$$\boldsymbol{W}(t+1) = \boldsymbol{W}(t) - 2\overline{\overline{\boldsymbol{u}}}(t)\boldsymbol{v}^T(t), \tag{3.21}$$

where $\boldsymbol{v}(t) = \boldsymbol{W}^T(t)\overline{\overline{\boldsymbol{u}}}(t),$, $\overline{\overline{\boldsymbol{u}}}(t) = \boldsymbol{u}(t)/\|\boldsymbol{u}(t)\|$, $\boldsymbol{u}(t) = (1-\vartheta(t))\hat{\boldsymbol{x}}(t)/(\eta\|\boldsymbol{y}(t)\|^2) + \vartheta(t)\boldsymbol{e}(t)$, $\hat{\boldsymbol{x}}(t) = \boldsymbol{W}(t)\boldsymbol{y}(t)$, $\boldsymbol{e}(t) = \boldsymbol{x}(t) - \hat{\boldsymbol{x}}(t)$, $\vartheta(t) = 1/\sqrt{1+\eta^2\|\boldsymbol{e}(t)\|^2\|\boldsymbol{y}(t)\|^2}$, $\boldsymbol{W}$ is initialized as an arbitrary orthogonal matrix and $\boldsymbol{y}$ is given by $\boldsymbol{y}(t) = \boldsymbol{W}^T(t)\boldsymbol{x}(t)$. The OOja is numerically very stable. In addition, the normalized Oja (NOja) [30], the normalized orthogonal Oja (NOOja) have been proposed [30]. OOja, NOja, and NOOja have less computation load.

- *Coupled MCA Algorithms* [32–35].

Among neural network-based MCA algorithms, most previously reviewed do not consider eigenvalue estimates in the update equations of the weights. In [32], Moller provided a framework for a special class of learning rules where eigenvectors and eigenvalues are simultaneously estimated in coupled update equations, and has proved that coupled learning algorithms are solutions for the speed stability problem that plagues most non-coupled learning algorithms. In non-coupled MCA learning algorithms, a larger learning rates may lead to the stability of the numerical procedure through faster convergence. Using small enough learning rates to still guarantee the stability of the numerical procedure, non-coupled MCA algorithms may converge very slowly.

Coupled MCA algorithms can resolve the speed stability problem. Several coupled MCA algorithms have been proposed in [32–35]. The online form of coupled MCA algorithm is given as in [34]

$$\begin{cases} \boldsymbol{w}(t+1) = \boldsymbol{w}(t) + \gamma(t)\left\{\frac{\boldsymbol{Q}(t)\boldsymbol{w}(t)}{\boldsymbol{w}^T(t)\boldsymbol{Q}(t)\boldsymbol{w}(t)} - \boldsymbol{w}(t)\right\} \\ \lambda(t+1) = \lambda(t) + \gamma(t)\left\{\frac{1}{\boldsymbol{w}^T(t)\boldsymbol{Q}(t)\boldsymbol{w}(t)} - \lambda(t)\right\} \end{cases}, \tag{3.22}$$

where $\gamma(t)$ is the learning rate, and $\boldsymbol{Q}(t)$ is updated by

$$\boldsymbol{Q}(t+1) = \frac{t+1}{\alpha t}\left[\boldsymbol{Q}(t) - \frac{\boldsymbol{Q}(t)\boldsymbol{x}(t+1)\boldsymbol{x}^T(t+1)\boldsymbol{Q}(t)}{t + \boldsymbol{x}^T(t+1)\boldsymbol{Q}(t)\boldsymbol{x}^T(t+1)}\right], \tag{3.23}$$

where $0 < \alpha < 1$ denotes the forgetting factor.

3.5.2 Neural Networks Methods for SVD

- *Cross-correlation Neural Networks Model and Deflation Method* [36]

In [36], the cross-coupled Hebbian rule and the APCA networks were proposed, which were used to extract the singular values of the cross-correlation matrix of two stochastic signal vectors or to implement the SVD of a general matrix. The cross-correlation APCA network consists of two sets of neurons that are laterally hierarchically connected. The algorithm extracts the first m principal singular values in the descending order and their corresponding left and right singular vectors. The APCA algorithm can incrementally add nodes without retraining the learned nodes.

- *Parallel SVD Learning Algorithms on Double Stiefel Manifol* [37–40]

Parallel SVD learning algorithms allow to simultaneously compute the SVD vectors. The considered neural algorithms were developed in [38, 40] and have been analyzed in [37]. These algorithms are utilized to train in an unsupervised way a three-layer neural network with the classical "buttery" topology [37–39]. The first layer has connection matrix $\boldsymbol{A}$, the second one has connection matrix $\boldsymbol{B}$, and the middle (hidden) layer provides network's output. It has been shown that when proper initial conditions are chosen, the associated learning trajectories lie on the double Stiefel manifold [37].

- *Double Generalized Hebbian Algorithm (DGHA) for SVD* [41]

In [41], Sanger presented two iterative algorithms for finding the SVD of a matrix $\boldsymbol{P}$ given only samples of the inputs $\boldsymbol{u}$ and outputs $\boldsymbol{y}$. The first algorithm is the double generalized Hebbian algorithm (DGHA), which finds the eigenvectors of the autocorrelation matrix of its inputs $\boldsymbol{y}$, and it is described by two coupled difference equations. The second algorithm is the orthogonal asymmetric encoder (OAE), which converges to the left and right singular vectors of the matrix $\boldsymbol{P}$.

- *Coupled SVD of a Cross-covariance Matrix* [42]

It is known that there exists the speed stability problem in most non-coupled learning algorithms. In order to resolve the speed stability problem that plagues most non-coupled learning algorithms, Kaiser et al. proposed a coupled online coupled learning algorithms for the SVD of a cross-covariance matrix [42]. In this algorithm, the singular value is estimated alongside the singular vectors, and the effective learning rates for the singular vector rules are influenced by the singular value estimates.

3.5.3 Neural Networks that Iterate Only in the TLS Hyperplane

- *The Hopfield-Like Neural Network of Luo, Li, and He* [43, 44].

This network is made up of $3(m + n) + 2$ neurons (m and n are the dimensions of the data matrix) grouped in a main network and in four subnetworks; the output of the main network gives the TLS solution, and the available data matrix and observation vector are taken directly as the interconnections and the bias current of the network. Thus, the principal limit of this network is the fact that it is linked to the dimensions of the data matrix and cannot be used without structural changes for other TLS problems. The authors demonstrate the stability of the network for batch operation: that is, working on all the equations together (as opposed to sequential or online operation, which updates at every equation presentation). The initial conditions cannot be null. The network is based on an analog circuit architecture which has continuous-time dynamics. The authors apply the network to the TLS linear prediction frequency estimation problem.

- *The linear Neuron of Gao, Ahmad, and Swamy* [45, 46].

This is a single linear neuron associated with a constrained anti-Hebbian learning law, which follows from the linearization of $E_{TLS}(\boldsymbol{x})$, so it is correct enough for small gains and, above all, for weight norms much smaller than 1. The weight vector gives the TLS solution after the learning phase. It has been applied to adaptive FIR and IIR parameter estimation problems. In the first problem, after learning, the output of the neuron gives the error signal of the adaptive filter; this is a useful property because in a great many adaptive filter applications the error signal has the same importance as the filter parameters and other signal quantities. From now on, this neuron will be termed *TLS GAO*.

- *The Linear Neurons of Cichocki and Unbehauen* [47, 48].

This is a linear neurons with different learning laws to deal with OLS, TLS, and DLS problems. For each algorithm, an appropriate cost energy to be minimized is devised, which contains the classical minimization [e.g., the minimization of E_{TLS} $(\boldsymbol{x})$ for the TLS] plus regularization terms for ill-conditioned problems and nonlinear functions for the robustness of the method. Then, the system of differential equations describing the gradient flow of this energy is implemented in an analog network for the continuous-time learning law and in a digital network for the discrete-time learning law. The neurons for the TLS do not work on the exact gradient flow, but on its linearization. It gives the same learning law of TLS GAO for a particular choice of the independent signals. The DLS learning law is introduced empirically, without justification.

- *The TLS EXIN Neural Network and its Modified Version* [49, 50].

The TLS EXIN neuron is presented in [49] as a new neural approach for linear parameter estimation. In particular, it is able to implement the TLS technique, thus coping with noisy observation vector and data matrix. With respect to other techniques, both classical and neuronal, it is superior in performance and also, unlike them, it permits the use of such powerful algorithms as SCG and BFGS. All these properties are proved mathematically and not only numerically as is often the case for other neuronal approaches. Furthermore, the use of a sigmoidal activation function would improve the TLS EXIN robustness to outliers. Later, a neural approach for the parameter estimation of adaptive FIR filters for linear system identification is presented in [50]. It is based on a linear neuron with a modified gradient algorithm, capable of resolving the total least-squares (TLS) problem present in this kind of estimation, where noisy errors affect not only the observation vector but also the data matrix.

- *The LS-TLS Neural Algorithm of Bruce and Williamson* [51, 52].

Steepest descent gradient algorithms for unbiased equation error adaptive IIR filtering are analyzed collectively for both the TLS and mixed LS-TLS framework in [51]. These algorithms have a monic normalization that allows for a direct filtering implementation. The algorithms can converge to the desired filter coefficient vector, and a region of initialization is given under which the algorithm converges to the desired solution. In addition, a mixed LS-TLS is introduced in [52] as a QR-decomposition-based algorithm for unbiased, equation error adaptive IIR filtering.

The algorithm is based on casting adaptive IIR filtering into a mixed LS-TLS framework. This formulation is shown to be equivalent to the minimization of the mean-square equation error subject to a unit norm constraint on the denominator parameter vector. An efficient implementation of the mixed LS-TLS solution is achieved through the use of back substitution and inverse iteration.

3.6 Other Methods of Extended TLS Problems

The statistical model that corresponds to the basic TLS approach is the EIV model with the restrictive condition that the measurement errors are zero mean independent and identically distributed. In order to relax these restrictions, several extensions of the TLS problem have been developed, i.e., the mixed LS-TLS problem, the weighted TLS problem, the constrained TLS problem, the structured TLS, the TLS in nonlinear EIV models, TLS under non-Gaussian noises, etc. The methods in Sects. 3.2–3.5 are mainly applied to classical TLS problem. The methods specially for the extended TLS problems are briefly discussed as follows.

3.6.1 Weighted TLS Algorithms

Similarly to the classical TLS estimator, the generalized TLS estimator is computed reliably using the SVD. This is no longer the case for more general weighted TLS problems. Consistency of the weighted total least-squares estimator is proven and an iterative procedure for its computation is proposed in [53]. Special optimization methods for the weighted TLS problem are proposed in [54–58]. The Riemannian SVD framework of De Moor [54] is derived for the structured TLS problem but includes the weighted TLS problem with diagonal weight matrix and $d = 1$ as a special case. De Moor proposed an algorithm resembling the inverse power iteration algorithm for computing the solution. The maximum likelihood PCA method of Wentzell et al. [55] is an alternating LS algorithm. It applies to the general weighted TLS problems and is globally convergent with linear convergence rate. The method of Premoli and Rastello [56] is a heuristic for solving the first-order optimality condition of the weighted TLS. A solution of a nonlinear equation is sought instead of a minimum point of the original optimization problem. The method is locally convergent with superlinear convergence rate. The weighted low-rank approximation framework of Manton et al. [58] proposed specialized optimization methods on a Grassmann manifold.

3.6.2 Constrained TLS Algorithm

Arun [59] addressed the unitarily constrained TLS problem, i.e., $\boldsymbol{AX}\approx\boldsymbol{B}$, subject to the constraint that the solution matrix $\boldsymbol{X}$ is unitary. He proved that this solution is the same as the solution to the orthogonal Procrustes problem [60, p. 582]. Abatzoglou et al. [61] considered another constrained TLS problem, which extends the classical TLS problem to the case where the errors $[\widetilde{\boldsymbol{A}};\tilde{\boldsymbol{b}}]$ are algebraically related. In this case, the TLS solution is no longer statistically optimal (e.g., maximum likelihood in the case of normal distribution). Effective algorithm is presented and it has been applied into harmonic superresolution. More general problem formulations, such as restricted TLS, which also allow the incorporation of equality constraints, have been proposed, as well as equivalent problem formulations using other L_p norms and resulting in the so-called Total L_p approximations. The latter problems proved to be useful in the presence of outliers. Robustness of the TLS solution is also improved by adding regularization, resulting in the regularized TLS methods, or by exploiting the matrix structure (e.g., Toeplitz, Hankel) in the problem formulation, leading to the so-called structured TLS problems. The latter problems are also studied and fast algorithms are presented, by exploiting the matrix structure in the computations [3].

3.6.3 Structured TLS Solution

The structured TLS solution, in general, has no closed form expression in terms of the SVD. An important exception is the circulant structured TLS, which can be solved using the fast Fourier transform, see [62]. In the general case, a structured TLS solution is searched via numerical optimization methods. In [54], Aoki and Yue consider a SISO TLS system identification, they use classical nonlinear LS minimization methods for solving an equivalent unconstrained problem [54]. Cadzow, et al. propose heuristic solution methods that turn out to be suboptimal with respect to the ℓ_2—optimality criterion, see [63, 64]. Tufts and Shah propose a non-iterative method for Hankel structured TLS approximation that is based on perturbation analysis and provides nearly optimal solution for high SNR [64].

Abatzoglou et al. [61] are considered to be the first who formulated a structured TLS problem, and his solution approach is closely related to the one of Aoki and Yue. An equivalent optimization problem is derived, but it is solved numerically using a Newton-type optimization method. De Moor [54] outlines a new framework of structured TLS problem for deriving analytical properties and numerical methods. His approach is based on the Lagrange multipliers and the basic result is an equivalent problem, called Riemannian SVD, which can be considered as a "nonlinear" extension of the classical SVD. An iterative solution method based on the inverse power iteration is proposed.

Another algorithm for solving the structured TLS problem (even with ℓ_1 and ℓ_∞ norm in the cost function), called structured total least norm, is proposed by Rosen et al. [65]. Rosen et al. solve the problem in its original formulation. The constraint is linearized around the current iteration point, which results in a linearly constrained least-squares problem. In the algorithm of Rosen et al., the constraint is incorporated in the cost function by adding a multiple of its residual norm.

In addition, efficient algorithms for problems with block-Hankel/Toeplitz structure and rank reduction with more than one are proposed by Markovsky et al. [66–68]. Regularized structured TLS solution methods are also proposed in [69, 70]. Solution methods for nonlinearly structured TLS methods are also developed in [71, 72].

3.6.4 TLS in Nonlinear ElV Models

Different from the linear EIV model, the nonlinear EIV model has many types. The relation between the input variables and output variables of these nonlinear EIV models is more complex. For different nonlinear models, specific TLS methods or EIV models need to study. Fortunately, various types of nonlinear EIV models, including bilinear, polynomial, nonlinear functional, semi-linear, Volterra series, and Cox's proportional Hazards models, etc. have been studied, and some consistent estimators are derived. The advantages of using Bayesian approaches have also been investigated. The statistical properties of the estimators of the unknown model parameters are discussed and algorithms are derived for computing these estimators in a reliable way [3].

3.6.5 TLS Under Non-Gaussian Noises

In recent years, the TLS methods under non-Gaussian noises are researched. In order to resolve the parameter estimation problem under non-Gaussian noises, Hu et al. proposed a learning theory approach to minimum error entropy criterion (EEC) in [73]. Based on the EEC criterion, Shen et al. [74] proposed a minimum total error entropy method for parameter estimation, which is a non-Gaussian noise TLS (NG-TLS) algorithm. In the Shen's algorithm, the parameters are estimated by minimizing the quadratic Renyi's entropy of the total error, and the batch processing mode and online mode application methods were obtained. Theory analysis shows that NG-TLS algorithm outperform the traditional TLS under the non-Gaussian noises. Recently, some estimation methods of the TLS under the non-Gaussian noises in EIV model have been proposed. The mixed LS-TLS parameter estimation method was proposed under the non-Gaussian noises.

3.7 Conclusions

In this chapter, different typical TLS methods are discussed and important formulas and steps are presented, the focus of which is on the classical TLS methods. This chapter is an introduction to methods addressed in other chapters.

References

1. S. Van Huffel, J. Vandewalle, in *The Total Least Squares Problems: Computational Aspects and Analysis. Frontiers in Applied Mathematics* (SIAM, Philadelphia, 1991)
2. G. Cirrincione, M. Cirrincione, *Neural-Based Orthogonal Data Fitting-the EXIN Neural Networks* (John Wiley & Sons, Inc., Publication, 2010)
3. S. Van Huffel, P. Lemmerling (eds.), *Total Least Squares and Errors-in-variables Modeling: Analysis* (Algorithms and Applications. Kluwer Academic Publishers, Dordrecht, 2002)
4. G.H. Golub, C.F. Van Loan, An analysis of the total least squares problem. SIAM J. Numer. Anal. **17**, 883–893 (1980)
5. J.A. Cadzow, Spectral estimation: an overdetermined rational model equation approach. Proceed. IEEE **70**, 907–938 (1982)
6. G.H. Golub, C.F. Van Loan, *Matrix Computations*, 2nd edn. (Johns Hopkins University Press, Baltimore, 1989)
7. S. Van Huffel, J. Vandewalle, A. Haegemans, An efficient and reliable algorithm for computing the singular subspace of a matrix, associated with its smallest singular values. J. Comput. Appl. Math. **19**, 313–330 (1987)
8. S. Van Huffel, J. Vandewalle, The partial total least squares algorithm. J. Comput. Appl. Math. **21**, 333–341 (1988)
9. B.N. Parlett, *The Symmetric Eigenvalue Problem* (Englewood Cliffs, NJ, Prentice Hall, 1980)
10. J.H. Wilkinson, *The Algebraic Eigenvalue Problem* (Clarendon Press, Oxford, UK, 1965)
11. B. De Moor, J. Vandewalle, An adaptive singular value decomposition algorithm based on generalized Chebyshev recursion, in *Proceedings of the Conference on Mathematics in Signal Processing*, ed. by T.S. Durrani, J.B. Abbiss, J.E. Hudson, R.W. Madan, J.G. McWhirter, and T.A. Moore, (Clarendon Press, Oxford, UK, 1987), pp. 607–635
12. Y. Saad, Chebyshev acceleration techniques for solving nonsymmetric eigenvalue problems. Math. Comput. **42**, 567–588 (1984)
13. B.N. Parlett, The Rayleigh quotient iteration and some generalizations for nonormal matrices. Math. Comput. **28**, 679–693 (1974)
14. P. Comon, G.H. Golub, Tracking a few extreme singular values and vectors in signal processing. Proc. IEEE **78**, 1327–1343 (1990)
15. C.E. Davila, An efficient recursive total least squares algorithm for FIR adaptive filtering. IEEE Trans. Sign. Process. **42**, 268–280 (1994)
16. N.K. Bose, H.C. Kim, H.M. Valenzuela, Recursive total least squares algorithm for image reconstruction. Multidimens. Syst. Sign. Process. **4**, 253–268 (1993)
17. W. Zhu, Y. Wang, Y. Yao, J. Chang, L. Graber, R. Barbour, Iterative total least squares image reconstruction algorithm for optical tomography by the conjugate gradient method. J. Optic. Soc. Am. A:Optics **14**(4), 799–807 (1997)
18. D.Z. Feng, X.D. Zhang, D.X. Chang, W.X. Zheng, A fast recursive total least squares algorithm for adaptive FIR filtering. IEEE Trans. Signal Process. **52**(10), 2729–2737 (2004)
19. D.X. Chang, D.Z. Feng, W.X. Zheng, A fast recursive total least squares algorithm for adaptive IIR filtering. IEEE Trans. Signal Process. **53**(3), 975–965 (2005)
20. D.Z. Feng, W.X. Zheng, Fast approximate inverse power iteration algorithm for adaptive total least-squares FIR filtering. IEEE Trans. Signal Process. **54**(10), 4032–4039 (2006)

21. L. Xu, E. Oja, C. Suen, Modified Hebbian learning for curve and surface fitting. Neural Netw. **5**(3), 441–457 (1992)
22. E. Oja, Principal components, minor components and linear neural networks. Neural Netw. **5**(6), 927–935 (1992)
23. F. Luo, R. Unbehauen, A. Cichocki, A minor component analysis algorithm. Neural Netw. **10**(2), 291–297 (1997)
24. G. Cirrincione, M. Cirrincione, The MCA EXIN neuron for the minor component analysis. IEEE Trans. Neural Netw. **13**(1), 160–187 (2002)
25. R. Möller, A self-stabilizing learning rule for minor component analysis. Int. J. Neural Syst. **14**(1), 1–8 (2004)
26. S.C. Douglas, S.Y. Kung, S. Amari, A self-stabilized minor subspace rule. IEEE Sign. Process. Lett. **5**(12), 328–330 (1998)
27. D.Z. Feng, W.X. Zheng, Y. Jia, Neural network Learning algorithms for tracking minor subspace in high-dimensional data stream. IEEE Trans. Neural Netw. **16**(3), 513–521 (2005)
28. X.Y. Kong, C.H. Hu, C.Z. Han, On the discrete-time dynamics of a class of self-stabilizing MCA extraction algorithms. IEEE Trans. Neural Netw. **21**(1), 175–181 (2010)
29. K. Abed-Meraim, S. Attallah, A. Chkeif, Y. Hua, Orthogonal Oja algorithm. IEEE Sign. Process. Lett. **7**(5), 116–119 (2000)
30. S. Attallah, K. Abed-Meraim, Fast algorithms for subspace tracking. IEEE Sign. Process. Lett. **8**(7), 3–206 (2001)
31. S. Attallah, The generalized Rayleigh's quotient adaptive noise subspace algorithm: a householder transformation-based implementation. IEEE Trans. Circ. Syst. II **53**(1), 3–7 (2006)
32. R. Moller, A. Konies, Coupled principal component analysis. IEEE Trans. Neural Netw. **15**(1), 214–222 (2004)
33. L. Hou, T.P. Chen, Online algorithm of coupled principal (minor) component analysis. J. Fudan Univ. **45**(2), 158–169 (2006)
34. X.W. Feng, X.Y. Kong, H.G. Ma, H.M. Liu, Unified and coupled self-stabilizing algorithm for minor and principal eigen-pair extraction. Neural Process. Lett. (2016). https://doi.org/10.1007/s11063-016-9520-3
35. X.Y. Kong, C.H. Hu, Z.S. Duan. *Principal Component Analysis Networks and Algorithms* (Springer, 2017)
36. K.I. Diamantaras, S.Y. Kung, Cross-correlation neural network models. IEEE Trans. Sign. Process. **42**(11), 3218–3223 (1994)
37. S. Fiori, Singular value decomposition learning on double Stiefel manifolds. Int. J. Neural Syst. **13**(2), 1–16 (2003)
38. A. Weingessel, K. Hornik, SVD algorithms: APEX-like versus subspace methods. Neural Process. Lett. **5**(3), 177–184 (1997)
39. A. Weingessel, *An analysis of learning algorithms in PCA and SVD Neural Networks*. PhD thesis (Technische University at Wien, 1999)
40. U. Helmke, J.B. Moore, Singular-value decomposition via gradient and self-equivalent flows. Linear Algebra Appl. **169**(92), 223–248 (1992)
41. T.D. Sanger, Two iterative algorithms for computing the singular value decomposition from input/output samples, in *Advances in Neural Information Processing Systems*, vol. 6, ed. by J.D. Cowan, G. Tesauro, J. Alspector (Morgan Kaufmann, 1994), pp. 144–151
42. A. Kaiser, M. Schenck, R. Moller, Coupled singular value decomposition-of a cross-covariance matrix. Int. J. Neural Syst. **20**(4), 293–318 (2010)
43. F. Luo, Y. Li, C. He, Neural network approach to the TLS linear prediction frequency estimation problem. Neurocomputing **11**, 31–42 (1996)
44. F. Luo, R. Unbehauen, *Applied Neural Networks for Signal Processing* (Cambridge University Press, New York, 1997)
45. K. Gao, M.O. Ahmad, M.N. Swamy, Learning algorithm for total least-squares adaptive signal processing. Electron. Lett. **28**(4), 430–432 (1992)
46. K. Gao, M.O. Ahmad, M.N. Swamy, A constrained anti-Hebbian learning algorithm for total least-squares estimation with applications to adaptive FIR and IIR filtering. *IEEE Transact. Circ. Syst.-II*. **41**(11), 718–729 (1994)

47. A. Cichocki, R. Unbehauen, *Neural Networks for Optimization and Signal Processing* (Wiley, New York, 1993)
48. A. Cichocki, R. Unbehauen, Simplified neural networks for solving linear least squares and total least squares problems in real time. IEEE Trans. Neural Netw. **5**(6), 910–923 (1994)
49. G. Cirrincione, M. Cirrincione, Linear system identification using the TLS EXIN neuron. Neurocomputing **28**, 53–74 (1999)
50. X.Y. Kong, C.Z. Han, R.X. Wei, Modified gradient algorithm for total least square filtering. Neurocomputing **70**, 568–576 (2006)
51. B.E. Dunne, G.A. Williamson, Analysis of gradient algorithms for TLS-based adaptive IIR filter. IEEE Trans. Sign. Process. **52**(12), 3345–3356 (2004)
52. B.E. Dunne, G.A. Williamson, QR-based TLS and mixed LS-TLS algorithms with applications to adaptive IIR filtering. IEEE Trans. Sign. Process. **51**(2), 386–394 (2003)
53. A. Kukush, S. Van Huffel, Consistency of elementwise weighted total least squares estimator in a multivariate errors-in-variables model AX=B. Metrika **59**(1), 75–97 (2004)
54. B. De Moor, Structured total least squares and L_2 approximation problems. Linear Algebra Appl. **188–189**, 163–207 (1993)
55. P. Wentzell, D. Andrews, D. Hamilton, K. Faber, B. Kowalski, Maximum likelihood principle component analysis. J. Chemom. **11**, 339–366 (1997)
56. A. Premoli, M.-L. Rastello, The parametric quadratic form method for solving TLS problems with elementwise weighting, in *Total Least Squares Techniques and Errors-in-Variables Modeling: Analysis, Algorithms and Applications*, ed. by S. Van Huffel, P. Lemmerling (Kluwer Academic Publishers, Dordrecht, 2002), pp. 67–76
57. I. Markovsky, M.-L. Rastello, A. Premoli, A. Kukush, S. Van Huffel, The element-wise weighted total least squares problem. Comput. Stat. Data Anal. **50**(1), 181–209 (2005)
58. J. Manton, R. Mahony, Y. Hua, The geometry of weighted low-rank approximations. IEEE Trans. Sign. Process. **51**(2), 500–514 (2003)
59. K. Arun, A unitarily constrained total least-squares problem in signal-processing. SIAM J. Matrix Anal. Appl. **13**, 729–745 (1992)
60. G. Golub, C. Van Loan, *Matrix Computations*, 3rd edn. (Johns Hopkins University Press, Baltimore, MD, 1996)
61. T. Abatzoglou, J. Mendel, G. Harada, The constrained total least squares technique and its application to harmonic superresolution. IEEE Trans. Signal Process. **39**, 1070–1087 (1991)
62. A. Beck, A. Ben-Tal, A global solution for the structured total least squares problem with block circulant matrices. SIAM J. Matrix Anal. Appl. **27**(1), 238–255 (2006)
63. B. De Moor, Total least squares for affinely structured matrices and the noisy realization problem. IEEE Trans. Sign. Process. **42**(11), 3104–3113 (1994)
64. D. Tufts, A. Shah, Estimation of a signal waveform from noisy data using low-rank approximation to a data matrix. IEEE Trans. Sign. Process. **41**(4), 1716–1721 (1993)
65. J. Rosen, H. Park, J. Glick, Total least norm formulation and solution of structured problems. SIAM J. Matrix Anal. Appl. **17**, 110–126 (1996)
66. I. Markovsky, S. Van Huffel, R. Pintelon, Block-Toeplitz/Hankel structured total least squares. SIAM J. Matrix Anal. Appl. **26**(4), 1083–1099 (2005)
67. I. Markovsky, S. Van Huffel, A. Kukush, On the computation of the structured total least squares estimator. Numeric. Lin. Algebra Appl. **11**, 591–608 (2004)
68. I. Markovsky, S. Van Huffel, On weighted structured total least squares, in *Large-Scale Scientific Computing, Lecture Notes in Computer Science*, vol. 3743, ed. by I. Lirkov, S. Margenov, J. Was´ niewski (Springer, Berlin, 2006), pp. 695–702
69. N. Younan, X. Fan, Signal restoration via the regularized constrained total least squares. Sign. Process. **71**, 85–93 (1998)
70. N. Mastronardi, P. Lemmerling, S. Van Huffel, Fast regularized structured total least squares algorithm for solving the basic deconvolution problem. Numeric. Linear Algebra Appl. **12**(2–3), 201–209 (2005)
71. J. Rosen, H. Park, J. Glick, Structured total least norm for nonlinear problems. SIAM J. Matrix Anal. Appl. **20**(1), 14–30 (1998)

72. P. Lemmerling, S. Van Huffel, B. De Moor, The structured total least squares approach for nonlinearly structured matrices. Numeric. Linear Algebra Appl. **9**(1–4), 321–332 (2002)
73. T. Hu, J. Fan, Q. Wu et al., Learning theory approach to minimum error entropy criterion. J. Mach. Learn. Res. **14**(1), 377–397 (2013)
74. P. Shen, C. Li, Minimum total error entropy method for parameter estimation. IEEE Trans. Sign. Process. **63**(15), 4079–4090 (2015)

Chapter 4
Fast Recursive TLS Algorithms

4.1 Introduction

The recursive least squares (RLS) algorithm has been applied extensively in adaptive signal processing areas including the adaptive filtering, online system identification, adaptive equalization, adaptive spectrum estimation, adaptive noise canceling, and so on [1]. The RLS algorithm has many desirable properties, for example, it can track variation of system parameters, and if only the system output includes a white Gaussian noise sequence, it can get unbiased estimates of system parameters. However, if both the system input and output are corrupted by white Gaussian noise, then the RLS algorithm can only provide biased estimates of the system parameters, which decreases the performance of adaptive filtering.

In [2–4], the TLS problems and their basic performances have been extensively studied. However, the study on TLS solutions is still insufficient, and their application in signal processing is limited, perhaps due to a lack of efficient algorithms for solving a TLS problem online and/or offline. In general, the solution of a TLS problem can be obtained by the SVD of a matrix [3, 5]. Since the computational complexity $O(N^3)$ is large, the application of the TLS methods is limited in practice, especially in real-time signal processing. To adaptively compute the generalized eigenvector associated with the smallest eigenvalue of an autocorrelation matrix, a number of algorithms have been proposed in the context of Pisarenko spectral estimation [6]. These algorithms fall into two broad categories.

The first category involves stochastic-type adaptive algorithms. Thompson [7] proposed an adaptive algorithm that is used to extract a single minor Eigen component and can be applied to find the TLS solutions of adaptive filtering and online system identification. Other similar algorithms have also been proposed by many authors in [8–17], all leading to an adaptive implementation of Psiarenko's harmonic retrieval estimator. However, the above stochastic-type adaptive algorithms have no equilibrium point under the persistent excitation condition and with the constant learning rate, as shown in [18]. In contrast, the total least mean squares (TLMS)

X. Kong and D. Feng, *Efficient Online Learning Algorithms for Total Least Square Problems*, Engineering Applications of Computational Methods 21,
https://doi.org/10.1007/978-981-97-1765-1_4

algorithm developed in [19] has an equilibrium point under the persistent excitation condition [20]. In general, the stochastic-type algorithms have simple structure and require only $O(N)$ multiplication per iteration but have a relatively slow convergence speed compared with the following second category of algorithms.

A large variety of the second-category algorithms are called the recursive total least squares (RTLS) algorithms, which usually have $O(N^2)$ computational complexity per iteration. In particular, for online solution of the TLS problems in adaptive FIR filtering, Davila [21] proposed a fast RTLS algorithm that can fast track the eigenvector associated with the smallest eigenvalue of the augmented correlation matrix. Since the Kalman gain vector can be fast estimated by utilizing the shift structure of the input data vector [22], the computational complexity of Davila's algorithm is $O(N)$ per iteration. When both the system input and output are corrupted by white Gaussian noise, the RTLS algorithms yield unbiased estimates, and thus, their performances are better than the well-known RLS algorithm [1].

It should be pointed out that computation of the Kalman gain vector may be potentially unstable [23]. Some efficient solution approaches [24, 25] were developed to overcome the potential instability of the Kalman gain vector, which results in the more complex structure and increases the computational complexity. By using the fast gain vector (FGV), the numerically stable fast transversal filter algorithms were established in [24] and [25].

The aim of this chapter is to introduce and develop two RTLS algorithms by using the fast computation of the FGV and the adaptation minimization of the constrained Rayleigh quotient (c-RQ), the one is to recursively compute the TLS solution for adaptive FIR filtering, the other is to recursively compute the TLS solution for adaptive IIR filtering. The proposed RTLS algorithms are based on the minimization of the c-RQ in which the last/first entry of the parameter vector is constrained to the negative one. The high computational efficiency of the new algorithms depends on the efficient computation of the FGV and the adaptation of the c-RQ. The computational load of the FGV algorithms is smaller than that of the fast Kalman algorithm. The global convergence of the new algorithms is studied by LaSalle's invariance principle. Moreover, the new algorithms are numerically stable since they do not use the matrix inversion lemma.

In this chapter, the existing RTLS algorithms will be reviewed and discussed. Two algorithms proposed by us will be analyzed in detail. The remainder of this chapter is organized as follows. An overview of the existing RTLS algorithms is presented in Sect. 4.2. A fast recursive TLS algorithm for adaptive FIR filtering is presented in Sect. 4.3. A fast recursive TLS algorithm for adaptive IIR filtering is proposed in Sect. 4.4, followed by summary in Sect. 4.5.

4.2 Review of Davila's RTLS Algorithm for FIR Adaptive Filtering

Among the non-neural methods, the first to use such an idea in a recursive TLS algorithm was Davila [21]; here, an algorithm for recursively computing the TLS solution to the adaptive filtering problem is described. The desired N-dimensional eigenvector associated with the minimum eigenvalue of an augmented sample covariance matrix was updated with a correction vector chosen as a Kalman filter gain vector, and the scalar step size was determined by minimizing RQ.

4.2.1 RLS Filter Bias and Mean Squared Error for IR Estimation

Consider the IR estimation configuration in Fig. 4.1.

The unknown system impulse response is assumed to be the $M \times 1$ vector

$$\boldsymbol{\theta}^* = [\, b_0^* \; b_1^* \; \cdots \; b_{M-1}^* \,]^T . \tag{4.1}$$

These quantities may be time-varying, but will be assumed constant here. The desired signal at time t is given by

$$d(t) = \phi_t^T \boldsymbol{\theta}^* + n_o(t), \tag{4.2}$$

where the observation noise $n_0(t)$ is a zero-mean white noise process, independent of the input vector with variance σ_o^2. The input vector ϕ_t at time t is given by

$$\phi_t = [\, x(t) \; x(t-1) \; \cdots \; x(t-M+1) \,]^T . \tag{4.3}$$

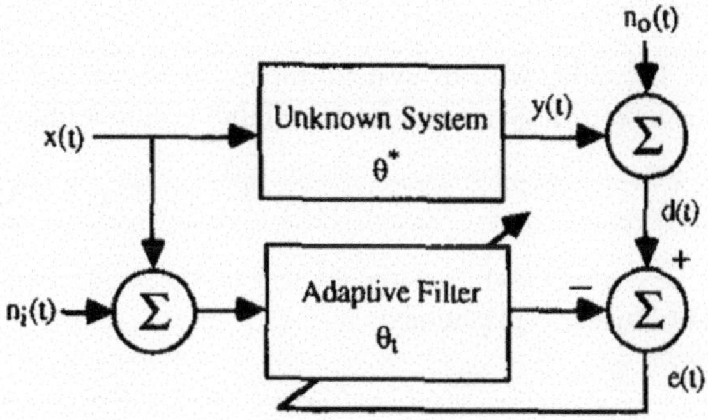

Fig. 4.1 Impulse response estimation structure showing input vector noise $n_i(t)$

First consider the bias in the RLS filter coefficients resulting from input vector noise. The adaptive filter is assumed to be engaged in impulse response (IR) estimation as in Fig. 4.1. It is assumed that the input vector noise $n_i(t)$ is zero-mean white noise having variance σ_i^2, independent of the observation noise $n_o(t)$, and the unknown system input ϕ_t. The observation noise $n_o(t)$ is also assumed to be independent, white, and zero mean with variance σ_o^2. Further assume that the input data is drawn from a persistently exciting random process of at least order M. This latter assumption is common for most adaptive filtering applications. The resulting RLS normal equation at time t is given by

$$\boldsymbol{R}_t \boldsymbol{\theta}_t = \boldsymbol{p}_t, \tag{4.4}$$

where

$$\boldsymbol{R}_t = \frac{1}{t} \sum_{j=1}^{t} \boldsymbol{\gamma}_j \boldsymbol{\gamma}_j^T, \tag{4.5}$$

and

$$\boldsymbol{p}_t = \frac{1}{t} \sum_{j=1}^{t} d(j) \boldsymbol{\gamma}_j, \tag{4.6}$$

where $\boldsymbol{\gamma}_j$ is the noisy adaptive filter input at time j

$$\boldsymbol{\gamma}_j = \left[x(j) + n_i(j) \; x(j-1) + n_i(j-1) \cdots x(j-M+1) + n_i(j-M+1)^T \right], \tag{4.7}$$

and $d(j)$ is the desired response at time j

$$d(j) = \phi_j^T \boldsymbol{\theta}^* + n_o(j). \tag{4.8}$$

The unknown FIR system impulse response is $\boldsymbol{\theta}^* = [b_0^* \; b_1^* \cdots b_{M-1}^*]^T$, while the unknown system (noise-free) input vector at time j is

$$\phi_j = [x(j) \; x(j-1) \cdots x(j-M+1)]^T. \tag{4.9}$$

Assuming that t is large enough to replace the deterministic quantities (4.5) and (4.6) by the expectations

$$E[\boldsymbol{\gamma}_1 \boldsymbol{\gamma}_t^T] = \boldsymbol{R}_\phi + \sigma_i^2 \boldsymbol{I}_{M \times M}, \tag{4.10}$$

and

$$E[d(t)\boldsymbol{\gamma}_t] \equiv p, \tag{4.11}$$

respectively, the normal Eq. (4.4) may be written

$$(\boldsymbol{R}_\phi + \sigma_i^2 \boldsymbol{I}_{M\times M})\boldsymbol{\theta}_t = \boldsymbol{R}_\phi \boldsymbol{\theta}^*, \tag{4.12}$$

where

$$\boldsymbol{R}_\phi = E[\phi_t \phi_t^T], \tag{4.13}$$

and the facts that $n_\mathrm{i}(t)$ and $n_\mathrm{o}(t)$ are independent and independent of ϕ_t over all t were used. The persistently exciting assumption on $\phi(t)$ insures that $\boldsymbol{R}_\phi$ is positive definite. Applying the matrix inversion lemma to $\boldsymbol{R}_\phi + \sigma_i^2 \boldsymbol{I}_{M\times M}$ gives

$$[\boldsymbol{R}_\phi + \sigma_i^2 \boldsymbol{I}_{M\times M}]^{-1} = \boldsymbol{R}_\phi^{-1} - \boldsymbol{R}_\phi^{-1}(\boldsymbol{R}_\phi^{-1} + \sigma_i^{-2}\boldsymbol{I}_{M\times M})^{-1}\boldsymbol{R}_\phi^{-1}. \tag{4.14}$$

The resulting bias term is then seen to be

$$\theta_{\mathrm{bias}} = \boldsymbol{R}_\phi^{-1}(\boldsymbol{R}_\phi^{-1} + \sigma_i^{-2}\boldsymbol{I}_{M\times M})^{-1}\boldsymbol{\theta}^*. \tag{4.15}$$

Given the same assumptions on noise terms $n_\mathrm{i}(t)$ and $n_\mathrm{o}(t)$ made above, it is found that the asymptotic mean squared error associated with the normal equation in (4.12) is given by

$$\tilde{e} = \sigma_d^2 - \boldsymbol{p}^T(\boldsymbol{R}_\phi + \sigma_i^2\boldsymbol{I}_{M\times M})^{-1}\boldsymbol{p}. \tag{4.16}$$

Since the mean squared error resulting from no input vector noise is

$$e = \sigma_d^2 - \boldsymbol{p}^T\boldsymbol{R}_\phi^{-1}\boldsymbol{p}. \tag{4.17}$$

From (4.14), the mean squared error term becomes

$$\tilde{e} = e + \boldsymbol{\theta}^{*T}(\boldsymbol{R}_\phi^{-1} + \sigma_i^{-2}\boldsymbol{I}_{M\times M})^{-1}\boldsymbol{\theta}^*. \tag{4.18}$$

The second term in (4.18) is positive, consequently, $\tilde{e} > e$

4.2.2 *TLS for Adaptive FIR Filters*

Consider the classical linear regression problem: given a set of t data points each of which consists of M possibly noisy independent variables $\boldsymbol{\gamma}_j$ (see Eq. (7.7), $j = 1$, $2,\ldots,t$) and 1 associated dependent variable $d(j)$, $j = 1, 2, \ldots, t$, possibly containing a noise term $n_o(j)$, the problem is one of fitting t data points, denoted by

$$\bar{\boldsymbol{\gamma}}_j = [d(j), \boldsymbol{\gamma}_j^T]^T, \ j = 1, 2, ..., t \tag{4.19}$$

in $(M + 1)$-dimensional Euclidean vector space with a hyperplane containing the points

$$\bar{\boldsymbol{\theta}}_j = [\boldsymbol{\gamma}_j^T \boldsymbol{\theta}, \boldsymbol{\gamma}_j^T]^T, \ j = 1, 2, ..., t. \tag{4.20}$$

The $M \times 1$ vector

$$\boldsymbol{\theta} = [\, b_0 \ b_1 \ \cdots \ b_{M-1} \,]^T \tag{4.21}$$

constitutes the sought-after regression coefficients.

Define the vector $\bar{\boldsymbol{\theta}}_\perp = [1, -\boldsymbol{\theta}^T]^T$ which is seen to be normal to the sought-after hyperplane. The generalized linear regression problem is then given by

$$\min_{\boldsymbol{\theta}} \frac{1}{t} \sum_{j=1}^{t} \left\| \boldsymbol{P}_\theta \bar{\boldsymbol{\gamma}}_j \right\|^2 = \min_{\boldsymbol{\theta}} \frac{1}{t} \sum_{j=1}^{t} \frac{\bar{\boldsymbol{\theta}}_\perp^T \bar{\boldsymbol{\gamma}}_j \bar{\boldsymbol{\gamma}}_j^T \bar{\boldsymbol{\theta}}_\perp}{\bar{\boldsymbol{\theta}}_\perp^T \bar{\boldsymbol{\theta}}_\perp}, \tag{4.22}$$

which is recognized as the minimum generalized eigenvalue $\lambda_{\min}^*(t)$ of the sample extended covariance matrix

$$\bar{\boldsymbol{R}}_t = \frac{1}{t} \sum_{j=1}^{t} \bar{\boldsymbol{\gamma}}_j \bar{\boldsymbol{\gamma}}_j^T. \tag{4.23}$$

Theorem 4.1 *The TLS solution given by (4.22) results in asymptotically unbiased, consistent, FIR parameter estimates when the filter input contains additive, zero mean, independent white noise. The total mean squared error associated with the TLS estimate given by (4.22) is* $\lambda_{\min}^*(t) = \sigma_i^2$.

Proof See reference [21].

4.2.3 Recursive TLS Algorithm

The algorithm is based on the following update at time t for the estimated $N = M + 1$ dimensional generalized eigenvector $\bar{\boldsymbol{q}}_t$:

$$\bar{\boldsymbol{q}}_t = \bar{\boldsymbol{q}}_{t-1} + \alpha(t) \bar{\boldsymbol{\Psi}}_t, \tag{4.24}$$

where $\bar{\boldsymbol{\Psi}}_1$ is a given $N \times 1$ vector at time t and the scalar $\alpha(t)$ is chosen so that $\bar{\boldsymbol{q}}_t$ minimizes the generalized Rayleigh quotient

$$\mu(\bar{q}_t) = \frac{\bar{q}_t^T \bar{R}_t \bar{q}_t}{\bar{q}_t^T \bar{D} \bar{q}_t}, \tag{4.25}$$

where $\bar{D}$ is an identity matrix with $(N+1) \times (N+1)$ dimension.

Substituting (4.24) into (4.25) and differentiating with respect to $\alpha(t)$, it is straightforward to show that the value of $\alpha(t)$ for which (4.25) is minimized at time t is the smallest root of the quadratic polynomial

$$a\alpha(t)^2 + b\alpha(t) + c = 0, \tag{4.26}$$

where

$$a = \bar{q}_{t-1}^T \bar{R}_t \bar{\psi}_t \bar{\psi}_t^T \bar{D} \bar{\psi}_t - \bar{\psi}_t \bar{R}_t \bar{\psi}_t \bar{q}_{t-1}^T \bar{D} \bar{\psi}_t, \tag{4.27}$$

$$b = \bar{q}_{t-1}^T \bar{R}_t \bar{q}_{t-1} \bar{\psi}_t^T \bar{D} \bar{\psi}_t - \bar{\psi}_t^T \bar{R}_t \bar{\psi}_t \bar{q}_{t-1}^T \bar{D} \bar{q}_{t-1}, \tag{4.28}$$

$$c = \bar{q}_{t-1}^T \bar{R}_t \bar{q}_{t-1} \bar{q}_{t-1}^T \bar{D} \bar{\psi}_t - \bar{q}_{t-1}^T \bar{R}_t \bar{\psi}_t \bar{q}_{t-1}^T \bar{D} \bar{q}_{t-1}. \tag{4.29}$$

Choosing the update direction $\bar{\Psi}_1$ to be the Kalman gain vector

$$\bar{k}_t = \bar{R}_t^{-1} \bar{\gamma}_t. \tag{4.30}$$

The Kalman gain vector $\bar{k}_t$ can also be computed in $O(N)$ multiplications using a multichannel fast transversal filters (FTF) algorithm. The choice of $\bar{k}_t$ as the update direction reduces the following quadratic forms in (4.27)–(4.29) to the inner products

$$\bar{q}_{t-1}^T \bar{R}_t \bar{\Psi}_t = \bar{q}_{t-1}^T \bar{\gamma}_t, \tag{4.31}$$

$$\bar{\Psi}_t^T \bar{R}_t \bar{\Psi}_t = \bar{k}_t^T \bar{\gamma}_t. \tag{4.32}$$

Since $\bar{R}_t = ((t-1)/t)\xi \bar{R}_{t-1} + (1/t)\bar{\Phi}_t \bar{\Phi}_t^T$, it follows that the remaining quadratic form $\bar{q}_{t-1}^T \bar{R}_t \bar{q}_{t-1}$ can be efficiently computed

$$\bar{q}_{t-1}^T \bar{R}_t \bar{q}_{t-1} = \frac{(t-1)}{t} \xi \lambda_{\min}(t-1) \bar{q}_{t-1}^T \bar{D} \bar{q}_{t-1} + \frac{1}{t} (\bar{q}_{t-1}^T \bar{\Phi}_t)^2 + \frac{\delta}{t}, \tag{4.33}$$

where $\lambda_{\min}(t-1)$ is the minimum Rayleigh quotient estimate at time t-1 and δ is a small positive constant that keeps $\bar{R}_t$ positive definite during the first few iterations. The constant δ plays the same role as the initial sample covariance matrix used in matrix inversion lemma-based RLS algorithms. The parameter vector estimate can then be obtained from

$$\boldsymbol{\theta}_t = \frac{-[\bar{\boldsymbol{q}}_t]_{2,N}}{[\bar{\boldsymbol{q}}_t]_1}. \tag{4.34}$$

The RTLS algorithm is as follows:

Initialize: $\boldsymbol{q}_1 = [N^{-0.5}, N^{-0.5}, \ldots, N^{-0.5}]^T$, $\lambda_{\min}(1) = 100$, $\delta = 0.001$, $N = M + 1$, initialize FTF.

For $t = 1, \ldots$

1. update extended data vector $\bar{\boldsymbol{\Phi}}_t$
2. update Kalman gain vector $\bar{\boldsymbol{k}}_t$ using stabilized FTF
3. $\mathring{\lambda}_{\min}(t) = ((t-1)/t)\xi\lambda_{\min}(t-1) + (1/t)(\bar{\boldsymbol{q}}_{t-1}^T\bar{\boldsymbol{\Phi}}_t)^2 + (\delta/t)$
4. $a = (\bar{\boldsymbol{k}}_t^T\bar{\boldsymbol{\Phi}}_t)(\bar{\boldsymbol{q}}_{t-1}^T\bar{\boldsymbol{D}}\bar{\boldsymbol{k}}_t) - (\bar{\boldsymbol{q}}_{t-1}^T\bar{\boldsymbol{\Phi}}_t)(\bar{\boldsymbol{k}}_t^T\bar{\boldsymbol{D}}\bar{\boldsymbol{k}}_t)$
5. $b = (\bar{\boldsymbol{k}}_t^T\bar{\boldsymbol{\Phi}}_t) - \mathring{\lambda}_{\min}(t)(\bar{\boldsymbol{k}}_t^T\bar{\boldsymbol{D}}\bar{\boldsymbol{k}}_t)$
6. $c = (\bar{\boldsymbol{q}}_{t-1}^T\bar{\boldsymbol{\Phi}}_t) - \mathring{\lambda}_{\min}(t)(\bar{\boldsymbol{q}}_{t-1}^T\bar{\boldsymbol{D}}\bar{\boldsymbol{k}}_t)$
7. $\alpha(t) = ((-b + \sqrt{b^2 - 4ac})/2a)$
8. $\lambda_{\min}(t) = \frac{\bar{\boldsymbol{q}}_{t-1}^T\boldsymbol{R}_t\bar{\boldsymbol{q}}_{t-1} + 2\alpha(t)\bar{\boldsymbol{q}}_{t-1}^T\boldsymbol{R}_t\bar{\boldsymbol{\psi}}_t + \alpha^2(t)\bar{\boldsymbol{\psi}}_t^T\boldsymbol{R}_t\bar{\boldsymbol{\psi}}_t}{\bar{\boldsymbol{q}}_{t-1}^T\bar{\boldsymbol{D}}\bar{\boldsymbol{q}}_{t-1} + 2\alpha(t)\bar{\boldsymbol{q}}_{t-1}^T\bar{\boldsymbol{D}}\bar{\boldsymbol{\psi}}_t + \alpha^2(t)\bar{\boldsymbol{\psi}}_t^T\bar{\boldsymbol{D}}\bar{\boldsymbol{\psi}}_t}$
9. $\tilde{\boldsymbol{q}}_t = \tilde{\boldsymbol{q}}_{t-1} + \alpha(t)\bar{\boldsymbol{k}}_t$
10. $\bar{\boldsymbol{q}}_t = \tilde{\boldsymbol{q}}_t/\sqrt{(\tilde{\boldsymbol{q}}_t^T\tilde{\boldsymbol{q}}_t)}$
11. $\boldsymbol{\theta}_t = -[\bar{\boldsymbol{q}}_t]_{2,N}/[\bar{\boldsymbol{q}}_t]_1$

4.3 A Fast Recursive TLS Algorithm for Adaptive FIR Filtering

This section proposes a new fast recursive total least squares (N-RTLS) algorithm to recursively compute the TLS solution for adaptive FIR filtering. The N-RTLS algorithm is based on the minimization of the c-RQ in which the last entry of the parameter vector is constrained to the negative one. The high computational efficiency of the new algorithm depends on the efficient computation of the FGV and the adaptation of the c-RQ. Since the last entry of the parameter vector in the c-RQ has been fixed as the negative one, a minimum point of the c-RQ is searched only along the input data vector, and a more efficient N-RTLS algorithm is obtained by using the FGV. As compared with Davila's RTLS algorithms, the N-RTLS algorithm saves the 6 number of multiplies, divides, and square roots (MADs). The global convergence of the new algorithm is studied by LaSalle's invariance principle.

4.3.1 TLS Problems in Signal Processing

A. *Signal Model*

Consider an unknown system with FIR, and assume that both the input and output are corrupted by the additive white Gaussian noise (AWGN). We use an adaptive FIR filter to estimate the FIR system from noisy observations of the input and output, as shown in Fig. 4.2.

The FIR vector of the unknown system is described by

$$\boldsymbol{h} = [h_0, h_1, ..., h_{M-1}]^T \in \Re^{M\times 1}, \tag{4.35}$$

where $\boldsymbol{h}$ may be time-varying. The desired output is given by

$$d(t) = \boldsymbol{x}^T(t)\boldsymbol{h} + n_o(t), \tag{4.36}$$

where $n_o(t)$ is an AWGN with zero mean and variance σ_o^2 at the output and independent of the input signal. Moreover, the noise-free signal vector $\boldsymbol{x}(t) \in \Re^{M\times 1}$ is defined as

$$\boldsymbol{x}(t) = [x(t), x(t-1), ..., x(t-M+1)]^T, \tag{4.37}$$

while the noisy input vector of the adaptive FIR filter $\tilde{\boldsymbol{x}}(t) \in \Re^{M\times 1}$ is given by

$$\begin{aligned} \tilde{\boldsymbol{x}}(t) &= [\tilde{x}(t), \tilde{x}(t-1), ..., \tilde{x}(t-M+1)]^T \\ &= [x(t)+n_i(t), x(t-1)+n_i(t-1), ..., x(t-M+1)+n_i(t-M+1)]^T \\ &= \boldsymbol{x}(t)+\boldsymbol{n}_i(t), \end{aligned} \tag{4.38}$$

where $\boldsymbol{n}_i(t) = [n_i(t), n_i(t-1), ..., n_i(t-M+1)]^T$, and $n_i(t)$ is an AWGN with zero mean and variance σ_i^2. Notice that the input noise may originate from the measured error, interference, quantized noise, and so on. Hence, we adopt the more general

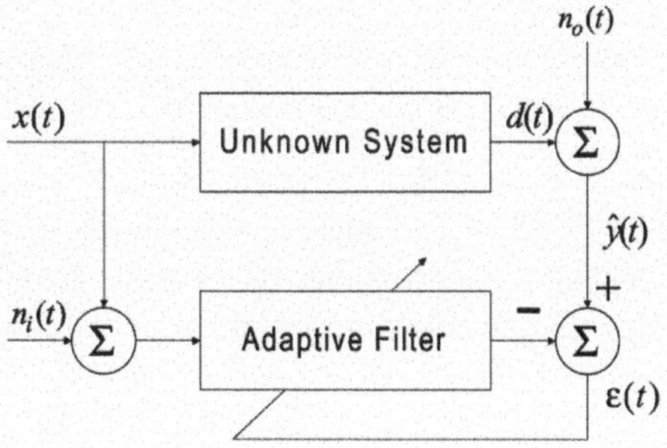

Fig. 4.2 Identification of unknown system $h(k)$ ($k = 0, 1, ..., N-1$) by using adaptive FIR filter

signal model than the adaptive LS-based filtering [1]. Moreover, the augmented data vector is defined as

$$\bar{\boldsymbol{x}}(t) = [\tilde{\boldsymbol{x}}^T(t), d(t)]^T \in \Re^{(M+1)\times 1}. \tag{4.39}$$

For convenience of analysis, the following several matrices is defined. The autocorrelation matrix of the noise-free input vector is given by

$$\boldsymbol{R} = E\{\boldsymbol{x}(t)\boldsymbol{x}^T(t)\}, \tag{4.40}$$

and the autocorrelation matrix of the noisy input vector is described by

$$\tilde{\boldsymbol{R}} = E\{\tilde{\boldsymbol{x}}(t)\tilde{\boldsymbol{x}}^T(t)\} = \boldsymbol{R} + \sigma_i^2 \boldsymbol{I}. \tag{4.41}$$

Then, the autocorrelation matrix of the augmented data vector can be represented as

$$\bar{\boldsymbol{R}} = E\{\bar{\boldsymbol{x}}(t)\bar{\boldsymbol{x}}^T(t)\} = \begin{bmatrix} \tilde{\boldsymbol{R}} & \boldsymbol{b} \\ \boldsymbol{b}^T & c \end{bmatrix}, \tag{4.42}$$

where $\boldsymbol{b} = E\{\tilde{\boldsymbol{x}}(t)d(t)\}$ and $c = E\{d^2(t)\}$. It is easy to show that

$$\boldsymbol{b} = E\{[\boldsymbol{x}(t) + \boldsymbol{n}_i(t)][\boldsymbol{x}^T(t)\boldsymbol{h} + n_o(t)]\} = \boldsymbol{Rh}, \tag{4.43}$$

$$c = E\{[\boldsymbol{h}^T\boldsymbol{x}(t) + n_o(t)][\boldsymbol{x}^T(t)\boldsymbol{h} + n_o(t)]\} = \boldsymbol{h}^T\boldsymbol{Rh} + \sigma_o^2. \tag{4.44}$$

It is seen from (4.43) that if the autocorrelation matrix of the noise-free input vector can be directly estimated, we can obtain an unbiased estimate of the FIR. However, when the input vector contains the additive noise, we cannot estimate the noise-free autocorrelation matrix.

B. *Rayleigh Quotient for Tracking TLS Solution*

In order to find the TLS solution for adaptive FIR filtering, Davila [21] established the following RQ:

$$J(\boldsymbol{q}) = (\boldsymbol{q}^T\bar{\boldsymbol{R}}\boldsymbol{q})/(\boldsymbol{q}^T\bar{\boldsymbol{D}}\boldsymbol{q}), \tag{4.45}$$

where $\boldsymbol{q} \in \Re^{(M+1)\times 1}$ is the parameter vector, and $\bar{\boldsymbol{D}} = \mathrm{diag}(1, 1, ..., 1, \beta) \in \Re^{(M+1)\times(M+1)}$ is a diagonal weighting matrix with $\beta = \sigma_o^2/\sigma_i^2$. It was shown in [21] that if the parameter vector associated with the minimization of $J(\boldsymbol{q})$ is $\boldsymbol{q}^*$, then the unbiased solution $\boldsymbol{w}_{\mathrm{TLS}}$ for adaptive FIR filtering is given by

$$\boldsymbol{w}_{\mathrm{TLS}} = -[\boldsymbol{q}^*]_{1,M}/q^*_{M+1}, \tag{4.46}$$

where $[\boldsymbol{q}^*]_{1,M}$ denotes the vector constructed by the first M entries of $\boldsymbol{q}^*$, and q^*_{M+1} is the last element of $\boldsymbol{q}^*$.

Let

$$\boldsymbol{q} = [\boldsymbol{w}^T, -1]^T. \tag{4.47}$$

Substituting (4.47) into (4.45), we have the cost function

$$\hat{J}(\boldsymbol{w}) = \{[\boldsymbol{w}^T, -1]\bar{\boldsymbol{R}}[\boldsymbol{w}^T, -1]^T / [\boldsymbol{w}^T, -1]\bar{\boldsymbol{D}}[\boldsymbol{w}^T, -1]^T\}. \tag{4.48}$$

Obviously, the unbiased TLS solution can also be obtained by minimizing $\hat{J}(\boldsymbol{w})$. Since the last entry of is constrained to -1, we refer to the above cost function as the c-RQ.

4.3.2 *Landscape of Criterion*

In this section, the saddle points of the c-RQ is studied.

Put $\boldsymbol{v} = \mathrm{diag}(1, 1, ..., 1, \sqrt{\beta})[\boldsymbol{w}^T, -1]^T = [\boldsymbol{w}^T, -\sqrt{\beta}]^T$. Then,

$$\hat{J}(\boldsymbol{w}) = \{[\boldsymbol{w}^T, -\sqrt{\beta}]\breve{\boldsymbol{R}}[\boldsymbol{w}^T, -\sqrt{\beta}]^T / (\boldsymbol{w}^T\boldsymbol{w} + \beta)\}, \tag{4.49}$$

where $\breve{\boldsymbol{R}} = \mathrm{diag}(1, 1, ..., 1, \sqrt{\beta})\bar{\boldsymbol{R}}\mathrm{diag}(1, 1, ..., 1, \sqrt{\beta})$. Let the EVD of $\breve{\boldsymbol{R}}$ be given by $\breve{\boldsymbol{R}} = \boldsymbol{Q}\boldsymbol{\Gamma}\boldsymbol{Q}^T$, where $\boldsymbol{Q} = [\boldsymbol{q}_1, ..., \boldsymbol{q}_{M+1}]$, and $\boldsymbol{\Gamma} = \mathrm{diag}(\gamma_1, ..., \gamma_{M+1})$ with the eigenvector $\boldsymbol{q}_i$ corresponding to the eigenvalue γ_i being arranged such that $\gamma_1 \geq \gamma_2 \geq \cdots \geq \gamma_M > \gamma_{M+1}$.

Differentiating $\hat{J}(\boldsymbol{w})$ with respect to, it holds that

$$\nabla\hat{J}(\boldsymbol{w}) = \left(\left[\boldsymbol{R} + \sigma_i^2\boldsymbol{I}, \frac{1}{\sqrt{\beta}}\boldsymbol{R}\boldsymbol{h}\right]\right)[\boldsymbol{w}^T, -\sqrt{\beta}]^T - \hat{J}(\boldsymbol{w})\boldsymbol{w}\Big) \Big/ (\beta + \boldsymbol{w}^T\boldsymbol{w}). \tag{4.50}$$

It is easy to show that the stationary points of are given by

$$\begin{aligned} &\boldsymbol{w}_j = -\sqrt{\beta}[\boldsymbol{q}_j]_{1,M}/q_{j,(M+1)} \\ &\text{for } q_{j,(M+1)} \neq 0 \text{ and } j = 1, 2, ..., M+1, \end{aligned} \tag{4.51}$$

where $q_{j,(M+1)}$ is the last element of $\boldsymbol{q}_j$. It is easily known that $\boldsymbol{w}_{M+1} = \boldsymbol{h}$.

Theorem 4.2 *If $\gamma_M > \gamma_{M+1}$ and $q_{(M+1),(M+1)} \neq 0$, then $\boldsymbol{w}_{M+1} = \boldsymbol{h}$ is a global minimum point of $\hat{J}(\boldsymbol{w})$. All the other stationary points are saddle (unstable) points of $\hat{J}(\boldsymbol{w})$.*

Proof See reference [26].

The above theorem shows that one can search for the global minimum point of by the gradient descent method.

4.3.3 New RTLS Algorithm

This section develops a new algorithm for finding the TLS solution of the adaptive filtering problem. The new RTLS algorithm is a special gradient search method with computational complexity $O(M)$. The parameter vector is updated by

$$\boldsymbol{w}(t) = \boldsymbol{w}(t-1) + \alpha(t)\tilde{\boldsymbol{x}}(t), \tag{4.52}$$

where $\alpha(t)$ can efficiently be determined in $O(M)$ multiplications by

$$\min_{\alpha(t)} \hat{J}(\boldsymbol{w}(t)) = \min_{\alpha(t)} \frac{[\boldsymbol{w}^T(t), -1]\bar{\boldsymbol{R}}(t)[\boldsymbol{w}^T(t), -1]^T}{\beta + \boldsymbol{w}^T(t)\boldsymbol{w}(t)}. \tag{4.53}$$

Notice that $\bar{\boldsymbol{R}}(t)$ can be computed via an iteration formula

$$\bar{\boldsymbol{R}}(t) = \begin{bmatrix} \tilde{\boldsymbol{R}}(t) & \boldsymbol{b}(t) \\ \boldsymbol{b}^T(t) & c(t) \end{bmatrix} = \mu\bar{\boldsymbol{R}}(t-1) + \bar{\boldsymbol{x}}(t)\bar{\boldsymbol{x}}^T(t), \tag{4.54}$$

where

$$\tilde{\boldsymbol{R}}(t) = \mu\tilde{\boldsymbol{R}}(t-1) + \tilde{\boldsymbol{x}}(t)\tilde{\boldsymbol{x}}^T(t), \tag{4.55}$$

$$\boldsymbol{b}(t) = \mu\boldsymbol{b}(t-1) + \tilde{\boldsymbol{x}}(t)d(t), \tag{4.56}$$

$$c(t) = \mu c(t-1) + d(t)d(t). \tag{4.57}$$

The parameter μ in (4.54)–(4.57) is the forgetting factor. Let the gradient of $\hat{J}(\boldsymbol{w}(t))$ with respect to $\alpha(t)$ be equal to zero. Then,

$$\frac{\partial \hat{J}(\boldsymbol{w}(t))}{\partial \alpha(t)} = 0, \tag{4.58a}$$

or equivalently

$$\begin{aligned} &\bar{\boldsymbol{R}}(t)[\boldsymbol{w}^T(t), -1]^T[\beta + \boldsymbol{w}^T(t)\boldsymbol{w}(t)] \\ &-[\boldsymbol{w}^T(t), -1]\bar{\boldsymbol{R}}(t)[\boldsymbol{w}^T(t), -1]^T[\tilde{\boldsymbol{x}}^T(t)\boldsymbol{w}(t)] = 0. \end{aligned} \tag{4.58b}$$

In order to efficiently solve (4.58), let

$$k(t) = \tilde{\boldsymbol{R}}(t)\tilde{\boldsymbol{x}}(t), \tag{4.59}$$

$$\lambda^0(t) = [\boldsymbol{w}^T(t-1), -1]\bar{\boldsymbol{R}}(t)[\boldsymbol{w}^T(t-1), -1]^T, \tag{4.60}$$

$$\lambda(t) = [\boldsymbol{w}^T(t), -1]\bar{\boldsymbol{R}}(t)[\boldsymbol{w}^T(t), -1]^T/[\beta + \boldsymbol{w}^T(t)\boldsymbol{w}(t)]. \tag{4.61}$$

By definition, and can efficiently be computed by

$$\begin{aligned}\lambda^0(t) &= [\boldsymbol{w}^T(t-1), -1]\{\mu\bar{\boldsymbol{R}}(t-1) + \tilde{\boldsymbol{x}}(t)\tilde{\boldsymbol{x}}^T(t)\}[\boldsymbol{w}^T(t-1), -1]^T \\ &= \mu\lambda(t-1)\{\beta + \boldsymbol{w}^T(t-1)\boldsymbol{w}(t-1)\} + [\boldsymbol{w}^T(t-1)\tilde{\boldsymbol{x}}(t) - d(t)]^2,\end{aligned} \tag{4.62}$$

$$\begin{aligned}\lambda(t) &= \{[\boldsymbol{w}^T(t-1), -1] + \alpha(t)[\tilde{\boldsymbol{x}}^T(t), 0]\}\bar{\boldsymbol{R}}(t) \\ &\quad \times \{[\boldsymbol{w}^T(t-1), -1] + \alpha(t)[\tilde{\boldsymbol{x}}^T(t), 0]\}^T/[\beta + \boldsymbol{w}^T(t)\boldsymbol{w}(t)] \\ &= \{\lambda^0(t) + 2\alpha(t)[\boldsymbol{k}^T(t)\boldsymbol{w}(t-1) - \tilde{\boldsymbol{x}}^T(t)\boldsymbol{b}(t)] + \alpha^2(t)\tilde{\boldsymbol{x}}^T(t)\boldsymbol{k}(t)\}/[\beta + \boldsymbol{w}^T(t)\boldsymbol{w}(t)].\end{aligned} \tag{4.63}$$

Table 4.1 shows the FGV algorithm [28, 29] for computing the gain vector $\boldsymbol{k}(t)$. It can be shown that (4.58) can be rewritten as

$$a\alpha^2(t) + b\alpha(t) + c = 0, \tag{4.64}$$

where

$$a = [\boldsymbol{k}^T(t)\tilde{\boldsymbol{x}}(t)][\boldsymbol{w}^T(t-1)\tilde{\boldsymbol{x}}(t)] - \|\tilde{\boldsymbol{x}}(t)\|^2[\boldsymbol{k}^T(t)\boldsymbol{w}(t-1) - \tilde{\boldsymbol{x}}^T(t)\boldsymbol{b}(t)], \tag{4.65}$$

$$b = [\boldsymbol{k}^T(t)\tilde{\boldsymbol{x}}(t)][\beta + \|\boldsymbol{w}(t-1)\|^2] - \lambda^0(t)\|\tilde{\boldsymbol{x}}(t)\|^2, \tag{4.66}$$

$$\begin{aligned}c &= [\boldsymbol{k}^T(t)\boldsymbol{w}(t-1) - \tilde{\boldsymbol{x}}^T(t)\boldsymbol{b}(t)][\beta + \boldsymbol{w}^T(t-1)\boldsymbol{w}(t-1)] \\ &\quad - \lambda^0(t)[\boldsymbol{w}^T(t-1)\tilde{\boldsymbol{x}}(t)].\end{aligned} \tag{4.67}$$

Table 4.1 Fast algorithm for computing the gain vector

Initialize: $\boldsymbol{g}_M(0) = 0, \tilde{\boldsymbol{g}}_M(0) = 0, \pi(0) = 0$
$\boldsymbol{g}_M(t) = \mu\boldsymbol{g}_M(t-1) + \tilde{\boldsymbol{x}}_M(t-1)\tilde{x}(t)$
$\tilde{\boldsymbol{g}}_M(t) = \mu\tilde{\boldsymbol{g}}_M(t-1) + \tilde{\boldsymbol{x}}(t)\tilde{x}(t-M)$
$\pi(t) = \mu\pi(t-1) + \tilde{x}(t)\tilde{x}(t)$
$\boldsymbol{k}(t) = \begin{bmatrix} \pi(t)\tilde{x}(t) + \boldsymbol{g}_M^T(t)\boldsymbol{x}(t-1) \\ \tilde{x}(t)[\boldsymbol{g}_M(t)]_{1,M-1} + [\boldsymbol{k}(t-1)]_{1,M-1} \end{bmatrix} - \tilde{\boldsymbol{g}}_M(t)\tilde{x}(t-M)$

Table 4.2 N-RTLS algorithm

Initialize: $w(0) = [0, 0, ..., 0]^T$, $\lambda(0) = 0$, $\mu = 0.99 \sim 1.0$
For t = 1,2,...
1. Update the data vector $\tilde{x}(t)$
2. Update gain vector $k(t)$ as given in Table I
3. $\lambda^0(t) = \mu\lambda(t-1)[\beta + \|w(t-1)\|^2] + [w^T(t-1)\tilde{x}(t) - d(t)]^2$
4. $b(t) = \mu b(t-1) + \tilde{x}(t)d(t)$
5. $a = [k^T(t)\tilde{x}(t)][w^T(t-1)\tilde{x}(t)] - \|\tilde{x}(t)\|^2[k^T(t)w(t-1) - \tilde{x}^T(t)b(t)]$
6. $b = [k^T(t)\tilde{x}(t)][\beta + \|w(t-1)\|^2] - \lambda^0(t)\|\tilde{x}(t)\|^2$
7. $c = [k^T(t)w(t-1) - \tilde{x}^T(t)b(t)][\beta + w^T(t-1)w(t-1)] - \lambda^0(t)[w^T(t-1)\tilde{x}(t)]$
8. $\alpha(t) = (-b + \sqrt{b^2 - 4ac})/2a$
9. $w(t) = w(t-1) + \alpha(t)\tilde{x}(t)$
10. $\lambda(t) = \{\lambda^0(t) + 2\alpha(t)[k^T(t)w(t-1) - \tilde{x}^T(t)b(t)] + \alpha^2(t)\tilde{x}^T(t)k(t)\}/[\beta + w^T(t)w(t)]$

A solution of (4.64) is given by

$$\alpha(t) = (-b + \sqrt{b^2 - 4ac})/2a. \tag{4.68}$$

The new RTLS (N-RTLS) algorithm is summarized in Table 4.2.

4.3.4 *Convergence Analysis*

The convergence property of the N-RTLS algorithm is studied. Since the sequence is a discrete-time dynamical system, its convergence can be analyzed by LaSalle's invariance principle [30].

Lemma 4.1 *The following result is true:*

$$\nabla \hat{J}(w(t))^T x(t) = 0, \text{ for all } t \geq 1. \tag{4.69}$$

Proof See reference [26].

Theorem 4.3 *If t is large enough so that* $\bar{R}(t) \to \bar{R} = E\{\bar{x}(t)\bar{x}^T(t)\}$, *then* $w(t) \to h$ *as* $t \to \infty$.

Proof See reference [26].

4.3.5 Simulations and Conclusions

In this simulation, our N-RTLS algorithm is compared with other three algorithms. For convenience, Davila's RTLS algorithms [21] are referred to as the original RTLS (O-RTLS for short) and RLS and IP are used to represent the RLS algorithm [1] and the inverse power algorithm [8], respectively. Each estimation curve is the result averaged over 30 independent runs.

Example Adaptive Identification of Linear System: The unknown system impulse response is given by

$$\boldsymbol{h}^* = [-0.3, -0.9, 0.8, -0.7, 0.6]^T. \tag{4.70}$$

The noise-free input is a first-order AR process and given by

$$x(k) = -0.5 * x(k-1) + \eta(t), \tag{4.71}$$

where $\eta(t)$ is a white Gaussian noise sequence of zero mean and unit variance. Both the white Gaussian input noise and the white Gaussian output noise have the unit variance, i.e., $\sigma_i^2 = 1$, and $\sigma_o^2 = 1$. Figure 4.3 shows the averaged estimation errors for the linear time-invariant system, where the estimation error is as defined in [21].

In order to test the tracking behavior of the four algorithms in a non-stationary environment, we simulated a time-varying system whose parameters undergo a step at time $t = 500$ and used the forgetting factor $\mu = 0.995$. Moreover, in order to make a broader comparison of the relevant algorithms under different input signals, the noise-free input $x(k)$ is also changed into the following first-order AR process

$$x(k) = 0.5 * x(k-1) + \eta(t), \tag{4.72}$$

where $\eta(t)$ is a white Gaussian noise sequence of zero mean and unit variance. Both the white Gaussian input noise and the white Gaussian output noise have also the unit variance. The averaged estimation errors are shown in Fig. 4.4.

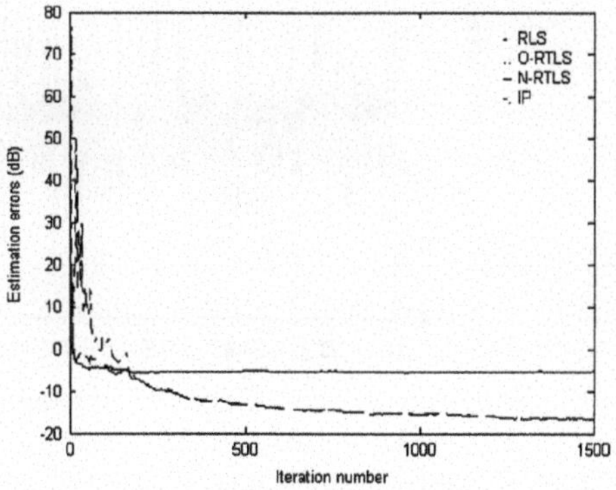

Fig. 4.3 Averaged estimation errors of the four methods for a time-invariant system, where the variances of the input noise and output noise are equal to 1

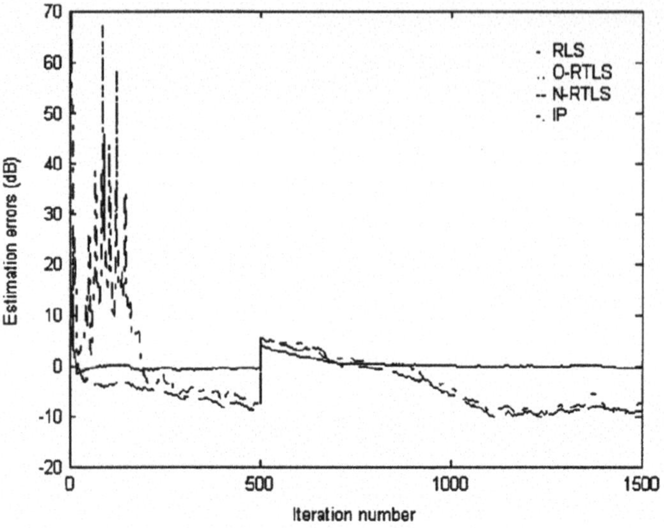

Fig. 4.4 Averaged estimation errors of the four methods for a time-varying system, where the variances of the input noise and output noise are equal to 1

Figure 4.5 shows the long-term numerical stability of the three TLS algorithms from $t = 1$ to $t = 2 \times 10^5$ for the time-invariant system, where $\mu = 0.995$, and the input signal and noises are as given in the above.

In this section, a fast RTLS algorithm has been introduced for adaptive FIR filtering. The proposed algorithm has been built on the adaptation of the c-RQ and

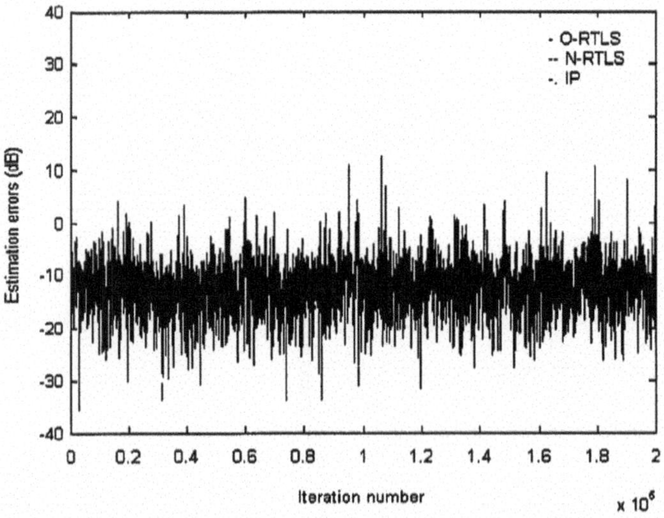

Fig. 4.5 Three TLS algorithms have the almost same long-time numerical stability

the efficient computation of the FGV. These represent its major differences with respect to Davila's RTLS algorithm. With the significantly reduced computational complexity, the proposed fast RTLS algorithm has demonstrated to be able to achieve the good performances that are closely consistent with those of Davila's algorithm.

4.4 A Fast Recursive TLS Algorithm for Adaptive IIR Filtering

This section develops a new fast recursive TLS (N-RTLS) algorithm to recursively compute the TLS solution for adaptive IIR filtering. The new algorithm is based on the minimization of the c-RQ in which the first entry of the parameter vector is fixed to the negative one. The highly computational efficiency of the proposed algorithm depends on the efficient computation of the gain vector and the adaptation of the RQ. Using the shift structure of the input data vectors, a fast algorithm for computing the gain vector is established, which is referred to as the FGV algorithm. The computational load of the FGV algorithm is smaller than that of the fast Kalman algorithm. Moreover, the new algorithm is numerically stable since it does not use the well-known matrix inversion lemma. The computational complexity of the new algorithm per iteration is also $O(L)$. The global convergence of the new algorithm is studied.

4.4.1 TLS Problems in Adaptive IIR Filtering and N-RTLS Algorithm

A. *Signal Model*

Consider an unknown system with IIR and assume that only the output is corrupted by the additive white Gaussian noise. We use an equation-error adaptive IIR filter to estimate the IIR system from the observations of the input and output, as shown in Fig. 4.6.

The IIR vector of the unknown system is described by

$$\boldsymbol{h} = [a_1, a_2, ..., a_{N-1}, b_0, b_1, ..., b_{M-1}]^T \in \Re^{L\times 1}, \tag{4.73}$$

where N is the autoregressive (AR) order, M is the moving average (MA) order, and $L = N + M-1$. $\boldsymbol{h}$ may be time-varying but is assumed constant here. Moreover, assume that the AR order and the MA order are known.

The desired output is given by

$$d(t) = \boldsymbol{r}^T(t)\boldsymbol{h} + n(t), \tag{4.74}$$

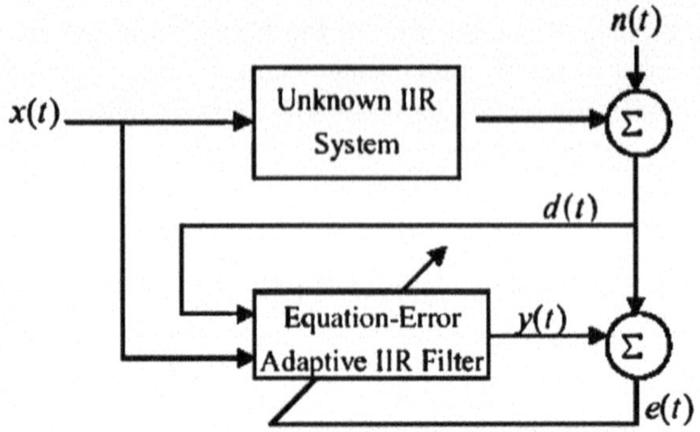

Fig. 4.6 Unknown adaptive IIR system with output noise $n(t)$

where the input vector $\boldsymbol{r}(t) \in \Re^{L\times 1}$ is given by

$$\begin{aligned}\boldsymbol{r}(t) = [&d(t-1), d(t-2), ..., d(t-N+1),\\ &x(t), x(t-1), ..., x(t-M+1)]^T,\end{aligned} \tag{4.75}$$

and the measurement noise $n(t)$ is a zero-mean Gaussian white noise with variance σ_o^2, independent of the input vector. The output for sufficient-order equation-error adaptive IIR filter is given by

$$y(t) = \sum_{m=1}^{N-1} a_m(t)d(t-m) + \sum_{m=0}^{M-1} b_m(t)x(t-m). \tag{4.76}$$

In vector notations, (4.76) can be written as $y(t) = \boldsymbol{r}^T(t)\boldsymbol{w}(t)$, where

$$\boldsymbol{w}(t) = [a_1(t), a_2(t), ..., a_{N-1}(t), b_0(t), b_1(t), ..., b_{M-1}(t)]^T \in \Re^{L\times 1}. \tag{4.77}$$

At time t, the augmented data vector is defined as

$$\bar{\boldsymbol{r}}(t) = [d(t), \boldsymbol{r}^T(t)]^T \tag{4.78}$$

For convenience of analysis, we define the following matrices. The autocorrelation matrix of the input vector is given by

$$\boldsymbol{R}(t) = E\{\boldsymbol{r}(t)\boldsymbol{r}^T(t)\}. \tag{4.79}$$

The autocorrelation matrix of the augmented data vector is described by

$$\bar{\boldsymbol{R}}(t) = E\{\bar{\boldsymbol{r}}(t)\bar{\boldsymbol{r}}^T(t)\} = \begin{bmatrix} c & \boldsymbol{b}^T \\ \boldsymbol{b} & \boldsymbol{R} \end{bmatrix}, \tag{4.80}$$

where $\boldsymbol{b} = E\{\boldsymbol{r}(t)d(t)\}$, $c = E\{d(t)d(t)\}$. We can further show that

$$\boldsymbol{b} = E\{\boldsymbol{r}(t)d(t)\} = E\{\boldsymbol{r}(t)[\boldsymbol{r}^T(t)\boldsymbol{h} + n(t)]\} = \boldsymbol{R}(t)\boldsymbol{h}, \tag{4.81}$$

$$c = E\{d(t)d(t)\} = E\{[\boldsymbol{h}^T\boldsymbol{r}(t) + n(t)][\boldsymbol{r}^T(t)\boldsymbol{h} + n(t)]\} = \boldsymbol{h}^T\boldsymbol{R}(t)\boldsymbol{h} + \sigma_o^2. \tag{4.82}$$

Therefore, the autocorrelation matrix of the augmented data vector can be written as

$$\bar{\boldsymbol{R}}(t) = \begin{bmatrix} \boldsymbol{h}^T\boldsymbol{R}\boldsymbol{h} + \sigma_o^2 & \boldsymbol{h}^T\boldsymbol{R} \\ \boldsymbol{R}\boldsymbol{h} & \boldsymbol{R} \end{bmatrix} = \begin{bmatrix} \boldsymbol{h}^T\boldsymbol{R}\boldsymbol{h} & \boldsymbol{h}^T\boldsymbol{R} \\ \boldsymbol{R}\boldsymbol{h} & \boldsymbol{R} \end{bmatrix} + \begin{bmatrix} \sigma_o^2\boldsymbol{I}_N & \boldsymbol{0} \\ \boldsymbol{0} & \boldsymbol{0} \end{bmatrix} = \boldsymbol{R}^* + \boldsymbol{R}_n. \tag{4.83}$$

B. *New RTLS Algorithm*

In order to find the TLS solution for adaptive IIR filtering, Davila [31] established the following RQ:

$$\bar{J}(\bar{\boldsymbol{w}}(t)) = \frac{\bar{\boldsymbol{w}}^T(t)\bar{\boldsymbol{R}}(t)\bar{\boldsymbol{w}}(t)}{\bar{\boldsymbol{w}}^T(t)\bar{\boldsymbol{D}}\bar{\boldsymbol{w}}(t)}, \tag{4.84}$$

where $\bar{\boldsymbol{w}}(t) \in \Re^{(L+1)\times 1}$ is the parameter vector, and $\bar{\boldsymbol{D}}$ is given

$$\bar{\boldsymbol{D}} = \begin{bmatrix} \boldsymbol{I}_{N\times N} & \boldsymbol{0}_{N\times M} \\ \boldsymbol{0}_{M\times N} & \boldsymbol{0}_{M\times M} \end{bmatrix}. \tag{4.85}$$

The algorithm given in [31] searches the minimum point of $\bar{J}(\boldsymbol{w})$ along the Kalman gain vector. Although there are some efficient approaches that may eliminate the instability of the Kalman gain vector [24, 25], the algorithm proposed in [31] has the potential instability caused by the Kalman gain vector.

It was shown in [31] that if the parameter vector $\boldsymbol{w}^*$ associated with the minimization of $\bar{J}(\boldsymbol{w})$ is obtained, then the unbiased solution $\boldsymbol{w}_{\mathrm{TLS}}$ for adaptive IIR filtering is given by

$$\begin{bmatrix} -1 \\ \boldsymbol{w}_{\mathrm{TLS}} \end{bmatrix} = -\boldsymbol{w}^*/w_1^*, \tag{4.86}$$

where w_1^* is the first element of $\boldsymbol{w}^*$. This shows that if the first element of $\bar{\boldsymbol{w}}(t)$ is constrained to the negative one, the scaling operation (4.86) can be avoided, which, at least, saves L multiplies, divides, and square roots (MADs). Thus, we adopt the following c-RQ:

$$\min_{\boldsymbol{w}(t)} \bar{J}(\boldsymbol{w}(t)) = \frac{[-1\ \boldsymbol{w}^T(t)]\bar{\boldsymbol{R}}(t)[-1\ \boldsymbol{w}^T(t)]^T}{[-1\ \boldsymbol{w}^T(t)]\bar{\boldsymbol{D}}[-1\ \boldsymbol{w}^T(t)]^T} = \frac{[-1\ \boldsymbol{w}^T(t)]\bar{\boldsymbol{R}}(t)[-1\ \boldsymbol{w}^T(t)]^T}{1 + \boldsymbol{w}^T(t)\boldsymbol{D}\boldsymbol{w}(t)}, \tag{4.87}$$

where

$$\boldsymbol{D} = \begin{bmatrix} \boldsymbol{I}_{(N-1)\times(N-1)} & \boldsymbol{0}_{(N-1)\times M} \\ \boldsymbol{0}_{M\times(N-1)} & \boldsymbol{0}_{M\times M} \end{bmatrix}.$$

Notice that the number of the unknown parameters in the cost function (4.87) is L, whereas the number of the variable parameters in the cost function (4.84) is $L+1$.

An algorithm for efficiently finding the TLS solution for the adaptive IIR filtering problem is now described. This algorithm is a special gradient search method. Moreover, the selection of the update direction to be the data vector will result in the computationally efficient algorithm with computational complexity $O(L)$. The parameter vector is updated by

$$\boldsymbol{w}(t) = \boldsymbol{w}(t-1) + \beta(t)\boldsymbol{r}(t). \tag{4.88}$$

$\beta(t)$ is determined by minimizing the Rayleigh quotient

$$\min_{\beta(t)} \bar{J}(\boldsymbol{w}(t)) = \frac{[-1\ \boldsymbol{w}^T(t)]\bar{\boldsymbol{R}}(t)[-1\ \boldsymbol{w}^T(t)]^T}{1+\boldsymbol{w}^T(t)\boldsymbol{D}\boldsymbol{w}(t)}. \tag{4.89}$$

Notice that $\bar{\boldsymbol{R}}(t)$ can be computed via an iteration formula

$$\bar{\boldsymbol{R}}(t) = \begin{bmatrix} c(t) & \boldsymbol{b}^T(t) \\ \boldsymbol{b}(t) & \boldsymbol{R}(t) \end{bmatrix} = \mu\bar{\boldsymbol{R}}(t-1) + \bar{\boldsymbol{r}}(t)\bar{\boldsymbol{r}}^T(t), \tag{4.90}$$

where

$$\boldsymbol{R}(t) = \mu\boldsymbol{R}(t-1) + \boldsymbol{r}(t)\boldsymbol{r}^T(t), \tag{4.91}$$

$$\boldsymbol{b}(t) = \mu\boldsymbol{b}(t-1) + \boldsymbol{r}(t)d(t), \tag{4.92}$$

$$c(t) = \mu c(t-1) + d(t)d(t). \tag{4.93}$$

The μ in (4.91)–(4.93) is a forgetting factor, and $0 < \mu \le 1$. Substituting (4.88) into (4.89), taking the derivative of the resulting equation with respect to $\beta(t)$, and forcing it to be equal to zero yields

$$\frac{\partial \bar{J}(\boldsymbol{w}(t))}{\partial \beta(t)} = 0, \tag{4.94a}$$

or equivalently

$$\begin{aligned}&\bar{\boldsymbol{R}}(t)[-1\ \boldsymbol{w}^T(t)]^T[1+\boldsymbol{w}^T(t)\boldsymbol{D}\boldsymbol{w}(t)]\\&-[-1\ \boldsymbol{w}^T(t)]\bar{\boldsymbol{R}}(t)[-1\ \boldsymbol{w}^T(t)]^T[\boldsymbol{r}^T(t)\boldsymbol{D}\boldsymbol{w}(t)]=0.\end{aligned} \tag{4.94b}$$

In order to solve (4.89) recursively, let

$$\boldsymbol{k}(t)=\boldsymbol{R}(t)\boldsymbol{r}(t), \tag{4.95}$$

$$\lambda^0(t)=[-1\ \boldsymbol{w}^T(t-1)]\bar{\boldsymbol{R}}(t)[-1\ \boldsymbol{w}^T(t-1)]^T, \tag{4.96}$$

$$\lambda(t)=[-1\ \boldsymbol{w}^T(t)]\bar{\boldsymbol{R}}(t)[-1\ \boldsymbol{w}^T(t)]^T/[1+\boldsymbol{w}^T(t)\boldsymbol{D}\boldsymbol{w}(t)]. \tag{4.97}$$

Notice that $\lambda^0(t)$ and $\lambda(t)$ can be efficiently computed by (4.98) and (4.99).

$$\begin{aligned}\lambda^0(t)&=[-1\ \boldsymbol{w}^T(t-1)]\{\mu\bar{\boldsymbol{R}}(t-1)+\bar{\boldsymbol{r}}(t)\bar{\boldsymbol{r}}^T(t)\}[-1\ \boldsymbol{w}^T(t-1)]^T\\&=\mu\lambda(t-1)\{1+\boldsymbol{w}^T(t-1)\boldsymbol{D}\boldsymbol{w}(t-1)\}+[\boldsymbol{w}^T(t-1)\boldsymbol{r}(t)-d(t)]^2,\end{aligned} \tag{4.98}$$

$$\begin{aligned}\lambda(t)&=\frac{\{[-1\ \boldsymbol{w}^T(t-1)]+\beta(t)[0\ \boldsymbol{r}^T(t)]\}\bar{\boldsymbol{R}}(t)[-1\ \boldsymbol{w}^T(t-1)]+\beta(t)[0\ \boldsymbol{r}^T(t)]\}^T}{[1+\boldsymbol{w}^T(t)\boldsymbol{D}\boldsymbol{w}(t)]}\\&=\{\lambda^0(t)+2\beta(t)[\boldsymbol{k}^T(t)\boldsymbol{w}(t-1)-\boldsymbol{r}^T(t)\boldsymbol{b}(t)]+\beta^2(t)\boldsymbol{r}^T(t)\boldsymbol{k}(t)\}/[1+\boldsymbol{w}^T(t)\boldsymbol{D}\boldsymbol{w}(t)].\end{aligned} \tag{4.99}$$

Hence, the gain vector can be efficiently computed, with its fast algorithm given in Table 4.3. The fast algorithm only requires $O(L)$ multiplications.

It is shown by burdensome operations that (4.94), in fact, is a quadratic polynomial of and can be written as

$$a\beta^2(t)+b\beta(t)+c=0, \tag{4.100}$$

where

Table 4.3 FGA for computing the gain vector

Initialize $\boldsymbol{B}_{L2}(0)=\boldsymbol{0}$, $\tilde{\boldsymbol{B}}_{L2}(0)=\boldsymbol{0}$, $\pi_{22}(0)=\boldsymbol{0}$
$\boldsymbol{B}_{L2}(t)=\boldsymbol{B}_{L2}(t-1)+\boldsymbol{r}(t-1)\xi_2^T(t)$
$\tilde{\boldsymbol{B}}_{L2}(t)=\tilde{\boldsymbol{B}}_{L2}(t-1)+\boldsymbol{r}(t)\rho_2^T(t)$
$\pi_{22}(t)=\pi_{22}(t-1)+\xi_2(t)\xi_2^T(t)$
$\bar{\boldsymbol{k}}(t)=\boldsymbol{S}_{pp}^T\begin{bmatrix}\pi_{22}\xi_2(t)+\boldsymbol{B}_{L2}^T(t)\boldsymbol{r}(t-1)\\\boldsymbol{B}_{L2}(t-1)\xi_2(t)+\boldsymbol{k}(t-1)\end{bmatrix}$
$\begin{bmatrix}\boldsymbol{m}(t)\\\eta(t)\end{bmatrix}=\boldsymbol{Q}_{pp}\bar{\boldsymbol{k}}(t)$
$\boldsymbol{k}(t)=\boldsymbol{m}(t)-\tilde{\boldsymbol{B}}_{L2}\rho_2(t)$

Table 4.4 Fast RTLS algorithm

Initialize: $\boldsymbol{w}(0) = [0, 0, ..., 0]^T$, $\lambda(0) = 0$, $\mu = 0.99 \sim 1.0$
For t = 1,2,...
1. Update the data vector $\boldsymbol{r}(t)$
2. Update gain vector $\boldsymbol{k}(t)$ as shown in Table III
3. $\lambda^0(t) = \mu\lambda(t-1)\{1 + \boldsymbol{w}^T(t-1)\boldsymbol{D}\boldsymbol{w}(t-1)\} + [\boldsymbol{w}^T(t-1)\boldsymbol{r}(t) - d(t)]^2$
4. $\boldsymbol{b}(t) = \mu\boldsymbol{b}(t-1) + \boldsymbol{r}(t)d(t)$
5. $a = \boldsymbol{k}^T(t)\boldsymbol{r}(t)\boldsymbol{w}^T(t-1)\boldsymbol{D}\boldsymbol{r}(t) - [\boldsymbol{k}^T(t)\boldsymbol{w}(t-1) - \boldsymbol{r}^T(t)\boldsymbol{b}(t)]\boldsymbol{r}^T(t)\boldsymbol{D}\boldsymbol{r}(t)$
6. $b = \boldsymbol{k}^T(t)\boldsymbol{r}(t)[1 + \boldsymbol{w}^T(t-1)\boldsymbol{D}\boldsymbol{w}(t-1)] - \lambda^0(t)\boldsymbol{r}^T(t)\boldsymbol{D}\boldsymbol{r}(t)$
7. $c = [1 + \boldsymbol{w}^T(t-1)\boldsymbol{D}\boldsymbol{w}(t-1)][\boldsymbol{k}^T(t)\boldsymbol{w}(t-1) - \boldsymbol{r}^T(t)\boldsymbol{b}(t)] - \lambda^0(t)\boldsymbol{r}^T(t)\boldsymbol{D}\boldsymbol{w}(t-1)$
8. $\beta(t) = (-b + \sqrt{b^2 - 4ac})/2a$
9. $1 + \boldsymbol{w}^T(t)\boldsymbol{D}\boldsymbol{w}(t) = 1 + \boldsymbol{w}^T(t-1)\boldsymbol{D}\boldsymbol{w}(t-1) + 2\beta(t)\boldsymbol{r}^T(t)\boldsymbol{D}\boldsymbol{w}(t-1) + \beta^2(t)\boldsymbol{r}^T(t)\boldsymbol{D}\boldsymbol{r}(t)$
10. $\lambda(t) = \{\lambda^0(t) + 2\beta(t)[\boldsymbol{k}^T(t)\boldsymbol{w}(t-1) - \boldsymbol{r}^T(t)\boldsymbol{b}(t)] + \beta^2(t)\boldsymbol{r}^T(t)\boldsymbol{k}(t)\}/[1 + \boldsymbol{w}^T(t)\boldsymbol{D}\boldsymbol{w}(t)]$
11. $\boldsymbol{w}(t) = \boldsymbol{w}(t-1) + \beta(t)\boldsymbol{r}(t)$

$$a = \boldsymbol{k}^T(t)\boldsymbol{r}(t)\boldsymbol{w}^T(t-1)\boldsymbol{D}\boldsymbol{r}(t) - [\boldsymbol{k}^T(t)\boldsymbol{w}(t-1) - \boldsymbol{r}^T(t)\boldsymbol{b}(t)]\boldsymbol{r}^T(t)\boldsymbol{D}\boldsymbol{r}(t), \tag{4.101}$$

$$b = \boldsymbol{k}^T(t)\boldsymbol{r}(t)[1 + \boldsymbol{w}^T(t-1)\boldsymbol{D}\boldsymbol{w}(t-1)] - \lambda^0(t)\boldsymbol{r}^T(t)\boldsymbol{D}\boldsymbol{r}(t), \tag{4.102}$$

$$c = [1 + \boldsymbol{w}^T(t-1)\boldsymbol{D}\boldsymbol{w}(t-1)][\boldsymbol{k}^T(t)\boldsymbol{w}(t-1) - \boldsymbol{r}^T(t)\boldsymbol{b}(t)] - \lambda^0(t)\boldsymbol{r}^T(t)\boldsymbol{D}\boldsymbol{w}(t-1). \tag{4.103}$$

A solution of (4.100) is given by

$$\beta(t) = (-b + \sqrt{b^2 - 4ac})/2a. \tag{4.104}$$

This root makes $\left[-1\ \boldsymbol{w}^T(t)\right]^T$ converge to the eigenvector associated with the smallest eigenvalue of $\bar{\boldsymbol{R}}(t)$. Another root $\beta(t) = (-b - \sqrt{b^2 - 4ac})/2a$ will make $\left[-1\ \boldsymbol{w}^T(t)\right]^T$ converge to the eigenvector associated with the largest eigenvalue of $\bar{\boldsymbol{R}}(t)$. Thus, (4.104) represents the expected root. The new RTLS algorithm is given in Table 4.4.

4.4.2 Algorithm Convergence

Several issues regarding the existence and uniqueness of the TLS solution applied to identifying pole-zero systems were discussed in [31]. Here, only the case is considered where the TLS solution exists and is unique. This is the case when the polynomials associated with the unknown pole-zero system

$$B(z) = \sum_{m=0}^{M-1} b_m^* z^m, \; A(z) = 1 - \sum_{m=1}^{N-1} a_m^* z^m \tag{4.105}$$

have no common factor [32].

Perform the following generalized eigenvalue decomposition (GEVD) of the matrix pairs $\bar{\boldsymbol{R}}$ and $\bar{\boldsymbol{D}}$:

$$\bar{\boldsymbol{R}}\bar{\boldsymbol{V}} = \bar{\boldsymbol{D}}\bar{\boldsymbol{V}}\,\mathrm{diag}(\bar{\lambda}_1, \bar{\lambda}_2, ..., \bar{\lambda}_{L+1}) \;\text{ or }\; \bar{\boldsymbol{R}}\bar{\boldsymbol{v}}_j = \bar{\lambda}_j \bar{\boldsymbol{D}}\,\bar{\boldsymbol{v}}_j, \tag{4.106}$$

where $\bar{\boldsymbol{V}}$ is the generalized eigen matrix, and $\bar{\boldsymbol{v}}_j$ and $\bar{\lambda}_j$ are the jth generalized eigenvector and eigenvalue, respectively. Note that the eigenvalues have been arranged in the descending order $\bar{\lambda}_1 \geq \bar{\lambda}_2 \geq \cdots \geq \bar{\lambda}_L > \bar{\lambda}_{L+1}$. Differentiating $\bar{J}(\boldsymbol{w})$ with respect to $\boldsymbol{w}$ yields

$$\nabla\bar{J}(\boldsymbol{w}) = \frac{[\boldsymbol{Rh}\;\boldsymbol{R}][-1\;\boldsymbol{w}^T]^T}{1+\boldsymbol{w}^T\boldsymbol{Dw}} - \frac{[-1\;\boldsymbol{w}^T]\bar{\boldsymbol{R}}[-1\;\boldsymbol{w}^T]^T}{(1+\boldsymbol{w}^T\boldsymbol{Dw})^2}\boldsymbol{Dw}. \tag{4.107}$$

The stationary points are obtained by solving the following equation:

$$\nabla\bar{J}(\boldsymbol{w}) = \frac{[\boldsymbol{Rh}\;\boldsymbol{R}][-1\;\boldsymbol{w}^T]^T}{1+\boldsymbol{w}^T\boldsymbol{Dw}} - \frac{[-1\;\boldsymbol{w}^T]\bar{\boldsymbol{R}}[-1\;\boldsymbol{w}^T]^T}{(1+\boldsymbol{w}^T\boldsymbol{Dw})^2}\boldsymbol{Dw} = 0, \tag{4.108}$$

i.e.,

$$[\boldsymbol{Rh}\;\boldsymbol{R}][-1\;\boldsymbol{w}^T]^T - \bar{J}(\boldsymbol{w})\boldsymbol{Dw} = 0. \tag{4.109}$$

By making the substitution (4.106) into (4.109), it holds that

$$\widehat{\boldsymbol{V}}(\bar{\boldsymbol{\Lambda}}\boldsymbol{v} - \bar{J}(\boldsymbol{v})\boldsymbol{Dv}) = 0 \text{ subject to } \tilde{\boldsymbol{v}}\boldsymbol{v} = -1, \tag{4.110}$$

where $\widehat{\boldsymbol{V}}$ consists of the last L rows of $\bar{\boldsymbol{V}}$, and $\tilde{\boldsymbol{v}}$ is the first row of $\bar{\boldsymbol{V}}$. Hence, $\bar{J}(\boldsymbol{w})$ can be denoted by the following equation:

$$\bar{J}(\boldsymbol{v}) = \frac{\boldsymbol{v}^T\bar{\boldsymbol{\Lambda}}\boldsymbol{v}}{\boldsymbol{v}^T\bar{\boldsymbol{D}}\boldsymbol{v}}. \tag{4.111}$$

Clearly, the solution of (4.110) can be written as

$$\boldsymbol{v}_j = -\bar{\boldsymbol{v}}_j/\bar{v}_{1,j} \quad \bar{v}_{1,j} \neq 0, \; j = 1, 2, ..., L+1\,, \tag{4.112}$$

where $\bar{v}_{1,j}$ is the jth element of the first row of $\widehat{\boldsymbol{V}}$. Therefore, the following stationary points of $\bar{J}(\boldsymbol{w})$ can be obtained:

$$w_j = -\hat{v}_j/\bar{v}_{1,j} \;\; \bar{v}_{1,j} \neq 0, \; j = 1, 2, ..., L + 1. \tag{4.113}$$

Lemma 4.2 *If is of full rank, then* $\bar{\lambda}_L > \bar{\lambda}_{L+1}$.

Proof See reference [27].

Theorem 4.4 *If* $\bar{\lambda}_L > \bar{\lambda}_{L+1}$ *and* $\bar{v}_{1,L+1} \neq 0$, *then* $w_{L+1} = -\hat{v}_{L+1}/\bar{v}_{1,L+1}$ *is the global minimum point of* $\bar{J}(w)$. *All the other stationary points are the saddle (unstable) points of* $\bar{J}(w)$.

Proof See reference [27].

The above theorem guarantees that we can search the global minimum point of $\bar{J}(w)$ by the gradient descent method.

Lemma 4.3 *For an arbitrary* $t \geq 1$, *there always exists*

$$\nabla \bar{J}(w(t))^T r(t) = 0. \tag{4.114}$$

Proof See reference [27].

Theorem 4.5 *Assuming that t is large enough so that* $\bar{R}(t) \rightarrow \bar{R} = E\{\bar{r}(t)\bar{r}^T(t)\}$, *then* $w(t) \rightarrow h$ *as* $t \rightarrow \infty$.

Proof See reference [27].

4.4.3 Simulations and Conclusions

Some simulation results are now presented to support the theoretical analysis of the new fast RTLS IIR algorithm. Here, the proposed algorithm and Davila's algorithm [31] are simply called N-RTLS and O-RTLS, respectively.

Example 4.1 *System Identification:* In this example, the RLS, O-RTLS, and the N-RTLS are applied to the system identification experiment. The unknown system impulse response is defined by

$$h^* = [0.5, -0.5, 0.9, -0.3, -0.9, 0.8, -0.7, 0.6]^T, \tag{4.115}$$

where $N = 4$, $M = 5$, and $L = N + M - 1 = 8$. For the convenience of comparison and computation, the system is only stimulated by the white noise with a unit variance, as shown in Fig. 7.6. Only the system output is contaminated by additive, zero mean, and white Gaussian noise that is statistically independent of the input signal. The forgetting factor μ in the evaluation was chosen to be 0.998. The estimate error is defined by

$$E(t) = \|w(t) - h\|_2. \tag{4.116}$$

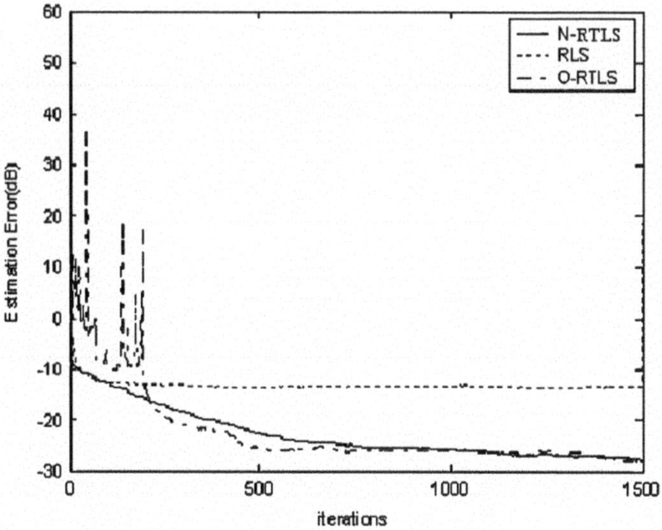

Fig. 4.7 N-RTLS, O-RTLS, and RLS are used for identifying linear time-varying system

When the system is linear time invariant, the estimation results are shown in Fig. 4.7 for the RLS, N-RTLS, and O-RTLS algorithms. Notice that although the performance of N-RTLS is similar to that of O-RTLS, the computational complexity of N-RTLS is significantly lower than that of O-RTLS. Here, all the results are averaged over 50 independent tests.

In order to test the tracking behavior of the relative algorithms in a non-stationary environment, the parameter estimation experiment is repeated, but the unknown system parameters undergo a step change at $t = 1000$. The obtained results are shown in Fig. 4.8.

Example 4.2 *Adaptive Line Enhancement:* The N-RTLS algorithm is then applied to an adaptive line enhancement (ALE) experiment. The adaptive line enhancers based on IIR were first discussed by Rao and Kung [33]. In Experiment 1, the periodic signal consists of a single sinusoid $s(t) = \cos(0.3\pi t + \phi) + n(t)$ in additive Gaussian white noise with variance of 0.25. The phase angle ϕ is a pseudo-random variable uniformly distributed on $[0, 2\pi]$ and is held constant for each experiment. To determine the relative performance of the ALE, periodograms are computed from successive output of the ALE corresponding to $t = 1$ to 1024. These periodograms are consequently averaged over 50 independent tests. The average periodograms $S(\omega)$ are shown in Fig. 4.9 for ALE lengths of $M = 2$.

In Experiment 2, the observed signal includes two sinusoids in additive white noise with variance of 0.25. It can be represented as $s(t) = \cos(0.4\pi t + \phi_1) + \cos(0.6\pi t + \phi_2) + n(t)$, where ϕ_1 and ϕ_2 are randomly produced on a uniform distribution of $[-\pi, \pi]$. The obtained results are shown in Fig. 4.10 for ALE lengths of $M = 4$.

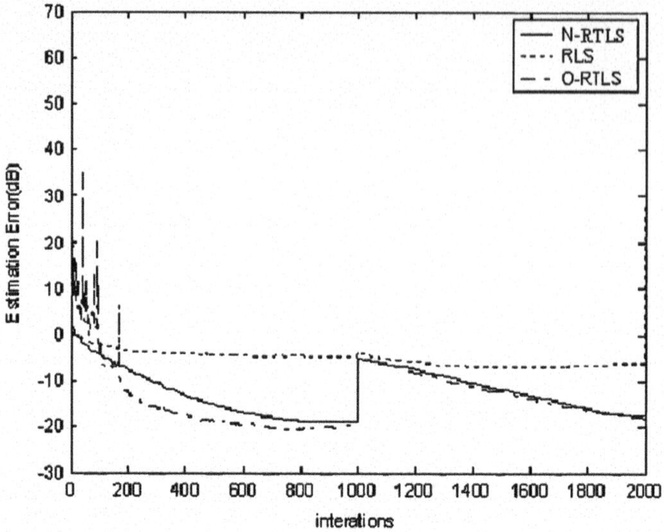

Fig. 4.8 N-RTLS, O-RTLS, and RLS are used linear time-invariant system

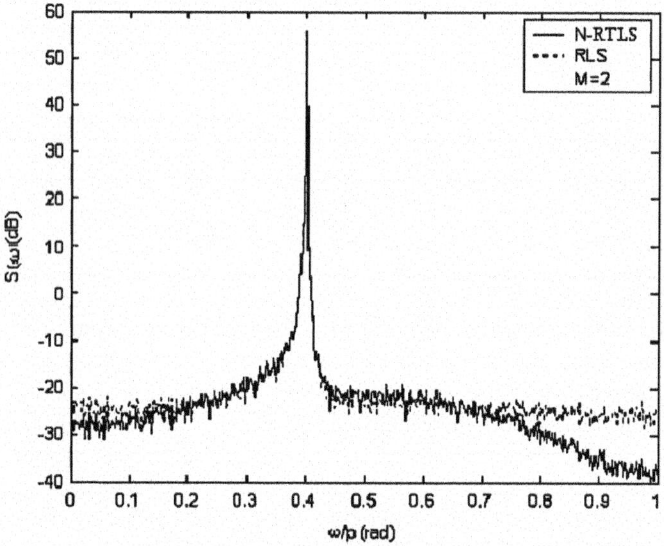

Fig. 4.9 Averaged periodograms of ALE outputs for N-RTLS and RLS for one sinusoid ALE lengths of 2

In this section, an algorithm for efficiently computing the eigenvector associated with the smallest eigenvalue of the sample covariance matrix has been described and applied to recursively solving the TLS solution to the adaptive IIR filtering problem. When only the output vector contains additive noise, it has been shown that the

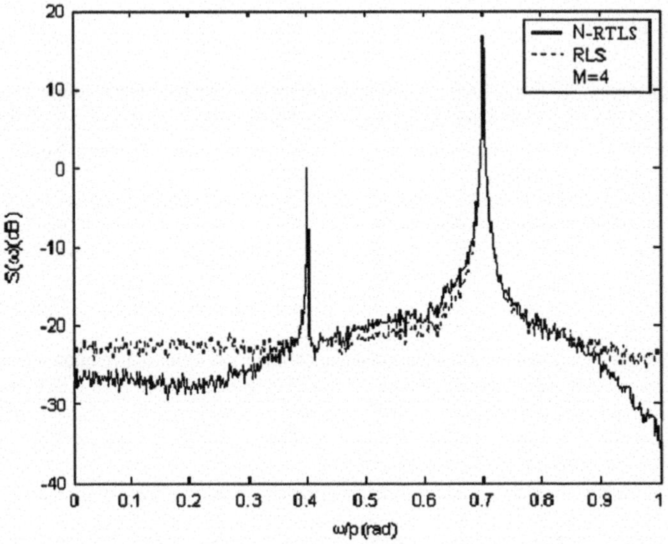

Fig. 4.10 Averaged periodograms of ALE outputs for N-RTLS and RLS for two sinusoids, ALE lengths of 4

filter coefficients produced by the N-RTLS algorithm are unbiased. The experiment results have been provided to confirm the efficiency of the proposed algorithm. The computational complexity of the N-RTLS algorithm is significantly lower than the existing algorithms.

4.5 Summary

In this chapter, the existing Davila's RTLS algorithms were reviewed. A new fast RTLS algorithm to recursively compute the TLS solution for adaptive FIR filtering is presented, and the algorithm is based on the minimization of the c-RQ in which the last entry of the parameter vector is constrained to the negative one. The high computational efficiency of the new algorithm depends on the efficient computation of the FGV and the adaptation of the c-RQ. In addition, a new fast RTLS algorithm to recursively compute the TLS solution for adaptive IIR filtering is also proposed, and the proposed algorithm is based on the minimization of the c-RQ in which the first entry of the parameter vector is fixed to the negative one. The new algorithm is numerically stable since it does not use the well-known matrix inversion lemma.

References

1. S. Haykin. *Adaptive Filter Theory* (PrenticeHall, Englewood Cliffs, NJ, 1996).
2. K. Pearson, On lines and planes of closest fit to points in space. Phil. Mag. **2**, 559–572 (1901)
3. G.H. Golub, C.F. Van Loan, An analysis of the total least squares problem. SIAM J. Number. Anal. **17**, 883–893 (1980)
4. S. Van Huffel, J. Vanderwalle, *The total least squares problem: computational aspects and analysis* (SIAM, Philadelphia, PA, 1991)
5. G.H. Golub, C.F. van Loan, *Matrix Computations* (Johns Hopkins University Press, Baltimore, MD, 1996)
6. V.F. Pisarenko, The retrieval of harmonics from a covariance function. Geophys. J. Royal Astonom. Soc. **33**, 347–366 (1973)
7. P. A. Thompson, Adaptive spectral analysis technique for unbiased frequency estimation in the presence white noise, in *Processing 13th Asilomar Conference on Circuits, Systems and Computing*, pp. 529–533 (1979)
8. R. Vaccaro, On adaptive implementations of Pisarenko's harmonic retrieval method, in *Proceeding of ICASSP* (1984), pp. 224–227
9. J.F. Yang, M. Kaveh, Adaptive eigen-subspace algorithms for direction or frequency estimation and tracking. IEEE Trans. Acoustic, Speech Signal Proc. **36**, 241–251 (1988)
10. N.L. Owsley, Adaptive data orthogonalization, in *Proceeding of ICASSP* (1978), pp. 109–112
11. E. Oja, Principal components, minor components, and linear neural networks. Neural Netw. **5**, 927–935 (1992)
12. L. Xu, E. Oja, C.Y. Suen, Modified Hebbian learning for curve and surface fitting. Neural Netw. **5**, 441–457 (1992)
13. L. Wang, J. Karhunen, A unified neural bi-gradient algorithm for robust PCA and MCA. Int. J. Neural Syst. **7**, 53–67 (1996)
14. F.L. Luo, R. Unbehauen, A. Cichocki, A minor component analysis algorithm. Neural Netw. **10**(2), 291–297 (1997)
15. K.Q. Gao, M.O. Ahmad, M.N.S. Swamy, A constrained anti-Hebbian learning algorithm for total least-squares estimation with applications to adaptive FIR and IIR filtering. IEEE Trans. Circ. Syst. **II**(41), 718–729 (1994)
16. C.S.T. Choy, W.C. Siu, Analysis of the convergence and divergence of a constrained anti-Hebbian learning algorithm. IEEE Trans. Circ. Syst. **II**(45), 1494–1502 (1998)
17. Q. Zhang, Y.W. Leung, Convergence of a Hebbian-type learning algorithm. IEEE Trans. Circ. Syst. II **46**(12), 1599–1601 (1999)
18. V. Solo, X. Kong, Performance analysis of adaptive eigen-analysis algorithms. IEEE Trans. Signal Proc. **46**, 636–646 (1998)
19. D.Z. Feng, Z. Bao, L.C. Jiao, Total least mean squares algorithm. IEEE T Trans. Sig. Proc. **46**, 2212–2220 (1998)
20. S. Ouyang, Z. Bao, G. Liao, Adaptive step-size minor component extraction algorithm. Electron. Lett. **35**(6), 443–444 (1999)
21. C.E. Davila, An efficient recursive total least squares algorithm for FIR adaptive filtering. IEEE Trans. Signal Process. **42**, 268–280 (1994)
22. L. Ljung, M. Morf, D. Falconer, Fast calculation of gain matrices for recursive estimation schemes. Int. J. Control. **7**(1), 1–19 (1978)
23. S. Ljung, L. Ljung, Error propagation properties of recursive leastsquares adaptation algorithms. Automatica **21**(2), 157–167 (1985)
24. J.L. Botto, G.V. Moustakides, Stabilization of fast Kalman algorithms. IEEE Trans. Acoustic, Speech Sig. Proc. **37**, 1342–1348 (1989)
25. D.T. Slock, T. Kailath, Numerically stable fast transversal filters for recursive least squares adaptive filtering. IEEE Trans. Signal Process. **39**, 92–114 (1991)
26. D.Z. Feng, X.D. Zhang, D.X. Chang, W.X. Zheng, A fast Recursive total least squares algorithm for adaptive FIR filtering. IEEE Trans. Signal Process. **52**(10), 2729–2737 (2004)

27. D.X. Chang, D.Z. Feng, W.X. Zheng, A fast recursive total least squares algorithm for adaptive IIR filtering. IEEE Trans. Signal Process. **53**(3), 975–965 (2005)
28. C.E. Davila, Line search algorithms for adaptive filtering. IEEE Trans. Signal Process. **41**, 2490–2494 (1993)
29. C.E. Davila, Efficient, high performance, subspace tracking for time-domain data. IEEE Trans. Signal Process. **48**, 3307–3315 (2000)
30. J.P. LaSalle, *The stability of dynamical systems* (SIAM, Philadelphia, PA, 1976)
31. C.E. Davila, An algorithm for efficient, unbiased, equation-error infinite impulse response adaptive filtering. IEEE Trans. Signal Process. **42**(5), 1221–1226 (1994)
32. T. Soderstrom, P. Stoica, *System Identification* (Prentice-Hall, Englewood Cliffs, NJ, 1989)
33. D.V.R. Rao, S.Y. Kung, Adaptive notch filtering for the retrieval of sinusoids in noise. *IEEE Transactions on Acoustic, Speech, and Signal Processing*, ASSP-32 (1984), pp. 791–802

Chapter 5
Approximate Inverse Power Iteration TLS Algorithm

5.1 Introduction

The RLS algorithms [1] may be one of the most powerful techniques that implement adaptive parameter estimation, which exhibit numerous attractive features, including the capability of tracking a variation of system parameters and provision of an unbiased estimate of system parameters in the case when only the system output is contaminated by white Gaussian noise. However, when there exists white Gaussian noise in both the system input and the system output or only within the system input, the RLS algorithms usually fail to yield an unbiased estimate of the system parameters, and cause a significantly degenerated performance of adaptive filtering. Instead, the TLS methods [2] may be the efficient technique used for achieving the unbiased estimate of the system parameters when the system input and the output are contaminated by white Gaussian noise.

Several recursive TLS (RTLS) algorithms have been established and are normally of computational complexity $O(M^2)$ per iteration. Some algorithms such as the inverse power (IP) method [3], the conjugate-gradient method [4], and the least squares-like method [5], also require $O(M^2)$ multiplication operations per eigenvector update. In order to solve online TLS problems in adaptive FIR filtering, a fast RTLS algorithm based on gradient search for the Rayleigh quotient (RQ) along Kalman gain vector [7] was firstly proposed by Davila [6]. This algorithm can fast-track the eigenvector associated with the smallest eigenvalue of the augmented correlation matrix, since Kalman gain vector can be fast estimated by the shift structure of the input data vector. Its computational complexity is $O(M)$ per iteration. When the system input and the output are contaminated by white Gaussian noise, the RTLS algorithms give an unbiased estimate, and their performance is superior to the famous RLS algorithms [1]. It is worth noting that Davila's RTLS algorithms are dependent on fast computation of Kalman gain vector.

In [8], it was shown that the computation of the Kalman gain vector might be potentially unstable. In order to overcome this potential instability, some efficient solution

X. Kong and D. Feng, *Efficient Online Learning Algorithms for Total Least Square Problems*, Engineering Applications of Computational Methods 21,
https://doi.org/10.1007/978-981-97-1765-1_5

approaches [9, 10] were developed, but such approaches have a more complex structure with increased computational complexity. Thus, to maintain numerical stability, Davila's RTLS algorithms would also have a more complex structure due to adoption of the Kalman gain vector. Later, the fast gain vector (FGV) was first given in [11], and a fast RTLS algorithm was established in [12]. The algorithms given in [6] and [12] find the TLS solution through searching the eigenvector associated with the smallest eigenvalue of the augmented data vector and by using gradient descent method, the computational complexity is decreased.

Later, several iterative TLS algorithms were proposed. In [13], the mixed LS-TLS method framework [13] was shown to be equivalent to minimization of the mean square equation error subject to a unit-norm constraint on the denominator parameter vector in ARMA model, and the mixed LS-TLS method framework provides an interesting way for unbiased adaptive filtering. Furthermore, steepest gradient-based algorithms based on the TLS or mixed LS-TLS cost function was established and analyzed in [14]. An extended least squares criterion (XLS) was introduced, and two iterative minimization algorithms for computing the XLS estimates were proposed in [15]. The error whitening criterion (EWC) methods, including the M instrumental variables, were established in [16].

The aim of this chapter is to develop more efficient fast recursive TLS algorithm for adaptive FIR filtering. In this chapter, we study the problem of adaptive FIR filtering in the TLS framework and develop a fast recursive algorithm by approximating inverse power (AIP) iteration along the input data vector. In addition, a novel and efficient algorithm to recursively extract multiple minor components from an N-dimensional vector sequence is developed. This algorithm is obtained by AIP iteration in conjunction with Galerkin method, and the convergence speed is faster than that of the stochastic gradient-based algorithms.

In this chapter, the existing IP iteration algorithms will be reviewed and discussed. A fast AIP iteration algorithm for adaptive TLS FIR filtering and an AIP algorithm for adaptive extraction of minor subspace proposed by us will be analyzed in detail. The remainder of this chapter is organized as follows. An overview of the existing inverse power iteration algorithms is presented in Sect. 5.2. Section 5.3 is devoted to a fast recursive algorithm by approximating the well-known IP iteration along the input data vector. In Sect. 5.4, an AIP algorithm for adaptive extraction of minor subspace is presented. Finally, the Sect. 5.5 is the summary of this chapter.

5.2 Review of Inverse Power Iteration

Given the fact that $\overline{\mathbf{R}}$ is a positive definite matrix and $\overline{\mathbf{D}}$ a positive semi-definite matrix, it follows from the matrix theory ([2], pp.468) that the matrix pair $(\overline{\mathbf{R}}, \overline{\mathbf{D}})$ has the generalized eigenvalue decomposition (GEVD):

$$\overline{V}^T\overline{DV} = \boldsymbol{\Gamma} \quad \text{and} \quad \overline{V}^T\overline{R}\,\overline{V} = \boldsymbol{\Phi}, \tag{5.1}$$

where $\boldsymbol{\Gamma} = \text{diag}(\rho_1, \rho_2, \ldots, \rho_{M+1})$, $\boldsymbol{\Phi} = \text{diag}(\varphi_1, \varphi_2, \ldots, \varphi_{M+1})$ and $\overline{\mathbf{V}} = [\mathbf{v}_1, \mathbf{v}_2, \ldots, \mathbf{v}_{M+1}] \in \Re^{(M+1)\times(M+1)}$ is a non-singular matrix.

Let all the generalized eigenvalues $\overline{\lambda}_i = \varphi_i/\rho_i \quad (i = 1, 2, \ldots, M+1)$ be arranged in a non-increasing order: $\overline{\lambda}_1 \geq \overline{\lambda}_2 \geq \overline{\lambda}_3 \geq \cdots \overline{\lambda}_{M+1}$. Then, we have

$$\overline{\mathbf{R}}\overline{\mathbf{v}}_i = \overline{\lambda}_i \overline{\mathbf{D}}\overline{\mathbf{v}}_i \quad (i = 1, 2, \ldots, M+1). \tag{5.2}$$

Moreover, it is apparent that

$$\overline{\boldsymbol{D}} = \overline{\boldsymbol{V}}^{-T} \boldsymbol{\Gamma} \overline{\boldsymbol{V}}^{-1} \quad \text{and} \quad \overline{\boldsymbol{R}} = \overline{\boldsymbol{V}}^{-T} \boldsymbol{\Phi} \overline{\boldsymbol{V}}^{-1} \tag{5.3}$$

It is easy to see that $\overline{\lambda}_{M+1} = \sigma_i^2$ and $\overline{\mathbf{v}}_{M+1} = [\mathbf{h}^T, -1]/\|\mathbf{h}^T, -1\|$. Since $\overline{\mathbf{v}}_{M+1}$ is associated with the unbiased solution for FIR filtering, an iterative algorithm is presented to find the desired generalized eigenvector $\overline{\mathbf{v}}_{M+1}$ corresponding to the smallest generalized eigenvalue $\overline{\lambda}_{M+1} = \sigma_i^2$.

The TLS solution for adaptive FIR filtering is usually related only to the single eigenvector corresponding to the smallest generalized eigenvalue of the matrix pair $(\overline{\mathbf{R}}, \overline{\mathbf{D}})$. The well-known IP iteration [2] provides an efficient technique for finding such a single eigenvector associated with the smallest generalized eigenvalue of the matrix pair $(\overline{\mathbf{R}}, \overline{\mathbf{D}})$, which is stated as follows.

Randomly produce the initial value $q(0)$; for $t = 1, 2, \ldots$, solve a set of linear equations

$$\overline{\mathbf{R}}\widehat{\mathbf{q}}(t) = \overline{\mathbf{D}}\mathbf{q}(t-1), \tag{5.4}$$

and perform the following normalization

$$\mathbf{q}(t) = \widehat{\mathbf{q}}(t)/\left\|\widehat{\mathbf{q}}(t)\right\|_2. \tag{5.5}$$

It is known from [2] that the IP iteration globally exponentially converges to the eigenvector associated with the smallest generalized eigenvalue of the matrix pair $(\overline{\mathbf{R}}, \overline{\mathbf{D}})$. Unfortunately, the IP iteration is an algorithm with computational complexity $O(M^3)$. Of course, if $\overline{\mathbf{R}}$ is updated via the matrix inversion formula like the RLS algorithms, then the IP iteration may become an algorithm with computational complexity $O(M^2)$. However, the potential instability of the matrix inverse update formula [8–10] may result in the potential instability of such IP iteration.

5.3 AIP Iteration Algorithm for Adaptive TLS FIR Filtering

The presence of contaminating noises at both the input and the output of an FIR system constitutes a major impediment to unbiased parameter estimation. The TLS method is known to be effective in achieving unbiased estimation. In this section, a fast recursive algorithm with a view to finding the TLS solution for adaptive FIR filtering is developed. Given the fact that the TLS solution is obtainable via IP iteration, a novel but AIP iteration is introduced in combination with Galerkin method so that the TLS solution can be updated adaptively at a lower computational cost. The regular form of the TLS solution is used to constrain the last element of the filter parameter vector to the negative one. The computational complexity of the developed algorithm is reduced by making efficient computation of the FGV defined in [11] and using rank-one update of the augmented autocorrelation matrix. Therefore, it has the simpler structure and may be more easily implemented. The global convergence of the developed algorithm is investigated carefully.

5.3.1 *Preliminaries*

A. *Signal Model*

As shown in Fig. 5.1, both the input and output of an unknown system are contaminated by white Gaussian noise. Let the FIR vector of the system be described by $\mathbf{h} = [h_0, h_1, \ldots, h_{M-1}]^T$. Then the so-called expectation output is given by $d(t) = \mathbf{x}^T(t)\mathbf{h} + n_o(t)$, where the output noise $n_o(t)$ is a white Gaussian noise with variance σ_o^2, and the noise-free input vector is represented by $\mathbf{x}(t) = [x(t), x(t-1), \ldots, x(t-M+1)]^T$. The noisy input vector of the adaptive FIR filter is given by $\tilde{\mathbf{x}}(t) = \mathbf{x}(t) + \mathbf{n}_i(t)$, where $\mathbf{n}_i(t) = [n_i(t), n_i(t-1), \ldots, n_i(t-M+1)]^T$ and the input noise $n_i(t)$ is white Gaussian noise with variance σ_i^2. Moreover, the augmented data vector is defined as $\bar{\mathbf{x}}(t) = [\tilde{\mathbf{x}}^T(t), d(t)]^T \in \Re^{(M+1)\times 1}$. Here, it is assumed that the input signal $x(t)$, the input noise $n_i(t)$, and the output noise $n_o(t)$ are mutually independent of each other.

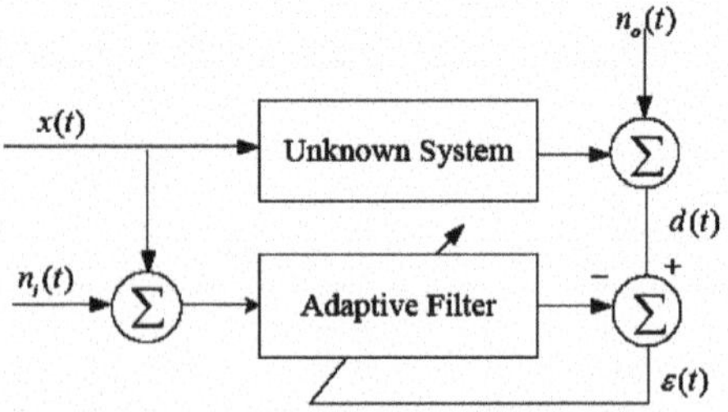

Fig. 5.1 Unknown system $h = [h_0, h_1, \ldots, h_{M-1}]$ is identified by an adaptive FIR filter with input noise

Let the autocorrelation matrix of the noise-free input vector be defined by $\mathbf{R} = E\{\mathbf{x}(t)\mathbf{x}^T(t)\}$. Then, the autocorrelation matrix of the noisy input vector is given by $\tilde{\mathbf{R}} = E\{\tilde{\mathbf{x}}(t)\tilde{\mathbf{x}}^T(t)\} = \mathbf{R} + \sigma_i^2\mathbf{I}$. The autocorrelation matrix of the augmented data vector is expressible as

$$\overline{\mathbf{R}} = E\{\overline{\mathbf{x}}(t)\overline{\mathbf{x}}^T(t)\} = \begin{bmatrix} \tilde{\mathbf{R}} & \mathbf{b} \\ \mathbf{b}^T & c \end{bmatrix}, \tag{5.6}$$

where $\mathbf{b} = E\{\tilde{\mathbf{x}}(t)d(t)\} = \mathbf{Rh}$ and $c = E\{d(t)d(t)\} = \mathbf{h}^T\mathbf{Rh} + \sigma_o^2$. Introduce a matrix

$$\mathbf{R}^* = \begin{bmatrix} \mathbf{R} & \mathbf{Rh} \\ \mathbf{h}^T\mathbf{R} & \mathbf{h}^T\mathbf{Rh} \end{bmatrix}. \tag{5.7}$$

Then, it is obvious that $\mathbf{R}^*[\mathbf{h}^T, -1]^T = 0$.

For the purpose of efficiently searching for the TLS solution for adaptive FIR filtering, the following RQ was established in [6]:

$$J(\mathbf{q}) = (\mathbf{q}^T\overline{\mathbf{R}}\mathbf{q})/(\mathbf{q}^T\overline{\mathbf{D}}\mathbf{q}), \tag{5.8}$$

where $\mathbf{q} \in \Re^{(M+1)\times 1}$ is the parameter vector, and $\overline{\mathbf{D}} = diag(\mathbf{I}_M, \xi)$ is a diagonal weighting matrix with $\xi = \sigma_o^2/\sigma_i^2$ denoting the ratio of the output noise variance to the input noise variance. It was shown in [6] that if $\mathbf{q}^*$ is the parameter vector associated with minimization of $J(\mathbf{q})$ and q_{M+1}^* denotes the last element of $\mathbf{q}^*$, then the unbiased solution $\boldsymbol{w}_{\text{TLS}}$ for adaptive FIR filtering is obtained by

$$\begin{pmatrix} \mathbf{w}_{TLS} \\ -1 \end{pmatrix} = -\mathbf{q}^*/q_{M+1}^*. \tag{5.9}$$

B. *The Proposed Inverse Power Iteration of TLS*

Although the IP iteration in Sect. 5.2 is usually efficient and robust, it is not especially appropriate for seeking the TLS solution of adaptive FIR filtering since an extra normalization operation is required, as shown in (5.5). It can be seen from (5.9) that the required non-normalized eigenvector associated with the smallest generalized eigenvalue of the matrix pair $(\overline{\mathbf{R}}, \overline{\mathbf{D}})$ should have the following regular form

$$\mathbf{q} = [\mathbf{w}^T, -1]^T, \ \mathbf{w} \in \Re^M, \tag{5.10}$$

i.e., the last entry of the required non-normalized eigenvector should be equal to -1. Thus, in order to maintain that the eigenvector associated with the smallest generalized eigenvalue of the matrix pair $(\overline{\mathbf{R}}, \overline{\mathbf{D}})$ has the form as shown in (5.10), we propose an alternative IP iteration as follows.

Randomly produce the initial value $\boldsymbol{w}(0)$; for $t = 1, 2, \ldots$, solve a set of linear equations

$$\overline{\mathbf{R}}\widehat{\mathbf{q}}(t) = \overline{\mathbf{D}}[\mathbf{w}^T(t-1), -1]^T \tag{5.11}$$

and perform the following normalization:

$$[\mathbf{w}^T(t-1), -1]^T = -\widehat{\mathbf{q}}(t)/[\widehat{\mathbf{q}}(t)]_{M+1,M+1}. \tag{5.12}$$

It can be shown that the following lemma holds for the above iteration.

Lemma 5.1 *Given the matrix pair* $(\overline{\mathbf{R}}, \overline{\mathbf{D}})$, *if the smallest generalized eigenvalue* $\overline{\lambda}_{M+1} = \sigma_i^2$ *is distinct from the other eigenvalues and* $[\overline{\mathbf{V}}^{-1}]_{M+1,M+1}\underset{\smile}{\mathbf{w}}(0) \neq 0$, *then* $\underset{\smile}{\mathbf{w}}(t) = [\mathbf{w}^T(t), -1]^T$ *in the iteration* (5.11) *and* (5.12) *globally exponentially converges to* $-\overline{\mathbf{v}}_{M+1}/[\overline{\mathbf{v}}_{M+1}]_{M+1,M+1} = [\mathbf{h}^T, -1]^T$ *as* $t \to \infty$.

Proof Substituting $\overline{\mathbf{D}} = \overline{\mathbf{V}}^{-T}\Gamma\overline{\mathbf{V}}^{-1}, \overline{\mathbf{R}} = \overline{\mathbf{V}}^{-T}\Phi\overline{\mathbf{V}}^{-1}, \underset{\smile}{\mathbf{u}}(t) = \overline{\mathbf{V}}^{-1}[\mathbf{w}^T(t), -1]^T$, $\widehat{\mathbf{u}}(t) = \overline{\mathbf{V}}^{-1}\widehat{\mathbf{q}}(t)$ and $[\widehat{\mathbf{q}}(t)]_{M+1,M+1} = [\overline{\mathbf{V}}\widehat{\mathbf{u}}(t)]_{M+1,M+1} = [\overline{\mathbf{V}}]_{M+1,M+1}\widehat{\mathbf{u}}(t)$ into (5.11) and (5.12) and conducting some operations give

$$\widehat{\mathbf{u}}(t) = \Lambda\underset{\smile}{\mathbf{u}}(t-1), \tag{5.13}$$

$$\underset{\smile}{\mathbf{u}}(t) = -\widehat{\mathbf{u}}(t)/(\tilde{\mathbf{v}}\,\widehat{\mathbf{u}}(t)) \tag{5.14}$$

where $\tilde{\mathbf{v}} = [\overline{\mathbf{V}}]_{M+1,M+1}$ and $\mathbf{\Lambda} = \mathbf{\Phi}^{-1}\mathbf{\Gamma} = diag(1/\overline{\lambda}_1, \ldots, 1/\overline{\lambda}_{M+1})$. Substituting (5.13) into (5.14) yields

$$\underset{\smile}{\mathbf{u}}(t) = -\Lambda\underset{\smile}{\mathbf{u}}(t-1)/[\tilde{\mathbf{v}}\,\Lambda\underset{\smile}{\mathbf{u}}(t-1)]. \tag{5.15}$$

Repeatedly using (5.15) produces

$$\begin{aligned}\underset{\smile}{\mathbf{u}}(t) &= -\frac{\Lambda\underset{\smile}{\mathbf{u}}(t-1)}{\tilde{\mathbf{v}}\,\Lambda\underset{\smile}{\mathbf{u}}(t-1)} = -\frac{\Lambda[-\Lambda\underset{\smile}{\mathbf{u}}(t-2)]/\tilde{\mathbf{v}}\,\Lambda\underset{\smile}{\mathbf{u}}(t-2)}{\tilde{\mathbf{v}}\,\Lambda[-\Lambda\underset{\smile}{\mathbf{u}}(t-2)]/\tilde{\mathbf{v}}\,\Lambda\underset{\smile}{\mathbf{u}}(t-2)]} \\ &= -\frac{\Lambda^2\underset{\smile}{\mathbf{u}}(t-2)}{\tilde{\mathbf{v}}\,\Lambda^2\underset{\smile}{\mathbf{u}}(t-2)} = \cdots = -\frac{\Lambda^t\underset{\smile}{\mathbf{u}}(0)}{\tilde{\mathbf{v}}\,\Lambda^t\underset{\smile}{\mathbf{u}}(0)}.\end{aligned} \tag{5.16}$$

Since $\overline{\lambda}_{M+1}^{t}\mathbf{\Lambda}^t$ globally exponentially converges to $\text{diag}(0, \ldots, 0, 1) \in \Re^{(M+1)\times(M+1)}$, as $t \to \infty$, we obtain

$$\underset{\smile}{\mathbf{u}}(t) = -\frac{\Lambda^t\underset{\smile}{\mathbf{u}}(0)}{\tilde{\mathbf{v}}\,\Lambda^t\underset{\smile}{\mathbf{u}}(0)} = -\frac{\overline{\lambda}_{M+1}^t\Lambda^t\underset{\smile}{\mathbf{u}}(0)}{\tilde{\mathbf{v}}\,\overline{\lambda}_{M+1}^t\Lambda^t\underset{\smile}{\mathbf{u}}(0)}$$

$$\rightarrow -\frac{\text{diag}(0,\ldots,0,1)\underset{\smile}{\mathbf{u}}(0)}{\tilde{\mathbf{v}}\text{diag}(0,\ldots,0,1)\underset{\smile}{\mathbf{u}}(0)}. \quad (5.17)$$

The above equation can be rewritten in the following component form:

$$u_i(t) \rightarrow 0 \quad \text{for} \quad i = 1, 2, \ldots, M, \quad (5.18)$$

$$u_{M+1}(t) \rightarrow -1/[\tilde{\mathbf{v}}]_{M+1,M+1}, \quad (5.19)$$

as $t \rightarrow \infty$. Then, we have

$$\underset{\smile}{\mathbf{w}}(t) = \overline{\mathbf{V}}\underset{\smile}{\mathbf{u}}(t) \rightarrow -\overline{\mathbf{v}}_{M+1}/[\overline{\mathbf{v}}_{M+1}]_{M+1,M+1} = [\mathbf{h}^T, -1]^T. \quad (5.20)$$

This completes the Proof of Lemma 5.1.

Combining (5.11) and (5.12) together, we get the following equivalent IP iteration.

Randomly produce the initial value $\mathbf{w}(0)$; for t = 1,2,..., solve a set of linear equations

$$\overline{\mathbf{R}}[\mathbf{w}^T(t), -1]^T = \gamma(t)\overline{\mathbf{D}}[\mathbf{w}^T(t-1), -1]^T, \quad (5.21)$$

where $\gamma(t)$ is an unknown scalar quantity that varies with time.

For convenience of convergence analysis, the cost function (5.8) in combination with (5.10) is redefined as

$$\hat{J}(\mathbf{w}) = ([\mathbf{w}^T, -1]\overline{\mathbf{R}}[\mathbf{w}^T, -1]^T)/([\mathbf{w}^T, -1]\overline{\mathbf{D}}[\mathbf{w}^T, -1]^T). \quad (5.22)$$

Since the last entry of $\boldsymbol{q}$ is constrained to -1, we refer to the above cost function as the constrained RQ (c-RQ). The following lemma holds for the c-RQ.

Lemma 5.2 [12]: *If $\boldsymbol{R}$ is of full rank, then the stationary point $\mathbf{w} = \boldsymbol{h}$ of $\hat{J}(\mathbf{w})$ is the unique global minimum point and the other stationary points are saddle or unstable.*

This lemma shows that the cost function $\hat{J}(\mathbf{w})$ can be taken as the energy function of a dynamical system [17].

5.3.2 *AIP Iteration Algorithm*

This section presents an alternative efficient algorithm for finding the TLS solution of the adaptive FIR filtering problem. The basic idea is to use Galerkin method for implementing an AIP iteration and to take the input data vector as the update direction. In this way, the developed algorithm is made with computational complexity $O(M)$, thus being computationally efficient.

First, the parameter vector in (5.21) is updated along the noisy input vector $\tilde{\mathbf{x}}(t)$ by the following rank-one updating formula:

$$\mathbf{w}(t) = \mathbf{w}(t-1) + \beta(t)\tilde{\mathbf{x}}(t), \tag{5.23}$$

where $\beta(t)$ is an unknown parameter. Substitution of (5.23) into (5.21) gives an approximate formula

$$\overline{\mathbf{R}}(t)[\mathbf{w}^T(t-1) + \beta(t)\tilde{\mathbf{x}}^T(t), -1]^T = \gamma(t)\overline{\mathbf{D}}[\mathbf{w}^T(t-1), -1]^T. \tag{5.24}$$

It can be seen that (5.24) represents an overdetermined system of linear equations with respect to two unknown parameters $\beta(t)$ and $\gamma(t)$, and it can be often solved by a least squares method. In order that the parameters $\beta(t)$ and $\gamma(t)$ can be efficiently determined in $O(M)$ multiplications, however, the Galerkin method [18] is used here. Pre-multiplying (5.24) with $[\tilde{\mathbf{x}}^T(t), 0]$ and $-[\mathbf{w}^T(t-1), -1]$, respectively, leads to

$$\begin{aligned}&[\tilde{\mathbf{x}}^T(t), 0]\{\bar{\mathbf{R}}(t)[\mathbf{w}^T(t-1) + \beta(t)\tilde{\mathbf{x}}^T(t), -1]^T\\&- \gamma(t)\bar{\mathbf{D}}[\mathbf{w}^T(t-1), -1]^T\} = 0,\end{aligned} \tag{5.25a}$$

$$\begin{aligned}&-[\mathbf{w}^T(t-1), -1]\{\bar{\mathbf{R}}(t)[\mathbf{w}^T(t-1) + \beta(t)\tilde{\mathbf{x}}^T(t), -1]^T\\&- \gamma(t)\bar{\mathbf{D}}[\mathbf{w}^T(t-1), -1]^T\} = 0.\end{aligned} \tag{5.25b}$$

Notice that $\overline{\mathbf{R}}(t)$ is computed by an iteration formula

$$\overline{\mathbf{R}}(t) = \begin{bmatrix}\tilde{\mathbf{R}}(t) & \mathbf{b}(t)\\ \mathbf{b}^T(t) & c(t)\end{bmatrix} = \mu\overline{\mathbf{R}}(t-1) + \overline{\mathbf{x}}(t)\overline{\mathbf{x}}^T(t), \tag{5.26}$$

where

$$\begin{cases}\tilde{\mathbf{R}}(t) = \mu\tilde{\mathbf{R}}(t-1) + \tilde{\mathbf{x}}(t)\tilde{\mathbf{x}}^T(t)\\ \mathbf{b}(t) = \mu\mathbf{b}(t-1) + \tilde{\mathbf{x}}(t)d(t)\\ c(t) = \mu c(t-1) + d(t)d(t).\end{cases} \tag{5.27}$$

The μ in (5.26) and (5.27) denotes the forgetting factor. By substituting (5.26) and (5.27) into (5.25a–5.25b), we have

$$\begin{aligned}&[\tilde{\mathbf{x}}^T(t)\tilde{\mathbf{R}}(t), \tilde{\mathbf{x}}^T(t)\mathbf{b}(t)][(\mathbf{w}(t-1) + \beta(t)\tilde{\mathbf{x}}(t))^T, -1]^T\\&- \gamma(t)[\tilde{\mathbf{x}}^T(t)\mathbf{w}(t-1)] = 0,\end{aligned} \tag{5.28a}$$

$$\begin{aligned}&-[\mathbf{w}^T(t-1), -1]\bar{\mathbf{R}}(t)[\mathbf{w}^T(t-1), -1]^T\\&- \beta(t)[\mathbf{w}^T(t-1), -1]\bar{\mathbf{R}}(t)[\tilde{\mathbf{x}}^T(t), 0]^T\end{aligned}$$

$$+\gamma(t)[\mathbf{w}^T(t-1)\mathbf{w}(t-1)+\xi]=0. \tag{5.28b}$$

By some manipulations, (5.28a–5.28b) can be further expressed as

$$\begin{aligned}&\beta(t)[\tilde{\mathbf{x}}^T(t)\tilde{\mathbf{R}}(t)\tilde{\mathbf{x}}(t)]-\gamma(t)[\tilde{\mathbf{x}}^T(t)\mathbf{w}(t-1)]\\&=-[\tilde{\mathbf{x}}^T(t)\tilde{\mathbf{R}}(t)\tilde{\mathbf{x}}(t-1)]+\tilde{\mathbf{x}}^T(t)\mathbf{b}(t),\end{aligned} \tag{5.29a}$$

$$\begin{aligned}&-\beta(t)[\mathbf{w}^T(t-1)\tilde{\mathbf{R}}(t)\widetilde{\mathbf{x}}(t)-\mathbf{b}^T(t)\widetilde{\mathbf{x}}(t)]\\&+\gamma(t)[\mathbf{w}^T(t-1)\mathbf{w}(t-1)+\xi]\\&=[\mathbf{w}^T(t-1),-1]\bar{\mathbf{R}}(t)[\mathbf{w}^T(t-1),-1]^T.\end{aligned} \tag{5.29b}$$

In order to efficiently solve (5.29a–5.29b), let

$$y(t)=\mathbf{w}^T(t-1)\tilde{\mathbf{x}}(t), \tag{5.30}$$

$$\mathbf{k}(t)=\tilde{\mathbf{R}}(t)\tilde{\mathbf{x}}(t), \tag{5.31}$$

$$\begin{cases}a_1(t)=\mathbf{b}^T(t)\tilde{\mathbf{x}}(t)\\a_2(t)=\tilde{\mathbf{x}}^T(t)\mathbf{k}(t)\\a_3(t)=\mathbf{w}^T(t-1)\mathbf{k}(t)\\a_4(t)=\mathbf{w}^T(t-1)\mathbf{w}(t-1)+\xi\end{cases}, \tag{5.32}$$

$$\begin{cases}\varsigma^0(t)=[\mathbf{w}^T(t-1),-1]\overline{\mathbf{R}}(t)[\mathbf{w}^T(t-1),-1]^T\\\varsigma(t)=[\mathbf{w}^T(t),-1]\overline{\mathbf{R}}(t)[\mathbf{w}^T(t),-1]^T\end{cases}. \tag{5.33}$$

The gain vector $\boldsymbol{k}(t)$ was first defined in [11], while its fast gain vector algorithm (FGA) was also obtained in [11]. A step-by-step scheme for the FGA was given in [12].

Note that in virtue of the gain vector $\boldsymbol{k}(t)$, the variable pairs $\varsigma^0(t)$ and $\varsigma(t)$ can be efficiently computed by

$$\begin{aligned}\varsigma^0(t)&=[\mathbf{w}^T(t-1),-1]\{\mu\overline{\mathbf{R}}(t-1)\\&\quad+\bar{\mathbf{x}}(t)\bar{\mathbf{x}}^T(t)\}[\mathbf{w}^T(t-1),-1]^T\\&=\mu\varsigma(t-1)+(y(t)-d(t))^2,\end{aligned} \tag{5.34}$$

$$\begin{aligned}&\varsigma(t)=[\mathbf{w}^T(t-1)+\beta(t)\widetilde{\mathbf{x}}^T(t),-1]\\&\overline{\mathbf{R}}(t)[\mathbf{w}^T(t-1)+\beta(t)\widetilde{\mathbf{x}}^T(t),-1]^T\\&=\varsigma^0(t)+2\beta(t)[\mathbf{k}^T(t)\mathbf{w}(t-1)-\mathbf{b}^T(t)\widetilde{\mathbf{x}}(t)]\\&\quad+\beta^2(t)\mathbf{k}^T(t)\widetilde{\mathbf{x}}(t).\end{aligned} \tag{5.35}$$

By using (5.30–5.34), (5.29a–5.29b) can be rewritten as

$$\beta(t)a_2(t) - \gamma(t)y(t) = -a_3(t) + a_1(t), \tag{5.36a}$$

$$-\beta(t)[a_3(t) - a_1(t)] + a_4(t)\gamma(t) = \varsigma^0(t). \tag{5.36b}$$

The above equation can be recast in matrix form as

$$\begin{bmatrix} a_2(t) & -y(t) \\ -a_3(t) + a_1(t) & a_4(t) \end{bmatrix} \begin{bmatrix} \beta(t) \\ \gamma(t) \end{bmatrix} = \begin{bmatrix} -a_3(t) + a_1(t) \\ \varsigma^0(t) \end{bmatrix}. \tag{5.37}$$

Solving (5.37) yields

$$\begin{aligned} \beta(t) &= \begin{vmatrix} -a_3(t) + a_1(t) & -y(t) \\ \varsigma^0(t) & a_4(t) \end{vmatrix} \Big/ \begin{vmatrix} a_2(t) & -y(t) \\ -a_3(t) + a_1(t) & a_4(t) \end{vmatrix} \\ &= [\varsigma^0(t)y(t) - a_4(t)(a_3(t) - a_1(t))]/[a_2(t)a_4(t) - y(t)(a_3(t) - a_1(t))]. \end{aligned} \tag{5.38}$$

Notice that $\gamma(t)$ is not interested for finding the TLS solution of adaptive FIR filtering.

On the basis of the above derivation, we can now summarize the developed algorithm in Table 5.1, which is termed the AIP algorithm.

It is worth mentioning that the MAD's of the developed AIP algorithms is 15 M + 12. Since the MADs of the algorithm in [6] is 22 M + 74 and the MADs of the algorithm given in [12] is 16 M + 20, it is clear that the computational complexity

Table 5.1 AIP algorithm for adaptive TLS FIR filtering

Initialize: $\mathbf{w}(0) = [0, 0, \ldots, 0]^T$, $\mathbf{b}(0) = \mathbf{0}$, $\varsigma(0) = 0$, $\mu = 0.99 \sim 1.0$
For $t = 1, 2, \ldots$
1. Update the data vector $\tilde{\mathbf{x}}(t)$ 2. Update gain $\boldsymbol{k}(t)$ (See Table 5.1 in [12])
3. $y(t) = \mathbf{w}^T(t-1)\tilde{\mathbf{x}}(t)$
4. $\varsigma^0(t) = \mu\varsigma(t-1) + (y(t) - d(t))^2$
5. $\mathbf{b}(t) = \mu\mathbf{b}(t-1) + \tilde{\mathbf{x}}(t)d(t)$
6. $\begin{cases} a_1(t) = \mathbf{b}^T(t)\tilde{\mathbf{x}}(t) \\ a_2(t) = \tilde{\mathbf{x}}^T(t)\mathbf{k}(t) \\ a_3(t) = \mathbf{w}^T(t-1)\mathbf{k}(t) \\ a_4(t) = \mathbf{w}^T(t-1)\mathbf{w}(t-1) + \xi \end{cases}$
7. $\beta(t) = [\varsigma^0(t)y(t) - a_4(t)(a_3(t) - a_1(t))]/[a_2(t)a_4(t) - y(t)(a_3(t) - a_1(t))]$
8. $\mathbf{w}(t) = \mathbf{w}(t-1) + \beta(t)\tilde{\mathbf{x}}(t)$
9. $\varsigma(t) = \varsigma^0(t) + \beta(t)[2(a_3(t) - a_1(t)) + \beta(t)a_2(t)]$

of the AIP algorithm is not only significantly lower than that in [6], but also lower than that in [12]. Moreover, in comparison with the algorithms in [6] and [12], the structure of the formulations associated with the AIP algorithm is much simpler and may be more easily implemented. For example, unlike the algorithms given in [6, 12], the developed algorithm does not deal with solving a one-variable quadratic equation and it avoids the square root operation (note that the square root operation sometimes results in an unexpected complexity). In addition, a division operation is also avoided.

5.3.3 Algorithm Convergence

This section is concerned with convergence property of the developed AIP algorithm. To this end, we consider the stochastic sequence *w*(t) generated by the AIP algorithm. Given the fact that the sequence *w*(t) is a discrete-time dynamical system, its convergence can be shown by the well-known energy function method [17].

Lemma 5.3 *If t is large enough such that* $\overline{\mathbf{R}}(t) \to \overline{\mathbf{R}}$, *then* $\hat{J}(\mathbf{w}(t))$ *is monotonically decreasing, namely* $\hat{J}(\mathbf{w}(t)) \leq \hat{J}(\mathbf{w}(t-1))$.

Proof Substitution of (5.23) into the expression $\hat{J}(\mathbf{w}(t)) \leq \hat{J}(\mathbf{w}(t-1))$ gives rise to (5.39). By means of Table 5.1 and (5.26), (5.30–5.32), and (5.37–5.38), (5.39) can be further written as (5.40).

$$
\begin{aligned}
&\hat{J}(\mathbf{w}(t)) - \hat{J}(\mathbf{w}(t-1)) \\
&= \frac{[\mathbf{w}^T(t-1) + \beta(t)\tilde{\mathbf{x}}^T(t), -1]\overline{\mathbf{R}}[\mathbf{w}^T(t-1) + \beta(t)\tilde{\mathbf{x}}^T(t), -1]^T}{[\mathbf{w}^T(t-1) + \beta(t)\tilde{\mathbf{x}}^T(t), -1]\bar{\mathbf{D}}[\mathbf{w}^T(t-1) + \beta(t)\tilde{\mathbf{x}}^T(t), -1]^T} \\
&\quad - \frac{[\mathbf{w}^T(t-1), -1]\overline{\mathbf{R}}[\mathbf{w}^T(t-1), -1]^T}{\mathbf{w}^T(t-1)\mathbf{w}(t-1) + \xi} \\
&= \frac{[\mathbf{w}^T(t-1), -1]\overline{\mathbf{R}}[\mathbf{w}^T(t-1), -1]^T + 2\beta(t)[\mathbf{w}^T(t-1), -1]\overline{\mathbf{R}}[\tilde{\mathbf{x}}^T(t), 0]^T + \beta^2(t)\tilde{\mathbf{x}}^T(t)\overline{\mathbf{R}}\tilde{\mathbf{x}}(t)}{[\mathbf{w}^T(t-1)\mathbf{w}(t-1) + \xi] + 2\beta(t)y(t) + \beta^2(t)\|\tilde{\mathbf{x}}(t)\|^2} \\
&\quad - \frac{[\mathbf{w}^T(t-1), -1]\overline{\mathbf{R}}[\mathbf{w}^T(t-1), -1]^T}{\mathbf{w}^T(t-1)\mathbf{w}(t-1) + \xi},
\end{aligned} \tag{5.39}
$$

$$
\begin{aligned}
\hat{J}(\mathbf{w}(t)) - \hat{J}(\mathbf{w}(t-1)) &= \frac{\varsigma^0(t) + 2\beta(t)(a_3(t) - a_1(t)) + \beta^2(t)a_2(t)}{a_4(t) + 2\beta(t)y(t) + \beta^2(t)\|\tilde{\mathbf{x}}(t)\|^2} - \frac{\varsigma^0(t)}{a_4(t)} \\
&= \frac{a_4(t)[\varsigma^0(t) + 2\beta(t)(a_3(t) - a_1(t)) + \beta^2(t)a_2(t)] - \varsigma^0(t)[a_4(t) + 2\beta(t)y(t) + \beta^2(t)\|\tilde{\mathbf{x}}(t)\|^2]}{a_4(t)[a_4(t) + 2\beta(t)y(t) + \beta^2(t)\|\tilde{\mathbf{x}}(t)\|^2]} \\
&= \frac{-2\beta(t)[\varsigma^0(t)y(t) - a_4(t)(a_3(t) - a_1(t))] + \beta^2(t)[a_2(t)a_4(t) - \varsigma^0(t)\|\tilde{\mathbf{x}}(t)\|^2]}{a_4(t)[a_4(t) + 2\beta(t)y(t) + \beta^2(t)\|\tilde{\mathbf{x}}(t)\|^2]} \\
&= \frac{-2\beta^2(t)[a_2(t)a_4(t) + y(t)(a_3(t) - a_1(t))] + \beta^2(t)[a_2(t)a_4(t) - \varsigma^0(t)\|\tilde{\mathbf{x}}(t)\|^2]}{a_4(t)[a_4(t) + 2\beta(t)y(t) + \beta^2(t)\|\tilde{\mathbf{x}}(t)\|^2]} \\
&= \frac{-\beta^2(t)[a_2(t)a_4(t) + 2y(t)(a_3(t) - a_1(t)) + \varsigma^0(t)\|\tilde{\mathbf{x}}(t)\|^2]}{a_4(t)[a_4(t) + 2\beta(t)y(t) + \beta^2(t)\|\tilde{\mathbf{x}}(t)\|^2]}.
\end{aligned} \tag{5.40}
$$

Let $\rho(t) = a_2(t)a_4(t)+2y(t)(a_3(t)-a_1(t))+\varsigma^0(t)\|\tilde{\mathbf{x}}(t)\|^2$. According to (5.40), in order to prove this lemma, it suffices to show that $\rho(t)$ is larger than or equal to zero. By virtue of Table 5.1, $\rho(t)$ is expressible as

$$\begin{aligned}\rho(t) &= [\tilde{\mathbf{x}}^T(t), 0]\overline{\mathbf{R}}[\tilde{\mathbf{x}}^T(t), 0]^T[\mathbf{w}^T(t-1), -1]\overline{\mathbf{D}}[\mathbf{w}^T(t-1), -1]^T \\ &- 2[\mathbf{w}^T(t-1), -1]\overline{\mathbf{R}}[\tilde{\mathbf{x}}^T(t), 0]^T[\mathbf{w}^T(t-1), -1]\overline{\mathbf{D}}[\tilde{\mathbf{x}}^T(t), 0]^T \\ &+ [\mathbf{w}^T(t-1), -1]\overline{\mathbf{R}}[\mathbf{w}^T(t-1), -1][\tilde{\mathbf{x}}^T(t), 0]\overline{\mathbf{D}}[\tilde{\mathbf{x}}^T(t), 0]^T.\end{aligned} \tag{5.41}$$

Introduce $\tilde{\mathbf{D}} = \overline{\mathbf{R}}^{-1/2}\overline{\mathbf{D}}\overline{\mathbf{R}}^{-1/2}$, $\tilde{\mathbf{u}} = \overline{\mathbf{R}}^{1/2}[\tilde{\mathbf{x}}^T(t), 0]^T$, $\tilde{\mathbf{v}} = \overline{\mathbf{R}}^{1/2}[\tilde{\mathbf{x}}^T(t), -1]^T$. Then, (5.41) can be transformed into

$$\begin{aligned}\rho(t) &= \tilde{\mathbf{u}}^T(t)\tilde{\mathbf{u}}(t)\tilde{\mathbf{v}}^T(t)\bar{\mathbf{D}}\tilde{\mathbf{v}}(t) \\ &- 2\tilde{\mathbf{v}}^T(t)\tilde{\mathbf{u}}(t)\tilde{\mathbf{v}}^T(t)\bar{\mathbf{D}}\tilde{\mathbf{u}}(t) + \tilde{\mathbf{v}}^T(t)\tilde{\mathbf{v}}(t)\tilde{\mathbf{u}}^T(t)\bar{\mathbf{D}}\tilde{\mathbf{u}}(t).\end{aligned} \tag{5.42}$$

Let the EVD of $\tilde{\mathbf{D}}$ be given by $\tilde{\mathbf{D}} = \mathbf{Q}\Psi\mathbf{Q}^T$, where $\Psi = \mathrm{diag}[\psi_1, \psi_2, \ldots, \psi_{M+1}]$ stands for the eigenvalue matrix and $\mathbf{Q}$ denotes the eigenvector matrix formed by all the corresponding eigenvectors. Furthermore, introduce $\mathbf{u} = \mathbf{Q}^T\tilde{\mathbf{u}}$ and $\mathbf{v} = \mathbf{Q}^T\tilde{\mathbf{v}}$, where $\mathbf{u} = [u_1(t), \ldots, u_{M+1}(t)]^T$ and $\mathbf{v} = [v_1(t), \ldots, v_{M+1}(t)]^T$. With these, (5.42) can be changed into

$$\begin{aligned}\rho(t) &= \mathbf{u}^T(t)\mathbf{u}(t)\mathbf{v}^T(t)\Psi\mathbf{v}(t) - 2\mathbf{v}^T(t)\mathbf{u}(t)\mathbf{v}^T(t)\Psi\mathbf{u}(t) + \mathbf{v}^T(t)\mathbf{v}(t)\mathbf{u}^T(t)\Psi\mathbf{u}(t) \\ &= \sum_{i=1}^{M+1}\sum_{j=1}^{M+1}[u_i^2(t)\psi_j v_j^2(t) - 2u_i(t)v_i(t)\psi_j v_j(t)u_j(t) + v_i^2(t)\psi_j u_j^2(t)] \\ &= \sum_{i=1}^{M+1}\sum_{j=1}^{M+1}\psi_j[u_i^2(t)v_j^2(t) - 2u_i(t)v_i(t)v_j(t)u_j(t) + v_i^2(t)u_j^2(t)] \\ &= \sum_{i=1}^{M+1}\sum_{j=1}^{M+1}\psi_j[u_i(t)v_j(t) - u_j(t)v_i(t)]^2 \ge 0.\end{aligned} \tag{5.43}$$

Finally, it follows from (5.40) and (5.43) that $\hat{J}(\mathbf{w}(t)) - \hat{J}(\mathbf{w}(t-1)) \le 0$, namely $\hat{J}(\mathbf{w}(t)) \le \hat{J}(\mathbf{w}(t-1))$. Hence, the Proof of Lemma 5.3 is completed.

Theorem 5.1 *If t is large enough such that* $\overline{\mathbf{R}}(t) \to \overline{\mathbf{R}}$, *then w(t) generated by the AIP algorithm converges to the true FIR vector* $\mathbf{h}$ *with probability one as t tends to infinity, namely,* $\lim_{t\to\infty}\mathbf{w}(t) = \mathbf{h}$, *w.p.1.*

Proof Clearly, the stochastic sequence *w*(t) is a discrete-time dynamical system. It follows from Lemma 5.3 that $\hat{J}(\mathbf{w})$ is an energy function of the sequence *w*(t). Moreover, it is apparent that the sequence *w*(t) is bounded and remains in $\Re^M$ for all

$t > 0$. Using the energy function method [17], it is deduced that $\boldsymbol{w}(t)$ converges to a point in the invariance set defined by

$$F = \left\{ \mathbf{w}(t) | \hat{J}(\mathbf{w}(t)) - \hat{J}(\mathbf{w}(t-1)) = 0 \quad \text{and} \quad \text{for any } t \right\}. \tag{5.44}$$

Next, we proceed to prove that

$$F = \{\mathbf{w} | \mathbf{w} = \mathbf{h}\}. \tag{5.45}$$

It can be seen from (5.40) that $\hat{J}(\mathbf{w}(t)) - \hat{J}(\mathbf{w}(t-1)) = 0$ implies $\beta(t) \equiv 0$, which leads to

$$-a_4(t)(a_3(t) - a_1(t)) + \varsigma^0(t)y(t) \equiv 0. \tag{5.46}$$

By virtue of Table 5.1 and (5.41), it holds that

$$\begin{aligned}
&\frac{\varsigma^0(t)}{a_4(t)} - \frac{a_3(t) - b_1(t)}{y(t)} \\
&= \frac{[\mathbf{w}^T(t-1), -1]\overline{\mathbf{R}}[\mathbf{w}^T(t-1), -1]^T}{[\mathbf{w}^T(t-1), -1]\overline{\mathbf{D}}[\mathbf{w}^T(t-1), -1]^T} - \frac{[\mathbf{x}^T(t-1), 0]\overline{\mathbf{R}}[\mathbf{w}^T(t-1), -1]^T}{[\mathbf{x}^T(t-1), 0]\overline{\mathbf{D}}[\mathbf{w}^T(t-1), -1]^T} \\
&= \frac{[\mathbf{w}^T(t-1), -1]\mathbf{R}^*[\mathbf{w}^T(t-1), -1]^T}{[\mathbf{w}^T(t-1), -1]\overline{\mathbf{D}}[\mathbf{w}^T(t-1), -1]^T} \\
&\quad + \sigma_i^2 - \frac{[\mathbf{x}^T(t-1), 0]\mathbf{R}^*[\mathbf{w}^T(t-1), -1]^T}{[\mathbf{x}^T(t-1), 0]\overline{\mathbf{D}}[\mathbf{w}^T(t-1), -1]^T} - \sigma_i^2 \\
&= \left\{ \frac{[\mathbf{w}^T(t-1), -1]}{[\mathbf{w}^T(t-1), -1]\overline{\mathbf{D}}[\mathbf{w}^T(t-1), -1]^T} - \frac{[\mathbf{x}^T(t-1), 0]}{[\mathbf{x}^T(t-1), 0]\overline{\mathbf{D}}[\mathbf{w}^T(t-1), -1]^T} \right\} \\
&\quad \times \mathbf{R}^*[\mathbf{w}^T(t-1), -1]^T \\
&\equiv 0.
\end{aligned} \tag{5.47}$$

Since $\boldsymbol{x}(t)$ is an arbitrary vector, it follows from (5.47) that $\mathbf{R}^*[\mathbf{w}^T(t-1), -1] = 0$. According to Lemma 5.2, we have

$$\mathbf{w}(t-1) = \mathbf{h}, \tag{5.48}$$

i.e., there holds (5.45). Moreover, as shown in [12], $\hat{J}(\mathbf{w})$ has the unique stable point $\boldsymbol{w}(t) = \boldsymbol{h}$. Since the saddle point set is unstable, we conclude that $\boldsymbol{w}(t)$ converges to $\boldsymbol{h}$ with probability one as t tends to infinity. Therefore, the proof of Theorem 5.1 is completed.

5.3.4 Simulations Examples and Conclusion

We here present a simulation example to illustrate the behavior of the developed AIP algorithm and compare it with the inverse power method [3], Davila's RTLS algorithm [6], and the algorithm given in [12]. These four adaptive algorithms are simply called AIP, IP, O-RTLS and C-RTLS, respectively, in the following simulation illustrations.

This example is concerned with identifying adaptively the unknown FIR system with the parameter vector $\mathbf{h}^* = [-0.3, -0.9, 0.8, -0.7, 0.6]^T$. For convenience of comparison and computation, the noise-free input $\boldsymbol{x}(k)$ is a first-order AR process and given by $x(t) = -0.5 * x(t-1) + \eta(t)$, where $\eta(t)$ is a white Gaussian noise sequence of zero mean and unit variance, as shown in Fig. 5.1. The additive input and output noises are white Gaussian pseudo-random noise sequences with variances $\sigma_i^2 = 0.25$ and $\sigma_o^2 = 0.25$, respectively. Thus, it should be $\xi = 1$. The estimation error is as defined in [6]. An estimate of the impulse response is computed from $t = 1$ to $t = 1500$ and subsequently averaged over 30 independent tests.

When the system is linear time invariant, the estimation results are shown in Fig. 5.2, and they represent the average squared error-vector norm $\|\mathbf{w}(t) - \mathbf{h}^*\|^2$ obtained from AIP, O-RTLS, C-RTLS, and IP. It is seen that the AIP algorithm exhibits more stable convergence than the O-RTLS and C-RTLS algorithms.

Now consider the case when the unknown system parameters undergo a step change at $t = 500$. In order to assess the tracking capability in time-varying circumstances, the relative adaptive algorithms are implemented with an exponential forgetting factor $\mu = 0.995$ and are applied to adaptive parameter estimation of this time-varying FIR system. The corresponding simulation results are displayed in Fig. 5.3. In terms of the performance, the consistency between the AIP and O-RTLS is verified in this short-term experiment.

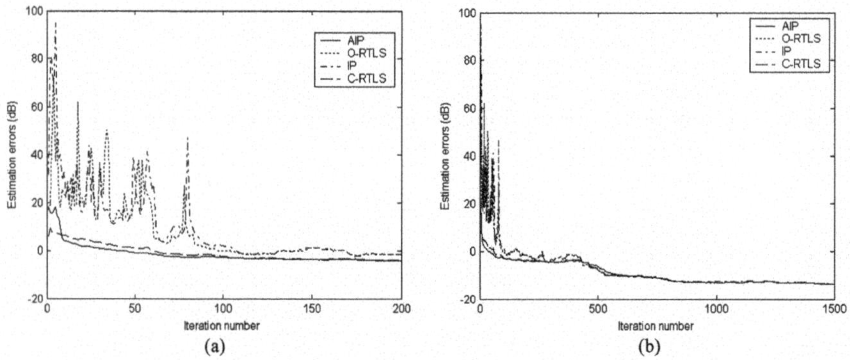

Fig. 5.2 AIP, O-RTLS, C-RTLS, and IP are used for identifying linear time-invariant system, where the variances of the input and output noises are 0.25. For clarity, **a** and **b** show the evolution curves from 0 to 200 iterations and from 0 to 1500 iterations, respectively

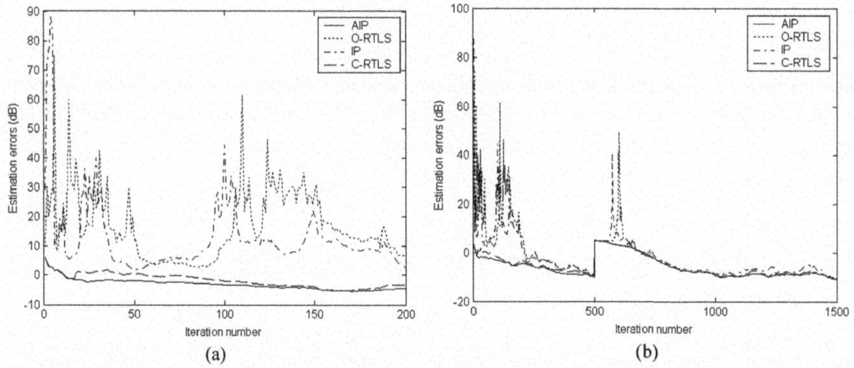

Fig. 5.3 AIP, O-RTLS, C-RTLS, and IP are used for identifying linear time-varying system, where there is a step change at $t = 500$, and the variances of the input and output noises are 0.25. For clarity, **a** and **b** show the evolution curves from 0 to 200 iterations and from 0 to 1500 iterations, respectively

The numerical stability of the relative adaptive algorithms has also been tested via a long-time experiment from $t = 1$ to $t = 2 \times 10^5$. Figure 5.4 shows the related results of the AIP algorithm, where the system is linear time invariant and an exponential forgetting factor $\mu = 0.995$ is used.

In this section, a low-complexity recursive algorithm is developed for computing the TLS solution for adaptive FIR filtering. At each recursion, the weight vector of the FIR filter is updated along the input data vector via AIP iteration. The algorithmic features of the developed AIP algorithm also include utilization of the regular form of the TLS solution for reducing the number of parameters being updated, efficient

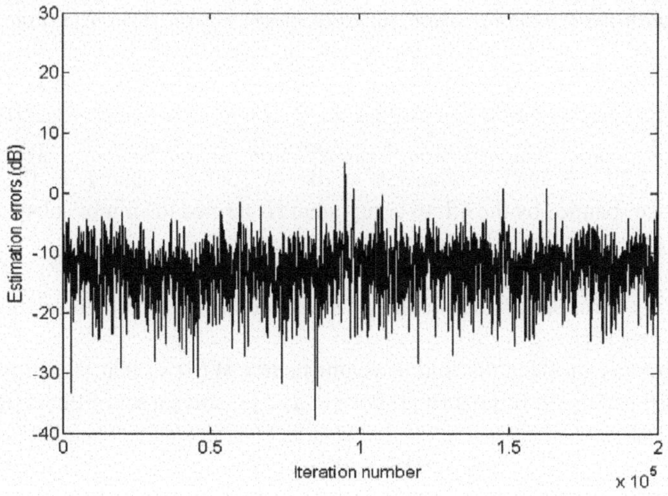

Fig. 5.4 Experiment for verifying the long-time numerical stability of the AIP algorithm

computation of the FGV, and rank-one update of the augmented autocorrelation matrix. All these lead to a significant decrease in the computational complexity, making the developed algorithm very effective in fast adaptation. The AIP algorithm not only has been shown to be globally convergent, but also has exhibited good long-term numerical stability, thanks to the use of the FGV.

5.4 An AIP Algorithm for Adaptive Extraction of Minor Subspace

This section develops a novel and efficient algorithm to recursively extract multiple minor components from an N-dimensional vector sequence. This algorithm is of computational complexity $O(N^2)$ and obtained by approximating the well-known IP iteration in conjunction with Galerkin method. Moreover, the convergence speed of the proposed algorithm is faster than that of the stochastic gradient-based algorithms with complexity $O(Nr)$, where r is the number of minor components. Global convergence of the proposed algorithm is established. The proposed algorithm may have good numerical stability over a very large data sequence due to no use of the matrix inversion lemma.

5.4.1 Proposed Adaptive Tracking Algorithm of MS

A. *Inverse Power Iteration*

Let $\mathbf{x}(t) \in \Re^{N\times 1}$ be an N-dimensional zero-mean, Gaussian stationary vector sequence with the positive definite autocorrelation matrix $\mathbf{R} = E\{\mathbf{x}(t)\mathbf{x}^T(t)\}$. Let λ_i and $\mathbf{l}_i (i = 1, 2, ..., N)$ denote the eigenvalues and the corresponding eigenvectors of $\boldsymbol{R}$, respectively. We shall arrange the orthonormal eigenvectors $\mathbf{l}_1, \mathbf{l}_2, \ldots, \mathbf{l}_N$ such that the corresponding eigenvalues are in a non-decreasing order: $0 < \lambda_1 \leq \lambda_2 \leq \cdots \leq \lambda_N$. The eigenvalue decomposition (EVD) of $\boldsymbol{R}$ is represented as $\mathbf{R} = \sum_{i=1}^{N} \lambda_i \mathbf{l}_i \mathbf{l}_i^T = \mathbf{L}\Lambda\mathbf{L}^T$, where $\Lambda = \text{diag}(\lambda_1, \ldots, \lambda_N)$ and $\mathbf{L} = [\mathbf{l}_1, \ldots, \mathbf{l}_N]$. The subspace spanned by the minor components is termed the minor subspace (MS). Denote $\mathbf{U}_{\text{MCA}} = [\mathbf{l}_1, \ldots, \mathbf{l}_r]$, where r is the dimension.

Given an $N \times r$ weight matrix $\boldsymbol{W}(t)$ which represents the estimate of the MS, the inverse power iteration for finding the MS associated with the autocorrelation matrix $\boldsymbol{R}$ is as follows [2].

Appropriately produce the initial weight matrix $\boldsymbol{W}(0)$ so that the initial weight matrix $\boldsymbol{W}(0)$ is column-orthonormal. For $t = 1, 2, \ldots$ solve a set of linear equations

$$\mathbf{R}\widehat{\mathbf{W}} = \mathbf{W}(t-1) \tag{5.49}$$

and perform the following orthonormalization

$$[\mathbf{W}(t), *] = QR(\hat{\mathbf{W}}(t)), \tag{5.50}$$

where * denotes that those terms are not interesting.

It can be seen that the core of the IP iteration is (5.49), while (5.50) is to orthonormalize the updated weight matrix and to avoid having some components converge to zero or infinity as $t \to \infty$. It is shown in [2] that the IP iteration globally exponentially converges to the subspace spanned by the eigenvectors associated with the MS of the autocorrelation matrix $\boldsymbol{R}$. Unfortunately, the IP iteration is an algorithm with computational complexity $O(N^3)$. Of course, if the inverse update formula of $\boldsymbol{R}$ is used like the RLS algorithms [8], then the IP iteration will become an algorithm with complexity $O(N^2r)$. However, the potential instability of the inverse update formula (see, e.g., [8]) of the autocorrelation matrix may cause the potential instability of such IP iteration. In order to improve the computational efficiency, an AIP iteration will be proposed.

B. *Development of the AIP Algorithm*

In this subsection, an efficient algorithm is developed for adaptively extracting the MS of the autocorrelation matrix $\boldsymbol{R}$ from a multiple dimensional data stream $\boldsymbol{x}(t)$. This algorithm is the AIP iteration obtained by Galerkin method [18, pp. 107–119]. Like the PAST [19], choosing the rank-one updating of the weight matrix to be $\boldsymbol{e}(t)\boldsymbol{g}^{\mathrm{T}}(t)$ will also result in the computationally efficient algorithm with complexity $O(N^2)$. Here, $\boldsymbol{g}(t)$ is the unknown parameter vector to be determined, and $\boldsymbol{e}(t)$ is the novel information vector which is defined as

$$\mathbf{y}(t) = \mathbf{W}^T(t-1)\mathbf{x}(t), \tag{5.51a}$$

$$\mathbf{e}(t) = \mathbf{x}(t) - \mathbf{W}(t-1)\mathbf{y}(t), \tag{5.51b}$$

where $\mathbf{W}(t-1)$ is a column-orthonormal matrix. Similar to the PAST [19], the weight matrix in (5.49) can be updated by the following rank-one updating formula:

$$\hat{\mathbf{W}}(t) = \mathbf{W}(t-1) + \mathbf{e}(t)\mathbf{g}^T(t). \tag{5.52}$$

Substituting (5.52) into (5.49) yields the approximate formula

$$\mathbf{R}(t)\left[\mathbf{W}(t-1) + \mathbf{e}(t)\mathbf{g}^T(t)\right] \approx \mathbf{W}(t-1)\mathbf{P}(t), \tag{5.53}$$

where $\boldsymbol{P}(t)$ is the $r \times r$ unknown parameter matrix that is not interesting for updating the weight matrix. Notice that $\boldsymbol{R}(t)$ is computed by an iteration formula

$$\mathbf{R}(t) = \mu\mathbf{R}(t-1) + \mathbf{x}(t)\mathbf{x}^T(t), \tag{5.54}$$

where μ is the forgetting factor. The unknown parameter vector $\boldsymbol{g}(t)$ in (5.53) can be usually determined by the Galerkin method [18]. It is easy to derive from (5.51a–5.51b) that

$$\mathbf{e}^T(t)\mathbf{W}(t-1) = 0. \tag{5.55}$$

Exploiting the well-known Galerkin method [18], considering (5.55), pre-multiplying (5.53) by $\boldsymbol{e}^{\mathrm{T}}(t)$, and forcing it to be equal to zero results in the following equation:

$$\mathbf{e}^T(t)\mathbf{R}(t)\big[\mathbf{W}(t-1) + \mathbf{e}(t)\mathbf{g}^T(t)\big] = 0. \tag{5.56}$$

Equation (5.56) includes only the single unknown parameter vector $\boldsymbol{g}(t)$, and it can be independently solved. It is easily shown that the solution to (5.56) is given by

$$\mathbf{v}(t) = \mathbf{R}(t)\mathbf{e}(t), \tag{5.57a}$$

$$\mathbf{g}(t) = \frac{-\mathbf{W}^T(t-1)\mathbf{v}(t)}{[\mathbf{v}^T(t)\mathbf{e}(t)]}. \tag{5.57b}$$

With $\boldsymbol{g}(t)$, a matrix $\boldsymbol{B}(t)$ should be found so as to make the matrix $[\mathbf{W}(t-1) + \mathbf{e}(t)\mathbf{g}^T(t)]\mathbf{B}(t)$ be orthonormal, namely

$$\mathbf{B}^T(t)[\mathbf{W}^T(t-1) + \mathbf{g}(t)\mathbf{e}^T(t)][\mathbf{W}(t-1) + \mathbf{e}(t)\mathbf{g}^T(t)]\mathbf{B}(t) = \mathbf{I}_r. \tag{5.58}$$

Considering (5.51a–5.51b) and (5.55), we further obtain from (5.58) that

$$\mathbf{B}^T(t)[\mathbf{I}_r + \|\mathbf{e}(t)\|^2\mathbf{g}(t)\mathbf{g}^T(t)]\mathbf{B}(t) = \mathbf{I}_r. \tag{5.59}$$

Equation (5.59) implies that $\boldsymbol{B}(t)$ should be the inverse-square root of the matrix $[\mathbf{I}_r + \|\mathbf{e}(t)\|^2\mathbf{g}(t)\mathbf{g}^T(t)]$. Thus, we have the following acceptable solution:

$$\begin{aligned}\mathbf{B}(t) &= \left\{[\mathbf{I}_r + \|\mathbf{e}(t)\|^2\mathbf{g}(t)\mathbf{g}^T(t)]^{1/2}\right\}^{-1}\\ &= \left\{\mathbf{I}_r + \xi(t)\mathbf{g}(t)\mathbf{g}^T(t)\right\}^{-1}\\ &= \mathbf{I}_r + \varsigma(t)\mathbf{g}(t)\mathbf{g}^T(t),\end{aligned} \tag{5.60}$$

where

$$\xi(t) = \frac{\left(\sqrt{1 + \|e(t)\|^2\|g(t)\|^2} - 1\right)}{\|g(t)\|^2}, \tag{5.61}$$

Table 5.2 AIP algorithm for adaptive extraction of MS

Initialize: $\mathbf{W}(0)$, $\mathbf{R}(0) = \mathbf{0}$ ($N \times N$ zero matrix) and $0 < \mu \leq 1$
For $t = 1,2,\ldots$
1. Update the data vector $\mathbf{x}(t)$
2. Compute autocorrelation matrix $\mathbf{R}(t) = \mu\mathbf{R}(t-1) + \mathbf{x}(t)\mathbf{x}^T(t)$
3. Novel information $\mathbf{e}(t) = \mathbf{x}(t) - \mathbf{W}(t-1)[\mathbf{W}^T(t-1)\mathbf{x}(t)]$
4. $\mathbf{v}(t) = \mathbf{R}(t)\mathbf{e}(t)$
5. $\mathbf{g}(t) = -\mathbf{W}^T(t-1)\mathbf{v}(t)/[\mathbf{v}^T(t)\mathbf{e}(t)]$
6. $a(t) = \|\mathbf{e}(t)\|^2$, $b(t) = \|\mathbf{g}(t)\|^2$
7. $\varsigma(t) = (1/\sqrt{1+a(t)b(t)} - 1)/b(t)$
8. $\mathbf{W}(t) = \mathbf{W}(t-1) + \{[1+\varsigma(t)b(t)]\mathbf{e}(t) + \mathbf{W}(t-1)[\varsigma(t)\mathbf{g}(t)]\}\mathbf{g}^T(t)$

$$\varsigma(t) = \frac{\left(\frac{1}{\sqrt{1+\|e(t)\|^2\|g(t)\|^2}} - 1\right)}{\|g(t)\|^2}. \tag{5.62}$$

The previous column orthonormalization process is adopted in [20] and can maintain the column orthonormality of the following updating weight matrix, which can now be represented as

$$\begin{aligned}\mathbf{W}(t) &= \left[\mathbf{W}(t-1) + \mathbf{e}(t)\mathbf{g}^T(t)\right]\mathbf{B}(t) \\ &= \left[\mathbf{W}(t-1) + \mathbf{e}(t)\mathbf{g}^T(t)\right]\left[\mathbf{I}_r + \varsigma(t)\mathbf{g}(t)\mathbf{g}^T(t)\right] \\ &= \mathbf{W}(t-1) + \left\{[1+\varsigma(t)\|\mathbf{g}(t)\|^2]\mathbf{e}(t) + \mathbf{W}(t-1)[\varsigma(t)\mathbf{g}(t)]\right\}\mathbf{g}^T(t).\end{aligned} \tag{5.63}$$

Table 5.2 gives a summary of the proposed adaptive tracking algorithm, which is called the AIP algorithm. It is also seen that the AIP algorithm is of computational complexity $O(N^2)$.

5.4.2 *Theoretical Analysis of Algorithm*

In this section, the globally asymptotic convergence of the AIP algorithm is analyzed. In order to establish our main result, a lemma is first stated.

Lemma 5.4 *Given a nonzero matrix* $\mathbf{C} \in \Re^{r \times N}$ *and two symmetric matrices* $\mathbf{A}_1 \in \Re^{N \times N}$ *and* $\mathbf{A}_2 \in \Re^{N \times N}$ *with full rank, let two subspaces spanned by all the columns of* $\mathbf{A}_2 - \mathbf{A}_1$ *and* $\mathbf{C}^T$ *be denoted by* $\mathbf{P}_a$ *and* $\mathbf{P}_c$*, respectively. If* $\mathbf{A}_2 - \mathbf{A}_1$ *is a positive semi-definite and nonzero matrix, and* $\mathbf{P}_a \cap \mathbf{P}_c \neq \emptyset$*, where* $\emptyset$ *denotes an empty set, then* $\mathrm{tr}(\mathbf{C}\mathbf{A}_1\mathbf{C}^T) < \mathrm{tr}(\mathbf{C}\mathbf{A}_2\mathbf{C}^T)$.

Proof Notice $\mathrm{tr}[\mathbf{B}] > 0$ for arbitrary positive semi-definite and nonzero matrix $\boldsymbol{B}$. Since

$$\mathrm{tr}\big[\mathbf{CA}_2\mathbf{C}^T\big] - \mathrm{tr}\big[\mathbf{CA}_1\mathbf{C}^T\big] = \mathrm{tr}\big[\mathbf{C}(\mathbf{A}_2 - \mathbf{A}_1)\mathbf{C}^T\big], \tag{5.64}$$

it suffices to prove that $\mathbf{C}(\mathbf{A}_2 - \mathbf{A}_1)\mathbf{C}^T$ is nonzero and positive semi-definite. It can be directly deduced from condition $\mathbf{P}_a \cap \mathbf{P}_c \neq \emptyset$ that $\mathbf{C}(\mathbf{A}_2 - \mathbf{A}_1)\mathbf{C}^T \neq \mathbf{0}$. Since $\mathbf{A}_2 - \mathbf{A}_1$ is nonzero and positive semi-definite, for arbitrary vector $\mathbf{x} \in \Re^{r\times 1}$, we have

$$\mathbf{x}^T\mathbf{C}(\mathbf{A}_2 - \mathbf{A}_1)\mathbf{C}^T\mathbf{x} = (\mathbf{C}^T\mathbf{x})^T(\mathbf{A}_2 - \mathbf{A}_1)(\mathbf{C}^T\mathbf{x}) \geq 0. \tag{5.65}$$

This shows that $\mathbf{C}(\mathbf{A}_2 - \mathbf{A}_1)\mathbf{C}^T$ is nonzero and positive semi-definite. Thus, Lemma 5.4 is proved.

The main result for the asymptotic convergence of the AIP algorithm is given in the following theorem.

Theorem 5.2 *If* $\mathbf{R}(t) \to \mathbf{R} = E\{\mathbf{x}(t)\mathbf{x}^T(t)\}$ *as t tends to a positive number large enough, then* $\mathbf{W}(t)$ *given by the AIP algorithm globally asymptotically converges to* $\mathbf{U}_{MCA}\mathbf{Q}$ *as* $t \to \infty$, *where* $\boldsymbol{Q}$ *denotes an* $r \times r$ *orthonormal matrix.*

Proof First, it is worth mentioning that the probability of even $\boldsymbol{e}(t) = 0$ is zero, since $\boldsymbol{x}(t)$ is not only spanned by noise subspace $\boldsymbol{W}$(t). Thus, the case e$(t) = 0$ will be not considered. Since $\mathbf{R}(t) \to \mathbf{R} = E\{\mathbf{x}(t)\mathbf{x}^T(t)\}$ as t tends to a positive number large enough, the state matrix $\boldsymbol{W}(t)$ for the AIP can be further represented as

$$\begin{aligned} J(\mathbf{W}(t)) &= \mathrm{tr}[\mathbf{W}^T(t)\mathbf{R}\mathbf{W}(t)] \\ &= \mathrm{tr}\{[(\mathbf{I} - \mathbf{P}_e(t))\mathbf{R}^{1/2}\mathbf{W}(t-1)\mathbf{B}(t)]^T \\ &[(\mathbf{I} - \mathbf{P}_e(t))\mathbf{R}^{1/2}\mathbf{W}(t-1)\mathbf{B}(t)]\} \\ &= \mathrm{tr}\{\mathbf{B}^T(t)\mathbf{W}^T(t-1)\mathbf{R}^{1/2}(\mathbf{I} - \mathbf{P}_e(t)) \\ &\mathbf{R}^{1/2}\mathbf{W}(t-1)\mathbf{B}(t)\}. \end{aligned} \tag{5.66}$$

where $\mathbf{P}_e(t) = [\mathbf{R}^{1/2}\mathbf{e}(t)\mathbf{e}^T(t)\mathbf{R}^{1/2}]/[\mathbf{e}^T(t)\mathbf{R}\mathbf{e}(t)]$ is a projection matrix. It is obvious that $\mathbf{I} - \mathbf{P}_e(t)$ is also a projection matrix. Again, since $\mathbf{R}(t) \to \mathbf{R} = E\{\mathbf{x}(t)\mathbf{x}^T(t)\}$ as t tends to a positive number large enough, inserting $\boldsymbol{W}(t)$ into $J(\boldsymbol{W})$ yields

$$\begin{aligned} J(\mathbf{W}(t)) &= \mathrm{tr}[\mathbf{W}^T(t)\mathbf{R}\mathbf{W}(t)] \\ &= \mathrm{tr}\{[(\mathbf{I} - \mathbf{P}_e(t))\mathbf{R}^{1/2}\mathbf{W}(t-1)\mathbf{B}(t)]^T[(\mathbf{I} - \mathbf{P}_e(t))\mathbf{R}^{1/2}\mathbf{W}(t-1)\mathbf{B}(t)]\} \\ &= \mathrm{tr}\{\mathbf{B}^T(t)\mathbf{W}^T(t-1)\mathbf{R}^{1/2}(\mathbf{I} - \mathbf{P}_e(t))\mathbf{R}^{1/2}\mathbf{W}(t-1)\mathbf{B}(t)\}. \end{aligned} \tag{5.67}$$

Since $\mathbf{R}^{1/2}(\mathbf{I}-\mathbf{P}_e(t))\mathbf{R}^{1/2} \leq \mathbf{R}$ and $\mathbf{R}-\mathbf{R}^{1/2}(\mathbf{I}-\mathbf{P}_e(t))\mathbf{R}^{1/2} = \mathbf{R}^{1/2}\mathbf{P}_e(t)\mathbf{R}^{1/2} \neq 0$, it is easily shown from Lemma 5.4 that

$$\begin{aligned} J(\mathbf{W}(t)) &< \mathrm{tr}\{\mathbf{B}^T(t)\mathbf{W}^T(t-1)\mathbf{R}\mathbf{W}(t-1)\mathbf{B}(t)\} \\ &= \mathrm{tr}\{\mathbf{W}^T(t-1)\mathbf{R}\mathbf{W}(t-1)\mathbf{B}(t)\mathbf{B}^T(t)\}. \end{aligned} \tag{5.68}$$

Equation (5.68) indicates that the weight matrix updating causes a decrease in the cost function. In addition

$$\begin{aligned}\mathbf{B}(t)\mathbf{B}^T(t) &= \left[\mathbf{I}_r + \varsigma(t)\mathbf{g}(t)\mathbf{g}^T(t)\right]\left[\mathbf{I}_r + \varsigma(t)\mathbf{g}(t)\mathbf{g}^T(t)\right] \\ &= \mathbf{I}_r + [2\varsigma(t) + \varsigma^2(t)\|\mathbf{g}(t)\|^2]\mathbf{g}(t)\mathbf{g}^T(t) \\ &= \mathbf{I}_r - \rho(t)\mathbf{g}(t)\mathbf{g}^T(t),\end{aligned} \tag{5.69}$$

where

$$r(t) = \frac{\left[\frac{1}{(1+\|\mathbf{e}(t)\|^2\|\mathbf{g}(t)\|^2)}\right]}{\|\mathbf{g}(t)\|^2} > 0. \tag{5.70}$$

Substituting (5.69) into (5.68) leads to

$$\begin{aligned}&J(\mathbf{W}(t)) < \mathrm{tr}\{\mathbf{W}^T(t-1)\mathbf{R}\mathbf{W}(t-1)\mathbf{B}(t)\mathbf{B}^T(t)\} \\ &= \mathrm{tr}\{\mathbf{W}^T(t-1)\mathbf{R}\mathbf{W}(t-1) - \rho(t)\mathbf{W}^T(t-1)\mathbf{R}\mathbf{W}(t-1)\mathbf{g}(t)\mathbf{g}^T(t)\} \\ &= J(\mathbf{W}(t-1) - \rho(t)\mathrm{tr}\{\mathbf{W}^T(t-1)\mathbf{R}\mathbf{W}(t-1)\mathbf{g}(t)\mathbf{g}^T(t)\} \\ &= J(\mathbf{W}(t-1) - \rho(t)\mathrm{tr}\{\mathbf{g}^T(t)\mathbf{W}^T(t-1)\mathbf{R}\mathbf{W}(t-1)\mathbf{g}(t)\} \\ &< J(\mathbf{W}(t-1).\end{aligned} \tag{5.71}$$

Furthermore, since $J(W)$ has the stable point set $\mathbf{W}(t) = \mathbf{U}_{\mathrm{MCA}}\mathbf{Q}$, and any saddle point set is unstable [21], we deduce that $W(t)$ globally asymptotically converges to $\mathbf{W}(t) = \mathbf{U}_{\mathrm{MCA}}\mathbf{Q}$ as $t \to \infty$. This completes the proof of Theorem 5.2.

5.4.2.1 Simulation Example and Conclusion

In this section, an example is given for testing the performance of the three algorithms, i.e., SS-MSR [22], AEE [23], and AIP. We first define two performance indices. The deviation of a state matrix from MS is defined as

$$\mathrm{dist}(\mathbf{W}(t)) = \left\|\mathbf{W}(t)[\mathbf{W}^T(t)\mathbf{W}(t)]^{-1}\mathbf{W}(t) - \mathbf{U}_{\mathrm{MCA}}\mathbf{U}_{\mathrm{MCA}}^T\right\|, \tag{5.72}$$

while the deviation of a state matrix from orthonormality is measured by

$$r(\mathbf{W}(t)) = \left\|\mathbf{W}^T(t)\mathbf{W}(t) - \mathbf{I}_r\right\|_F. \tag{5.73}$$

This experiment is run from time $t = 1$ to $t = 2000$. A 10-D vector stream $\boldsymbol{x}(t)$ is given by

$$\mathbf{x}(t) = \mathbf{A}^{1/2}\mathbf{y}(t), \tag{5.74}$$

where the 10-D vector sequence $\boldsymbol{y}(t)$ is a Gaussian stationary vector sequence, and $E\{\mathbf{y}(t)\mathbf{y}^T(t)\} = \mathbf{I}$. Here, A is a 10 10 matrix and is produced by the following equation:

$$\mathbf{A} = \mathbf{V}\mathrm{diag}(1, 1, 1, 1, 1, 0.01, 0.01, 0.001, 0.0001, 0.0001)\mathbf{V}^T, \tag{5.75}$$

where V is a 10×10 orthonormal matrix that is randomly generated. It is seen that the since the condition number of the autocorrelation matrix of x(t) is 10000, the autocorrelation matrix of $\boldsymbol{x}(t)$ is ill-conditioned and has a very small eigenvalue 0.0001 close to zero.

We simulate the algorithms starting from the unit-column-orthonormal initial matrix $\mathbf{W}(0) \in \Re^{10\times5}$ which is randomly produced in all the trials. The 5-D MS is adaptively extracted. Each result is averaged over 30 independent trials. The learning factors for the SS-MSR [22] are taken as 0.008, 0.012, 0.016, while the forgetting factors in the AIP are chosen as 0.996, 0.997, and 0.998. Figure 5.5a and b show the two performance indices of the AIP versus the iteration number, respectively. Figures 5.6 and 5.7 denote the two performance indices of the SS-MSR and the AEE versus the iteration number, respectively. The previous experimental results show that among the three algorithms the AEE converges fastest, while the convergence speed of the AIP is faster than that of the SS-MSR. It should be noticed that the AIP and the SS-MSR have a loss of the column orthonormality at each update, which also appears in the algorithms given in [20, 24]. To prevent $\boldsymbol{W}(t)$ from losing the column orthonormality, a column-orthonormal operation of $\boldsymbol{W}(t)$ may be performed by QR decomposition per multiple (for example, 1000) updates, which slightly increases the computational complexity of the proposed AIP algorithm since each QR decomposition only involves the computational complexity $O(Nr^2)$

In Fig. 5.8, it is shown by running on a long sequence that the AIP has favorable numerical stability. It can be seen from Fig. 5.8b that the loss of the column orthonormality of $\boldsymbol{W}(t)$ for the AIP is slow. Figure 5.9 denotes that the AEE may be numerically unstable, even though it has the fast convergence.

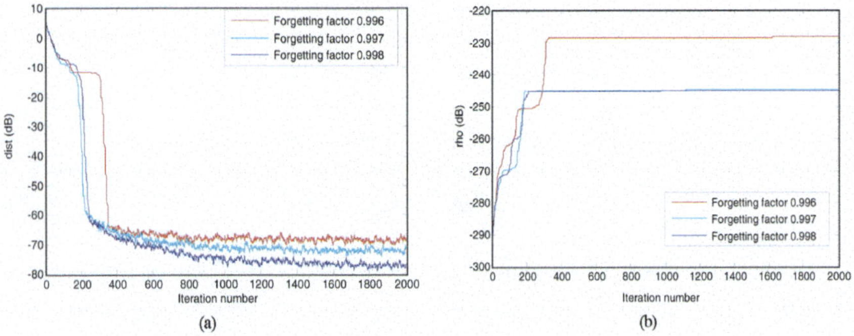

Fig. 5.5 Simulation results of AIP in Example 1. **a** and **b** denote the deviation of a state matrix from the MS and the orthonomality of a state matrix versus the iteration number, respectively

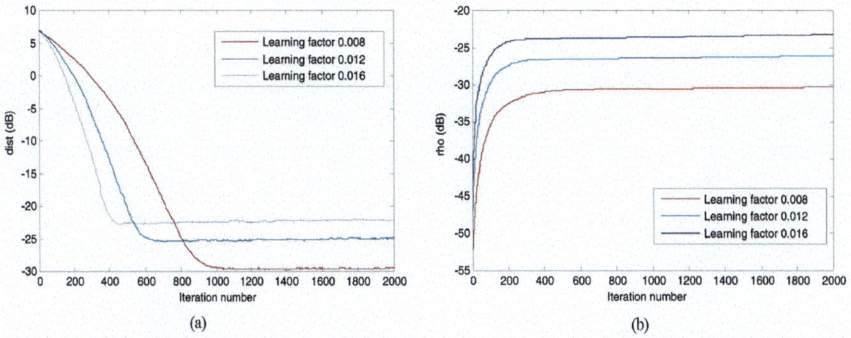

Fig. 5.6 Simulation results of SS-MSR in Example 1. **a** and **b** denote the deviation of a state matrix from the MS and the orthonomality of a state matrix versus the iteration number, respectively

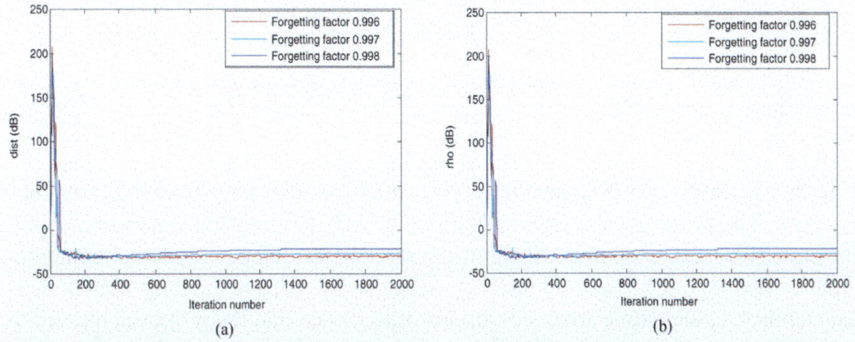

Fig. 5.7 Simulation results of AEE in Example 1. **a** and **b** denote the deviation of a state matrix from the MS and the orthonomality of a state matrix versus the iteration number, respectively. It can be seen that since no othonormalization is used in AEE, two performance indices are approximately consistent

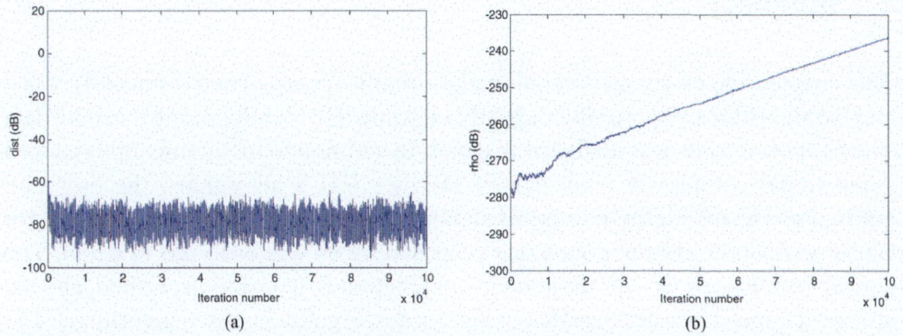

Fig. 5.8 Numerical stability of AIP is shown by an experiment run on a long sequence, where the forgetting factor is 0.998. **a** and **b** denote the deviation of a state matrix from the MS and the orthonomality of a state matrix versus the iteration number, respectively

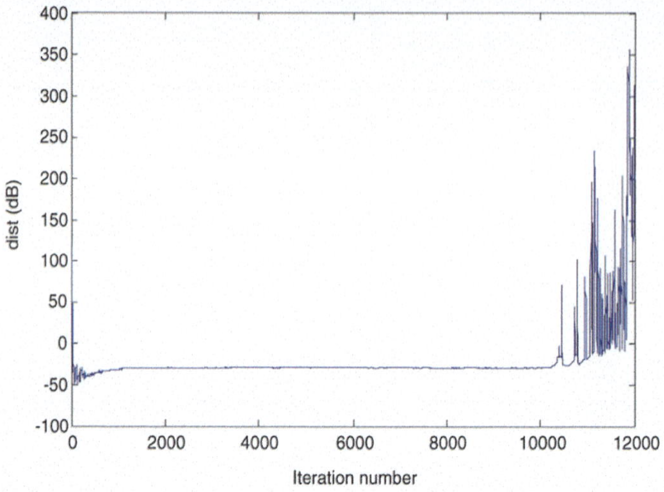

Fig. 5.9 Numerical instability of AEE is shown by deviation of a state matrix from the MS, where the forgetting factor is 0.998

In this section, an AIP algorithm is presented for adaptively tracking the minor subspace. The proposed algorithm is built upon AIP iteration and Galerkin method. The attractive features of the AIP algorithm include the comparably lower computational complexity, the relatively faster convergence speed, and the long-term numerical stability. Under some weak conditions, it is shown that the proposed algorithm globally asymptotically converges to the expected MS. The good performance of the proposed AIP algorithm is verified by simulations.

5.5 Summary

In this chapter, the existing IP iteration algorithm is reviewed and discussed. A fast recursive algorithm by approximating the well-known IP iteration along the input data vector proposed by us was analyzed in detail. In addition to its significantly reduced computational complexity from that of Davila's RTLS algorithms, the proposed algorithm possesses highly numerical stability. Furthermore, an AIP algorithm for adaptive extraction of minor subspace proposed by us was analyzed in detail. The algorithm is obtained by AIP iteration in conjunction with Galerkin method, and the convergence speed is faster than that of the stochastic gradient-based algorithms. The proposed algorithm have good numerical stability over a very large data sequence due to no use of the well-known matrix inversion lemma.

References

1. S. Haykin, *Adaptive Filter Theory*, 3rd edn. (Prentice-Hall, Englewood Cliffs, NJ, 1996)
2. G.H. Golub, C.F. van Loan, *Matrix Computations* (The Johns Hopkins Univ. Press, Baltimore, MD, 1983)
3. R. Vaccaro, On adaptive implementations of Pisarenko's harmonic retrieval method. in *Proceedings IEEE International Conference Acoustics, Speech, Signal Processing (ICASSP)*, San Diego, CA, March 1984, vol 9 (1984) pp. 224–227
4. H. Chen, T.K. Sarkar, S A. Dianat, J D. Brule, Adaptive spectral estimation by the conjugate gradient method. IEEE Trans. Acoustic Speech, and Signal Process. **ASSP-34**(2), 272–284 (1986)
5. V.U. Reddy, B. Egardt, T. Kailath, Least squares-type algorithm for adaptive implementations of Pisarenko's harmonic retrieval method. IEEE Trans. Acoustic Speech, and Signal Process. **ASSP-34**(3), 399–405 (1982)
6. C.E. Davila, An efficient recursive total least squares algorithm for FIR adaptive filtering. IEEE Trans. Signal Process. **42**(2), 268–280 (1994)
7. L. Ljung, M. Morf, D. Falconer, Fast calculation of gain matrices for recursive estimation schemes. Int. J. Control. **27**(1), 1–19 (1978)
8. S. Ljung, L. Ljung, Error propagation properties of recursive least-squares adaptation algorithms. Automatica **21**(2), 157–167 (1985)
9. J.L. Botto, G.V. Moustakides, Stabilization of fast Kalman algorithms. IEEE Trans. Acoustic Speech, and Signal Process. **ASSP-37**(9), 1342–1348 (1989)
10. D.T. Slock, T. Kailath, Numerically stable fast transversal filters for recursive least squares adaptive filtering. IEEE Trans. Signal Process. **39**(1), 92–114 (1991)
11. C.E. Davila, Line search algorithms for adaptive filtering. IEEE Trans. Signal Process. **41**(7), 2490–2494 (1994)
12. D.Z. Feng, X.D. Zhang, D.X. Chang, W.X. Zheng, A fast recursive total least squares algorithm for adaptive FIR filtering. IEEE Trans. Signal Process. **52**(10), 2729–2737 (2004)
13. B.E. Dunne, G.A. Williamson, QR-based TLS and mixed LS-TLS algorithms with applications to adaptive IIR filtering. IEEE Trans. Signal Process. **51**(2), 386–394 (2003)
14. B.E. Dunne, G.A. Williamson, Analysis of gradient algorithms for TLS-based adaptive IIR filters. IEEE Trans. Signal Process. **52**(12), 3345–3356 (2004)
15. A. Yeredor, The extended least squares criterion: minimization algorithms and applications. IEEE Trans. Signal Process. **49**(1), 74–86 (2001)
16. Y.N. Rao, D. Erdogmus, J.C. Principe, Error whitening criterion for adaptive filtering: theory and algorithms. IEEE Trans. Signal Process. **53**(3), 1057–1069 (2005)
17. J.J. Hopfield, Neural networks and physical systems with emergent collective computational abilities. Proc. Nat. Acad. Sci. USA **79**, 2554–2558 (1982)
18. K.H. Huebner, E.A. Thornton, *The Finite Element Method for Engineers* (Wiley, New York, 1982)
19. B. Yang, Projection approximation subspace tracking. IEEE Trans. Signal Process. **43**(1), 95–107 (1995)
20. K. Abed-Meraim, S. Attallah, A. Chkeif, Y. Hua, Orthonormal Oja algorithm. IEEE Signal Process. Letter **7**(5), 116–119 (2000)
21. D.Z. Feng, W.X. Zheng, An approximate inverse-power algorithm for adaptive extraction of minor subspace. IEEE Trans. Signal Process. **55**(7), 3937–3942 (2007)
22. S.C. Douglas, S.Y. Kung, S. Amari, A self-stabilized minor subspace rule. IEEE Signal Process. Letter **5**, 328–330 (1998)
23. G. Mathew, V.U. Reddy, S. Dasgupta, Adaptive estimation of eigensubspace. IEEE Trans. Signal Process. **43**(2), 401–411 (1995)
24. K. Abed-Meraim, A. Chkeif, Y. Hua, Fast orthonormal PAST algorithm. IEEE Signal Processing Letter **7**(3), 60–62 (2000)

Chapter 6
Neural-Based MCA Algorithms for Adaptive TLS

6.1 Introduction

From Chap. 3, it is known that the TLS solution $\mathbf{x}_{TLS}$ of an over-determined linear equations $\mathbf{AX} = \mathbf{b}$ can be obtained by $[\mathbf{x}_{\text{TLS}}, -1]^T = -\mathbf{v}_{n+1}/v_{n+1,n+1}$, where $\mathbf{v}_{n+1}$ is the right singular vector associated with the smallest singular value of $\mathbf{C} = [\mathbf{A}; \mathbf{b}]$. From the matrix theory, vector $\mathbf{v}_{n+1}$ is also equivalent to the eigenvector associated with the smallest eigenvalue of the autocorrelation matrix of the augmented input data $\mathbf{C} = [\mathbf{A}; \mathbf{b}]$. Thus, an adaptive algorithms for extracting the eigenvector associated with the smallest eigenvalue of the autocorrelation matrix of the augmented input data, in which a subsequent normalization (the last component of weight $\boldsymbol{x}_{n+1} = -1$) is needed, can be used to solve the TLS problem.

As known, neural networks on MCA pursue an effective online approach to update the eigendirection after each presentation of input vector. Compared with traditional algebraic approaches, neural-based algorithms have lower computational complexity and better suitability for data processing, especially for high-dimensional and non-stationary data processing. For over 30 years, many neural-based MCA algorithms [1–13], which can extract the eigenvector associated with the smallest eigenvalue of the correlation matrix, have been proposed. In neural-based MCA algorithms, the only nonlinear network is the Hopfield network in [12, 13], and all other existing neural networks are made up of one simple linear neuron. As pioneers in the MCA field, Xu and Oja [1] proposed the first learning algorithm based on the Hebbian learning rule. After that, many MCA learning algorithms [1–10, 14–17] were proposed in recent years, such as Luo's algorithm [3], the MCA EXIN algorithm [4], Ouyang's algorithm [18], Kong's algorithm [14], etc. This type of neural-based MCA algorithms can be divided into four types: (1) Oja's MCA algorithms [1–6]; (2) self-stabilizing MCA algorithms [8–10, 14]; (3) orthogonal Oja's MCA algorithms [19–24]; and (4) coupled MCA algorithms [15–17]. The four types MCA algorithms can be briefly reviewed as follows.

X. Kong and D. Feng, *Efficient Online Learning Algorithms for Total Least Square Problems*, Engineering Applications of Computational Methods 21,
https://doi.org/10.1007/978-981-97-1765-1_6

In [4], Cirrincione et al. distinguished between two classes of learning algorithms according to the time course of the length of the eigenvector estimate. In the first class, e.g., OJAn, Luo, and MCA EXIN, the length of the weight vector in the fixed point is undetermined. In an exact solution of the differential equation, the weight vector length would not deviate from its initial value. However, when a numerical procedure is applied, all these algorithms are plagued by "sudden divergence" of the weight vector length. In the second class, e.g., Doug, Chen, OJA+, the weight vector length converges toward a fixed value. Within the second class, some algorithms are self-stabilizing, which means that their weight vector length converges toward a fixed value independent of the presented input vector. All algorithms lacking self-stabilization are potentially prone to fluctuations and divergence in the weight vector length [10].

In addition to the self-stability, low complexity MCA algorithm are very beneficial for application. Due to the direct use of Gram–Schmidt orthonormalization, the computational complexity of DPM algorithm [25] is $O(np^2)$. FRANS algorithm [19] reduces its dominant complexity to $3np$. Afterward, several algorithms were proposed such as FDPM [20, 21], HFRANS [22], FOOJA [23], YAST [24], etc.

In the MCA field, most algorithms only focused on eigenvector(s) extraction or eigen-subspace tracking. This type of MCA algorithms are referred to as "non-coupled" algorithms in [15], and a speed-stability problem exists in most of them. The speed-stability problem is that the speed of convergence depends not only on the minor eigenvalue but also on all other eigenvalues of the covariance matrix. If these eigenvalues extend over a large interval, it is very difficult to select a suitable learning rate for a numerical solution that guarantees stability and a sufficient speed of convergence in all eigendirections [15]. To solve this problem, Möller [15] proposed a CMCA (coupled MCA) algorithm and also proved that the coupled algorithms can mitigate the speed-stability problem. Later, Hou [16] proposed a similar but easier information criterion and obtained the same CMCA algorithms as that of Möller's.

From the above discussion, it is evident that for neural-based MCA algorithms, these characteristics such as the self-stabilizing, coupling, parallel processing, etc., are beneficial to the completion of tasks. An algorithm that integrates multiple features is what we need. The aim of this chapter is to develop some new neural-based MCA algorithms to applied into the TLS parameter estimation. In this chapter, a novel parallel multiple minor components (MCs) extraction algorithm by diagonal matrix method [26] is presented. Then, a weighted information criterion for multiple MCs and its adaptive extraction algorithms [27] is introduced. Finally, based on a novel information criterion, a coupled self-stabilizing MCA algorithms [17] is proposed. The proposed coupled MCA algorithm can mitigate the speed-stability problem which exists in most non-coupled algorithms.

In this chapter, the existing neural-based MCA algorithms will be reviewed and discussed. Three algorithms proposed by us, including a novel parallel multiple MCs extraction algorithm by diagonal matrix, a weighted information criterion for multiple MCs and its adaptive algorithms and a coupled self-stabilizing MCA algorithm, will be analyzed in detail. The remainder of this chapter is organized as follows. An overview of the existing typical neural-based MCA algorithms is presented in

Sect. 5.2. A novel parallel multiple MCs extraction algorithm is presented in Sect. 5. 3. Section 5.4 is devoted to a weighted information criterion for multiple MCs and its adaptive algorithms. A coupled self-stabilizing MCA algorithm is proposed in in Sect. 5.5, followed by summary in Sect. 5.6.

6.2 Review of Neural-Network-Based MCA Algorithms

6.2.1 Oja's MCA Algorithms

Consider a linear unit with input $\mathbf{x}(t) = [x_1(t), x_2(t), \ldots, x_n(t)]^T$ and $y(t) = \sum_{i=1}^{n} w_i(t)x_i(t) = \mathbf{w}^T(t)\mathbf{x}(t)$, where $\mathbf{w}(t) = [w_1(t), w_2(t), \ldots, w_n(t)]^T$ is the weight vector. In the MCA analysis, $\mathbf{x}(t)$ is a bounded continuous-valued stationary ergodic data vector with finite second moments. The existing typical algorithms for the MCA of the autocorrelation matrix $\mathbf{R} = E[\mathbf{x}(t)\mathbf{x}^T(t)]$ of the input vector $\boldsymbol{x}(t)$ are listed below.

Oja's MCA algorithm is given by the following:

$$\mathbf{w}(t+1) = \mathbf{w}(t) - \alpha(t)y(t)[\mathbf{x}(t) - y(t)\mathbf{w}(t)], \tag{6.1}$$

where $\alpha(t)$ is a positive learning rate. The OJAn algorithm is as follows:

$$\mathbf{w}(t+1) = \mathbf{w}(t) - \alpha(t)y(t)\left[\mathbf{x}(t) - \frac{y(t)\mathbf{w}(t)}{\mathbf{w}^T(t)\mathbf{w}(t)}\right]. \tag{6.2}$$

Another Oja's learning rule (OJA+) is the following:

$$\mathbf{w}(t+1) = \mathbf{w}(t) - \alpha(t)[y(t)\mathbf{x}(t) - (y^2(t) + 1 - \|\mathbf{w}(t)\|_2^2)\mathbf{w}(t)]. \tag{6.3}$$

The TLMS algorithm [5] is given by the following:

$$\mathbf{w}(t+1) = \mathbf{w}(t) - \alpha(t)[\mathbf{w}^T(t)\mathbf{w}(t)y(t)\mathbf{x}(t) - \mathbf{w}(t)] \tag{6.4}$$

MCA EXIN algorithm [4] was proposed as follows:

$$\mathbf{w}(t+1) = \mathbf{w}(t) - \frac{\alpha(t)y(t)}{\|\mathbf{w}(t)\|_2^2}\left[\mathbf{x}(t) - \frac{y(t)\mathbf{w}(t)}{\|\mathbf{w}(t)\|_2^2}\right]. \tag{6.5}$$

The convergence of the above algorithms are indirectly proven by the convergence of their corresponding averaging ordinary differential equation (ODE). In [4], the analysis of the temporal behavior of all the above MCA algorithms are analyzed by using not only the ODE approximation, but especially the stochastic discrete laws. This study lead to the very important problem of sudden divergence [4].

6.2.2 Self-stabilizing MCA Algorithms

The concept of self-stability was presented in [10], which means that the norm of the weight vector in the algorithm converges toward a fixed value independent of the presented input vector. All algorithms lacking self-stability are prone to fluctuations and divergence in the weight vector norm, so the self-stability is an indispensable property for adaptive algorithms.

A self-stabilizing MCA algorithm was given in [8] as follows:

$$\mathbf{W}(t+1) = \mathbf{W}(t) - \eta[\mathbf{x}(t)\mathbf{y}^T(t)\mathbf{W}^T(t)\mathbf{W}(t)\mathbf{W}^T(t)\mathbf{W}(t) - \mathbf{W}(t)\mathbf{y}(t)\mathbf{y}^T(t)]. \quad (6.6)$$

Algorithm (6.6) is self-stabilizing, such that none of $\|\mathbf{w}_i(t)\| = 1$ deviates significantly from unity. Other self-stabilizing MCA algorithms, see [8–10, 14].

6.2.3 Orthogonal Oja's Algorithms

The orthogonal Oja (OOja) algorithm consists of Oja's MSA plus an orthogonalization of $\mathbf{W}(t)$ at each iteration [28] $\mathbf{W}^T(t)\mathbf{W}(t) = \mathbf{I}$. A Householder transform-based implementation of the MCA algorithm was given as [22]

$$\mathbf{W}(t+1) = \mathbf{W}(t) - 2\bar{\bar{\mathbf{u}}}(t)\mathbf{v}^T(t), \quad (6.7)$$

where W is initialized as an arbitrary orthogonal matrix and y is given by $\mathbf{y}(t) = \mathbf{W}^T(t)\mathbf{x}(t)$, $\mathbf{v}(t) = \mathbf{W}^T(t)\bar{\bar{\mathbf{u}}}(t)$, $\bar{\bar{\mathbf{u}}}(t) = \mathbf{u}(t)/\|\mathbf{u}(t)\|$, $\mathbf{u}(t) = (1 - \vartheta(t))\hat{\mathbf{x}}(t)/\eta\|\mathbf{y}(t)\|^2 + \vartheta(t)\mathbf{e}(t)$, $\vartheta(t) = 1/\sqrt{1+\eta^2\|\mathbf{e}(t)\|^2\|\mathbf{y}(t)\|^2}$, $\mathbf{e}(t) = \mathbf{x}(t) - \hat{\mathbf{x}}(t)$, $\hat{\mathbf{x}}(t) = \mathbf{W}(t)\mathbf{y}(t)$. The OOja is numerically very stable.

The normalized Oja (NOja) is derived by optimizing the MSE criterion subject to an approximation to the orthonormal constraint $\mathbf{W}^T(t)\mathbf{W}(t) = \mathbf{I}$ [29]. This leads to the optimal learning rate. The normalized orthogonal Oja (NOOja) is an orthogonal version of the NOja such that $\mathbf{W}^T(t)\mathbf{W}(t) = \mathbf{I}$ is perfectly satisfied [29]. Both algorithms offer a faster convergence, orthogonality, and a better numerical stability with a slight increase in the computational complexity. OOja, NOja, and NOOja have a complexity of $O(J_1 J_2)$ [28].

6.2.4 Coupled MCA Algorithms

Most neural-based MCA algorithms do not consider eigenvalue estimates in the update equations of the weights. In [15], Moller provided a framework for a special class of learning algorithms where eigenvectors and eigenvalues are simultaneously estimated in coupled update equations. Theoretical analysis has proved that coupled

MCA algorithms are solutions for the speed-stability problem that plagues most non-coupled MCA algorithms.

Several coupled MCA algorithms have been proposed in [15–17]. The online form of coupled MCA algorithm is given as in [17]

$$\mathbf{w}(t+1) = \mathbf{w}(t) + \gamma(t)\left\{\frac{\mathbf{Q}(t)\mathbf{w}(t)}{\mathbf{w}^T(t)\mathbf{Q}(t)\mathbf{w}(t)} - \mathbf{w}(t)\right\} \tag{6.8}$$

$$\lambda(t+1) = \lambda(t) + \gamma(t)\left\{\frac{1}{\mathbf{w}^T(t)\mathbf{Q}(t)\mathbf{w}(t)} - \lambda(t)\right\}, \tag{6.9}$$

where $\gamma(t)$ is the learning rate, and $\boldsymbol{Q}(t)$ is updated by

$$\mathbf{Q}(t+1) = \frac{t+1}{\alpha t}\left[\mathbf{Q}(t) - \frac{\mathbf{Q}(t)\mathbf{x}(t+1)\mathbf{x}^T(t+1)\mathbf{Q}(t)}{t + \mathbf{x}^T(t+1)\mathbf{Q}(t)\mathbf{x}^T(t+1)}\right], \tag{6.10}$$

where $0 < \alpha < 1$ denotes the forgetting factor.

6.3 Novel Parallel Multiple MCs Extraction Algorithm by Diagonal Matrix

The Douglas algorithm [8] is a self-stabilizing MCA algorithm. However, it is a MS tracking algorithm and cannot be used to extract multiple MCs. It is converted into a multiple MCs extraction algorithm through the time-oriented hierarchical learning method (TOHM) in [30]. The main drawback of this algorithm in [30] is the uncertain permutation of the estimated MCs. That is to say, for different values of the parameter and initial state matrix, the permutation of the extracted MCs is different. In this section, by adding a diagonal matrix into the Douglas algorithm, a novel algorithm is proposed, which can extract multiple MCs in parallel. This algorithm does not need to select parameters in advance and the permutation of estimated MCs is explicitly given in prior.

6.3.1 Preliminary and Novel Algorithm

Suppose that $\mathbf{x}(t) \in \Re^{n\times 1}$ is a stationary input signal of the linear neuron and its autocorrelation matrix is $\mathbf{R} = E\{\mathbf{x}(t)\mathbf{x}^T(t)\}$. Here, for the TLS application, $\boldsymbol{x}(t)$ can be written as $\mathbf{x}(t) = [\tilde{\mathbf{x}}^T(t), x_n(t)]^T = [x_1(t), x_2(t), \ldots, x_{n-1}(t), x_n(t)]^T = [x_1(t), x_2(t), \ldots, x_{n-1}(t), y(t)]^T$, where $\tilde{\mathbf{x}}(t)$ is the system input signal vector and $y(t)$ is the output signal of system at the time t. Let λ_i and $\mathbf{h}_i (i = 1, 2, ..., n)$ be the eigenvalue and corresponding eigenvector of $\boldsymbol{R}$, respectively, then $\boldsymbol{R}$ can be written as follows:

$$\mathbf{R} = \mathbf{H}^T \Lambda \mathbf{H}, \tag{6.11}$$

where $\boldsymbol{\Lambda} = \text{diag}[\lambda_{s(1)}, \lambda_{s(2)}, \ldots, \lambda_{s(n)}]$ is a diagonal matrix whose diagonal elements are the eigenvalues, $\mathbf{H} = [\mathbf{h}_{s(1)}, \mathbf{h}_{s(2)}, \ldots, \mathbf{h}_{s(n)}]$ consists of the corresponding eigenvectors, and s is a permutation of $\{1,2, \cdot,n\}$. For convenience, let's arrange the eigenvalues in an ascending order, i.e., $0 < \lambda_1 < \lambda_2 < \cdots < \lambda_n$. The eigenvectors $\boldsymbol{h}_1$, $\boldsymbol{h}_2,\ldots,\boldsymbol{h}_{\mathrm{r}}$ corresponding to $\lambda_1, \lambda_2, \cdots, \lambda_r$ are defined as the first r MCs of the input signal sequence $\{\boldsymbol{x}(k)\}$. Similarly, the subspace spanned by the first r MCs is called MS.

In order to track the MS, Douglas proposed the following neural-based algorithm in [8]

$$\mathbf{W}(t+1) = \mathbf{W}(t) - \eta\Big[\mathbf{W}(t)\mathbf{W}^T(t)\mathbf{R}\mathbf{W}(t) - \mathbf{R}\mathbf{W}(t)\big(\mathbf{W}^T(t)\mathbf{W}(t)\big)^2\Big], \tag{6.12}$$

where $\mathbf{W}(k)$ is the state matrix of the network, η is the learning rate and satisfies $0 < \eta \leq 1$. The ODE method is usually used to analyze the convergence of MCA algorithms. The ODE related to (6.12) is given by the following:

$$\frac{\mathrm{d}\mathbf{W}}{\mathrm{d}t} = \mathbf{W}\mathbf{W}^T\mathbf{R}\mathbf{W} - \mathbf{R}\mathbf{W}\big(\mathbf{W}^T\mathbf{W}\big)^2 \tag{6.13}$$

After the convergence, the state matrix in (6.12) just compose an orthonormal basis of the MS of $\mathbf{R}$, rather than the true MCs. In [30], by using the TOHM technique, a multiple MCs extraction algorithm is proposed, which has the following ODE

$$\frac{\mathrm{d}\mathbf{W}}{\mathrm{d}t} = \big[\mathbf{W}\mathbf{W}^T\mathbf{R}\mathbf{W} - \mathbf{R}\mathbf{W}(\mathbf{W}^T\mathbf{W})^2\big] - \gamma\big[\mathbf{R}\mathbf{W} - \mathbf{W}\text{diag}(\mathbf{W}^T\mathbf{R}\mathbf{W})\big], \tag{6.14}$$

where γ is a parameter and diag($\mathbf{B}$) is a $r \times r$ diagonal matrix, which is obtained by setting all the off-diagonal elements of $\mathbf{B}$ to zero. Herein, Eq. (6.14) is called TOHM-Douglas algorithm. Just as mentioned above, this algorithm has disadvantage on MCs permutation. In order to overcome this shortcoming, a novel multiple MCs extraction algorithm is proposed by us via adding a diagonal matrix into (6.13). The ODE of this proposed algorithm is given by the following:

$$\frac{\mathrm{d}\mathbf{W}}{\mathrm{d}t} = \mathbf{W}\mathbf{W}^T\mathbf{R}\mathbf{W}\mathbf{A} - \mathbf{R}\mathbf{W}\big(\mathbf{W}^T\mathbf{W}\big)^2, \tag{6.15}$$

where $\mathbf{A} = \text{diag}\{a_1, a_2, \ldots, a_r\}$ is an arbitrary diagonal matrix, whose diagonal elements only need to satisfy the requirement that $a_1 > a_2 > \cdots > a_r > 0$, and r means the number of desired MCs.

By discretizing (6.15), the following discrete algorithm can be obtained.

$$\mathbf{W}(t+1) = \mathbf{W}(t) - \eta\Big[\mathbf{W}(t)\mathbf{W}^T(t)\mathbf{R}\mathbf{W}(t)\mathbf{A} - \mathbf{R}\mathbf{W}(t)\big(\mathbf{W}^T(t)\mathbf{W}(t)\big)^2\Big] \tag{6.16}$$

Equation (6.16) is denoted as Diagonal-Douglas algorithm.

6.3.2 Fixed Points Analysis

Here, the convergence of the Diagonal-Douglas algorithm is investigated. From the optimization theory, the algorithm can only reach steady state at its fixed points. Thus, we firstly search all of the fixed points and the result is given by the following theorem.

Theorem 6.1 *Suppose that* $\boldsymbol{H}_r$ *is composed by any r distinct eigenvectors of* $\mathbf{R}$, then $\mathbf{H}_r\sqrt{\mathbf{A}}$ *is the fixed point of the Diagonal-Douglas algorithm.*

Proof
If $\mathbf{W} = \mathbf{H}_r\sqrt{\mathbf{A}}$, (6.15) can be written as follows:

$$\begin{aligned}\frac{\mathrm{d}\mathbf{W}}{\mathrm{d}t}\Big|_{\mathbf{W}=\mathbf{H}_r\sqrt{\mathbf{A}}} &= \mathbf{H}_r\sqrt{\mathbf{A}}\sqrt{\mathbf{A}}\mathbf{H}_r^T\mathbf{R}\mathbf{H}_r\sqrt{\mathbf{A}}\mathbf{A} - \mathbf{R}\mathbf{H}_r\sqrt{\mathbf{A}}\left(\sqrt{\mathbf{A}}\mathbf{H}_r^T\mathbf{H}_r\sqrt{\mathbf{A}}\right)^2 \\ &= \mathbf{H}_r\mathbf{\Lambda}_r(\sqrt{\mathbf{A}})^5 - \mathbf{H}_r\mathbf{\Lambda}_r(\sqrt{\mathbf{A}})^5 = 0\end{aligned} \tag{6.17}$$

According to definition, at the fixed point of (6.15), it holds that

$$\frac{\mathrm{d}\mathbf{W}}{\mathrm{d}t} = \mathbf{W}\mathbf{W}^T\mathbf{R}\mathbf{W}\mathbf{A} - \mathbf{R}\mathbf{W}\left(\mathbf{W}^T\mathbf{W}\right)^2 = 0 \tag{6.18}$$

Premultiplying both sides of (6.18) with $\boldsymbol{W}^T$, it holds that

$$\mathbf{W}^T\mathbf{W}\mathbf{W}^T\mathbf{R}\mathbf{W} = \mathbf{W}^T\mathbf{R}\mathbf{W}\left(\mathbf{W}^T\mathbf{W}\right)^2\mathbf{A}^{-1} \tag{6.19}$$

Suppose that the SVD of $\mathbf{W}$ is given by $\mathbf{W} = \mathbf{U}\mathbf{\Sigma}\mathbf{V}^T$, where $\mathbf{U}$ is a $n \times r$ matrix composed by r orthogonal columns, $\mathbf{\Sigma} = \mathrm{diag}(\sigma_1, \sigma_2, ..., \sigma_r)$ is a $r \times r$ diagonal matrix whose diagonal elements are denoted as the singular values of $\mathbf{W}$, and $\mathbf{V} \in \Re^{r\times r}$ is an orthogonal matrix. Since $\mathbf{W}$ is full column rank, then $\sigma_1, \sigma_2, ..., \sigma_r$ are positive.

Applying (6.19) into (6.18) and after some simplifications, we have

$$\mathbf{\Sigma}^2\mathbf{D}\mathbf{\Sigma} = \mathbf{D}\mathbf{\Sigma}^5\mathbf{V}^T\mathbf{A}^{-1}\mathbf{V}, \tag{6.20}$$

where $\mathbf{D} = \mathbf{U}^T\mathbf{R}\mathbf{U}$. Since $\mathbf{R}$ has nonzero eigenvalues, $\mathbf{D}$ must have full rank. Thus, the inverse of $\mathbf{D}$ exists, and then we have

$$\mathbf{\Sigma}^{-5}\mathbf{D}^{-1}\mathbf{\Sigma}^2\mathbf{D}\mathbf{\Sigma} = \mathbf{V}^T\mathbf{A}^{-1}\mathbf{V} \tag{6.21}$$

Due to the fact that $\mathbf{V}^T\mathbf{A}^{-1}\mathbf{V}$ is a symmetric matrix, $\mathbf{\Sigma}^{-5}\mathbf{D}^{-1}\mathbf{\Sigma}^2\mathbf{D}\Sigma$ must be a diagonal matrix, i.e.,

$$\mathbf{\Sigma}^2\mathbf{D}\Sigma^6\mathbf{D} = \mathbf{D}\Sigma^6\mathbf{D}\Sigma^2 \tag{6.22}$$

Considering the randomicity of $\mathbf{W}$ and the fact that $\mathbf{\Sigma}$ is a diagonal matrix, we have that $\mathbf{D} = \mathbf{U}^T\mathbf{R}\mathbf{U}$ is a diagonal matrix, which means that $\mathbf{U}$ is composed by r distinct eigenvectors of $\mathbf{R}$. Furthermore, from (6.21), we have

$$\mathbf{\Sigma} = \mathbf{V}^T\sqrt{\mathbf{A}}\mathbf{V} \tag{6.23}$$

Therefore, $\mathbf{V}^T\sqrt{\mathbf{A}}\mathbf{V}$ is also a diagonal matrix. Since $\mathbf{V}$ is an orthogonal matrix, $\mathbf{V}$ is also a permutation matrix.

Finally, it holds that

$$\mathbf{W} = \mathbf{U}\Sigma\mathbf{V}^T = \mathbf{U}\mathbf{V}^T\sqrt{\mathbf{A}}. \tag{6.24}$$

Furthermore, it holds that.

$$\mathbf{W}^T\mathbf{W} = \sqrt{\mathbf{A}}\mathbf{V}\mathbf{U}^T\mathbf{U}\mathbf{V}^T\sqrt{\mathbf{A}} = \mathbf{A}. \tag{6.25}$$

Denote $\mathbf{H}_r = \mathbf{U}\mathbf{V}^T$, then $\mathbf{H}_r$ contains r distinct eigenvectors of $\mathbf{R}$. That is to say, $\mathbf{H}_r\sqrt{\mathbf{A}}$ is a fixed point of the Diagonal-Douglas algorithm.

Theorem 6.1 lists all of the fixed points, yet it does not confirm that at which fixed point the Diagonal-Douglas algorithm is stable. Thus, the stable one must be picked out from all of the fixed points. The following theorem gives the main results.

Theorem 6.2 *The Diagonal-Douglas algorithm is stable if and only if* $\mathbf{W} = \mathbf{P}\sqrt{\mathbf{A}}$, *where* $\mathbf{P} \in \Re^{n\times r}$ *is composed by the first r MCs arranged by their corresponding eigenvalues in ascending order, i.e.,* $\mathbf{P} = [\mathbf{h}_1, \mathbf{h}_2, \ldots, \mathbf{h}_r]$.

Proof

Suppose that the eigenvectors in $\mathbf{H}$ is arranged in an arbitrary order. Denote $\mathbf{Z} = \mathbf{H}^T\mathbf{W}$. Then, (6.15) can be written as follows:

$$\frac{d\mathbf{Z}}{dt} = \mathbf{Z}\mathbf{Z}^T 3\mathbf{Z}\mathbf{A} - 3\mathbf{Z}(\mathbf{Z}^T\mathbf{Z})^2 \tag{6.26}$$

Denote that $\mathbf{H}'_r$ is composed by the first r columns of the matrix $\mathbf{H}$. If $\mathbf{W} = \mathbf{H}'_r\sqrt{\mathbf{A}}$, then $\overline{\mathbf{Z}} = \mathbf{H}^T\mathbf{H}'_r\sqrt{\mathbf{A}} = [\sqrt{\mathbf{A}}, \mathbf{0}_{r\times(n-r)}]^T$ is also a fixed point of (6.26).

Suppose that the ODE of (6.26) is perturbed by $\mathbf{E}(t)$ at the fixed point $\overline{\mathbf{Z}}$, then it holds that

$$\frac{d\mathbf{Z}}{dt}\Big|_{\mathbf{Z}=\overline{\mathbf{Z}}+\mathbf{E}} = (\overline{\mathbf{Z}}+\mathbf{E})(\overline{\mathbf{Z}}+\mathbf{E})^T\mathbf{\Lambda}(\overline{\mathbf{Z}}+\mathbf{E})\mathbf{A} - \mathbf{\Lambda}(\overline{\mathbf{Z}}+\mathbf{E})\left((\overline{\mathbf{Z}}+\mathbf{E})^T(\overline{\mathbf{Z}}+\mathbf{E})\right)^2$$
$$= \overline{\mathbf{Z}}\overline{\mathbf{Z}}^T 3\overline{\mathbf{Z}}\mathbf{A} + \overline{\mathbf{Z}}\mathbf{E}^T 3\overline{\mathbf{Z}}\mathbf{A} + \overline{\mathbf{Z}}\overline{\mathbf{Z}}^T 3\mathbf{E}\mathbf{A} + \mathbf{E}\overline{\mathbf{Z}}^T 3\overline{\mathbf{Z}}\mathbf{A} - \mathbf{\Lambda}\big(\overline{\mathbf{Z}}\mathbf{A}^2 + \overline{\mathbf{Z}}\mathbf{E}^T\overline{\mathbf{Z}}\mathbf{A} + \overline{\mathbf{Z}}\overline{\mathbf{Z}}^T\mathbf{E}\mathbf{A} +$$
$$\overline{\mathbf{Z}}\mathbf{A}\mathbf{E}^T\overline{\mathbf{Z}} + \overline{\mathbf{Z}}\mathbf{A}\overline{\mathbf{Z}}^T\mathbf{E} + \mathbf{E}\mathbf{A}^2\big) + O\left(\|\mathbf{E}\|^2\right), \tag{6.27}$$

where $O\left(\|\mathbf{E}\|^2\right)$ contains the second- and higher-order terms of the $\mathbf{E}$ and is negligibly small.

Rewrite $\mathbf{\Lambda}$ and $\mathbf{E}$ with block matrices, i.e.,

$$\mathbf{\Lambda} = \begin{bmatrix} \mathbf{\Lambda}_1 & \mathbf{0} \\ \mathbf{0} & \mathbf{\Lambda}_2 \end{bmatrix}, \mathbf{E} = \begin{bmatrix} \mathbf{E}_1 \\ \mathbf{E}_2 \end{bmatrix}, \tag{6.28}$$

where $\mathbf{\Lambda}_1, \mathbf{E}_1 \in \Re^{r\times r}$, $\mathbf{\Lambda}_2 \in \Re^{(n-r)\times(n-r)}$ and $\mathbf{E}_2 \in \Re^{(n-r)\times r}$. Inserting (6.28) into (6.27), it holds that

$$\frac{\mathrm{d}\mathbf{Z}}{\mathrm{d}t}\Big|_{\mathbf{Z}=\overline{\mathbf{Z}}+\mathbf{E}} = \begin{bmatrix} \sqrt{\mathbf{A}}\mathbf{E}_1^T\mathbf{\Lambda}_1(\sqrt{\mathbf{A}})^3 \\ \mathbf{0} \end{bmatrix} + \begin{bmatrix} \mathbf{E}_1\mathbf{A}^2\mathbf{\Lambda}_1 \\ \mathbf{E}_2\mathbf{A}^2\mathbf{\Lambda}_1 \end{bmatrix} - \begin{bmatrix} \sqrt{\mathbf{A}}\mathbf{\Lambda}_1\mathbf{E}_1^T(\sqrt{\mathbf{A}})^3 \\ \mathbf{0} \end{bmatrix}$$
$$- \begin{bmatrix} (\sqrt{\mathbf{A}})^3\mathbf{\Lambda}_1\mathbf{E}_1^T\sqrt{\mathbf{A}} \\ \mathbf{0} \end{bmatrix} - \begin{bmatrix} \mathbf{\Lambda}_1\mathbf{A}^2\mathbf{E}_1 \\ \mathbf{0} \end{bmatrix} - \begin{bmatrix} \mathbf{\Lambda}_1\mathbf{E}_1\mathbf{A}^2 \\ \mathbf{\Lambda}_1\mathbf{E}_2\mathbf{A}^2 \end{bmatrix} \tag{6.29}$$

Denote e_{ij} is the ith row and jth column entry of the matrix $\mathbf{E}$. Then, for $1 \le j \le r < i \le n$, it holds that

$$\frac{\mathrm{d}e_{ij}}{\mathrm{d}t} = (\lambda_{s(j)} - \lambda_{s(i)})a_j^2 e_{ij} \tag{6.30}$$

From (6.30), it can be concluded that if and only if $\lambda_{s(j)} < \lambda_{s(i)}$, we have $e_{ij} \to 0$. For $1 \le j, i \le r$, we have

$$\frac{\mathrm{d}e_{ij}}{\mathrm{d}t} = (\lambda_{s(j)}a_j^2 - \lambda_{s(i)}a_i^2 - \lambda_{s(i)}a_j^2)e_{ij}$$
$$+ \sqrt{a_i}\sqrt{a_j}(\lambda_{s(j)}a_j - \lambda_{s(i)}a_j - \lambda_{s(i)}a_i)e_{ji} \tag{6.31}$$

If $i = j$, it follows that

$$\frac{\mathrm{d}e_{ii}}{\mathrm{d}t} = -2a_i^2\lambda_{s(i)}e_{ii} \tag{6.32}$$

which means that $e_{ii} \to 0$ since $\lambda_{s(i)} > 0$.

If $i \ne j < r$, it can be also obtained that

$$\frac{\mathrm{d}e_{ji}}{\mathrm{d}t} = \sqrt{a_i}\sqrt{a_j}(\lambda_{s(i)}a_i - \lambda_{s(j)}a_i - \lambda_{s(j)}a_j)e_{ij}$$

$$+ (\lambda_{s(i)} a_i^2 - \lambda_{s(j)} a_j^2 - \lambda_{s(j)} a_i^2) e_{ji} \tag{6.33}$$

Combining (6.31) and (6.33), it follows that

$$\frac{d^2 e_{ij}}{dt^2} - (\alpha + \beta) \frac{de_{ij}}{dt} + \alpha\beta e_{ij} = 0, \tag{6.34}$$

where

$$\alpha + \beta = -\left(\lambda_{s(i)} a_j^2 + \lambda_{s(j)} a_i^2\right) \tag{6.35}$$

$$\alpha\beta = -(\lambda_{s(i)} - \lambda_{s(j)})(a_i - a_j)\left(\lambda_{s(i)} a_i^3 + a_j^3 \lambda_{s(j)}\right) \tag{6.36}$$

From (6.34), we can see that the two roots of (6.34) are α, β. According to matrix theory, if and only if the two roots are negative, e_{ij} will tend asymptotically to zero. Thus, it is expected that $\alpha + \beta < 0$ and $\alpha\beta > 0$. From (6.35), it holds that $\alpha + \beta < 0$, since λ_i and a_i are positive for all $i = 1, 2, ..., n$. In (6.36), the sign of $\alpha\beta$ is decided by the values of i and j. Then, the sign of (6.36) is analyzed in two scenarios. In the first scenario ($i > j$), since $a_i < a_j$, the expected result $\alpha\beta > 0$ can be satisfied if and only if $\lambda_{s(i)} > \lambda_{s(j)}$. On the contrary, when $i < j$, $\alpha\beta > 0$ if and only if $\lambda_{s(i)} < \lambda_{s(j)}$, since $a_i > a_j$.

From the above analysis, we have that $\mathbf{e}_{ij}$ will tend asymptotically to zero if and only if $s(i) = i$. It is to say, if and only if $\mathbf{H}'_r$ is equal to $\mathbf{P} = [\mathbf{h}_1, \mathbf{h}_2, ..., \mathbf{h}_r]$, the Diagonal-Douglas algorithm is stable.

Remark From the Theorem 6.2, it is obvious that the permutation of the MCs is arranged with an ascending order of the corresponding eigenvalues. No matter how you choose the initial conditions, this permutation does not change, which is different from the results obtained by the TOHM-Douglas algorithm. Besides, we only require the diagonal elements of the diagonal matrix to be positive and arranged in a descending order, which is in favor of real applications.

6.3.3 Stability on the Manifold

Here, the dynamic behavior of $\mathbf{W}^T\mathbf{W}$ on the manifold $\mathbf{W}^T\mathbf{W} = \mathbf{A}$ will be studied, which has been proven by (6.25). Herein, whether the state matrix on this manifold is stable or not, is our concern. The following theorem will answer this question.

Theorem 6.3 *For the Diagonal-Douglas algorithm* (6.15)*, the state matrix on the manifold* $\mathbf{W}^T\mathbf{W} = \mathbf{A}$ *is stable and any perturbation will finally decay to zero.*

Proof
From (6.15), it holds that.

$$\begin{aligned}\frac{\mathrm{d}(\mathbf{W}^T\mathbf{W})}{\mathrm{d}t} &= \frac{\mathrm{d}(\mathbf{W}^T)}{\mathrm{d}t}\mathbf{W} + \mathbf{W}^T\frac{\mathrm{d}(\mathbf{W})}{\mathrm{d}t} \\ &= \mathbf{A}\mathbf{W}^T\mathbf{R}\mathbf{W}\mathbf{W}^T\mathbf{W} - \left(\mathbf{W}^T\mathbf{W}\right)^2\mathbf{W}^T\mathbf{R}\mathbf{W} \\ &\quad + \mathbf{W}^T\mathbf{W}\mathbf{W}^T\mathbf{R}\mathbf{W}\mathbf{A} - \mathbf{W}^T\mathbf{R}\mathbf{W}\left(\mathbf{W}^T\mathbf{W}\right)^2\end{aligned} \tag{6.37}$$

Suppose that the state matrix $\hat{\mathbf{W}}$ is disturbed by $\delta\mathbf{W}(t)$ on the manifold. Then at the point $\mathbf{W}(t) = \hat{\mathbf{W}} + \delta\mathbf{W}(t)$, we have

$$\mathbf{W}^T(t)\mathbf{W}(t) = \mathbf{A} + \mathbf{W}^T\delta\mathbf{W}(t) + \delta^T\mathbf{W}(t)\mathbf{W}, \tag{6.38}$$

where the second-order term of $\delta W(t)$ is omitted. Denote $\delta(\mathbf{W}^T\mathbf{W})(t) = \mathbf{W}^T\delta\mathbf{W}(t) + \delta^T\mathbf{W}(t)\mathbf{W}$. We can see that the stability of the manifold relies on the behavior of the variation $\delta(\mathbf{W}^T\mathbf{W})(t)$. From (6.37), the ODE with respect to $\delta(\mathbf{W}^T\mathbf{W})(t)$ is given by the following:

$$\begin{aligned}\frac{\mathrm{d}\delta(\mathbf{W}^\mathbf{T}\mathbf{W})(t)}{\mathrm{d}t} &= \frac{d(\mathbf{W}^\mathbf{T}\mathbf{W})}{dt}\Big|_{\mathbf{W}=\hat{\mathbf{W}}+\delta\mathbf{W}} = 2\mathbf{A}\hat{\mathbf{W}}^\mathbf{T}\mathbf{R}\hat{\mathbf{W}}\mathbf{A} \\ &\quad + \mathbf{A}\hat{\mathbf{W}}^\mathbf{T}\mathbf{R}\hat{\mathbf{W}}(\delta\mathbf{W}) + (\delta\mathbf{W})\hat{\mathbf{W}}^\mathbf{T}\mathbf{R}\hat{\mathbf{W}}\mathbf{A} \\ &\quad - \left(\mathbf{A}^2\hat{\mathbf{W}}^\mathbf{T}\mathbf{R}\hat{\mathbf{W}} + \mathbf{A}(\delta\mathbf{W})\hat{\mathbf{W}}^\mathbf{T}\mathbf{R}\hat{\mathbf{W}} + (\delta\mathbf{W})\mathbf{A}\hat{\mathbf{W}}^\mathbf{T}\mathbf{R}\hat{\mathbf{W}}\right) \\ &\quad - \left(\hat{\mathbf{W}}^\mathbf{T}\mathbf{R}\hat{\mathbf{W}}\mathbf{A}^2 + \hat{\mathbf{W}}^\mathbf{T}\mathbf{R}\hat{\mathbf{W}}\mathbf{A}(\delta\mathbf{W}) + \hat{\mathbf{W}}^\mathbf{T}\mathbf{R}\hat{\mathbf{W}}(\delta\mathbf{W})\mathbf{A}\right)\end{aligned} \tag{6.39}$$

Considering the fact that $\hat{\mathbf{W}}^T\mathbf{R}\hat{\mathbf{W}}$ is a diagonal matrix, the Eq. (6.39) can be further simplified into

$$\frac{\mathrm{d}\delta(\mathbf{W}^T\mathbf{W})(t)}{\mathrm{d}t} = -\left[\mathbf{A}(\delta\mathbf{W})\hat{\mathbf{W}}^T\mathbf{R}\hat{\mathbf{W}} + \hat{\mathbf{W}}^T\mathbf{R}\hat{\mathbf{W}}(\delta\mathbf{W})\mathbf{A}\right] \tag{6.40}$$

Through the vectorization operator, (6.40) can be written as follows:

$$\frac{\mathrm{d}}{\mathrm{d}t}vec\left[\delta(\mathbf{W}^T\mathbf{W})(t)\right] = -\left[\mathbf{A}\otimes\hat{\mathbf{W}}^T\mathbf{R}\hat{\mathbf{W}} + \hat{\mathbf{W}}^T\mathbf{R}\hat{\mathbf{W}}\otimes\mathbf{A}\right]vec\left[\delta(\mathbf{W}^T\mathbf{W})(t)\right] \tag{6.41}$$

Since $\hat{\mathbf{W}}^T\mathbf{R}\hat{\mathbf{W}}$ and $\mathbf{A}$ are two positive definite matrices, $\mathbf{A}\otimes\hat{\mathbf{W}}^T\mathbf{R}\hat{\mathbf{W}} + \hat{\mathbf{W}}^T\mathbf{R}\hat{\mathbf{W}}\otimes\mathbf{A}$ is also positive definite. From (6.41), it concludes that $vec\left[\delta(\mathbf{W}^T\mathbf{W})(t)\right]$ will asymptotically approach to zero, which means that $\delta(\mathbf{W}^T\mathbf{W})(t) \to \mathbf{0}$ as $t \to \infty$. Thus, the state matrix in the Diagonal-Douglas algorithm is stable on the manifold $\mathbf{W}^T\mathbf{W} = \mathbf{A}$.

6.3.4 Simulation Experiments and Conclusion

Here, two simulation experiments are designed to explore the behaviors of the Diagonal-Douglas algorithm. The first experiment compare the Diagonal-Douglas algorithm and the TOHM-Douglas algorithm to test their convergence rate. The second experiment test the stability of the Diagonal-Douglas algorithm on the Manifold.

6.3.4.1 Experiment 1

Suppose that $\mathbf{x}(k)$ is a sequence of independent jointly Gaussian random vectors with the following autocorrelation matrix

$$\mathbf{R} = \begin{bmatrix} 1.0234 & 0.0773 & -0.4004 & 0.3871 & 0.0679 & 0.3410 \\ 0.0773 & 1.5168 & 0.0426 & -0.1241 & 0.1424 & -0.1442 \\ -0.4004 & 0.0426 & 1.2163 & 0.4420 & -0.1187 & -0.6513 \\ 0.3871 & -0.1241 & 0.4420 & 1.4153 & -0.3374 & 0.1466 \\ 0.0679 & 0.1424 & -0.1187 & -0.3374 & 2.2993 & 0.5359 \\ 0.3410 & -0.1442 & -0.6513 & 0.1466 & 0.5359 & 1.7522 \end{bmatrix}$$

Diagonal-Douglas algorithm and TOHM-Douglas algorithm are used to estimate the first three MCs of $\mathbf{R}$, i.e., $r = 3$. The initial parameters are set as follows: $\mathbf{A} = \text{diag}\{[1, 0.8, 0.6]\}$ and $\eta = 0.05$ for the Diagonal-Douglas algorithm, $\gamma = -0.05$ and $\eta = 0.05$ for the TOHM-Douglas algorithm. The norm of the initial state matrix $\mathbf{W}(0)$, which is randomly generated, is normalized to 1. In order to evaluate the estimation results and compare their characteristic, the direction cosine(DC) between the state matrix and the true MCs is calculated, which is given by the following:

$$\text{DC}_i(k) = \frac{\left|\mathbf{W}_i^T(k)\mathbf{h}_i\right|}{|\mathbf{W}_i(k)||\mathbf{h}_i|}, \tag{6.42}$$

where $\mathbf{W}_i(k)$ is the ith column of $\mathbf{W}(k)$ and $i = 1,2,3$. Clearly, if the values of DC tend to unity one, the columns of state matrix have the same directions with the MCs. Figure 6.1 illustrates the behaviors of $\text{DC}_i(k)$ obtained by the two algorithms. From Fig. 6.1, we can see that the DC curves of the Diagonal-Douglas algorithm tend to unity one after about 600 iterations, which means that the state matrix have converged to the directions of the true MCs. That is to say, the state matrix can be seen as the desired MCs after 600 iteration steps. Compared with the TOHM-Douglas algorithm, the Diagonal-Douglas algorithm has the faster convergence speed. Besides, it is easily found that the permutation of the MCs obtained by the TOHM-Douglas algorithm is ruleless, which means some extra work are needed to rearrange these MCs.

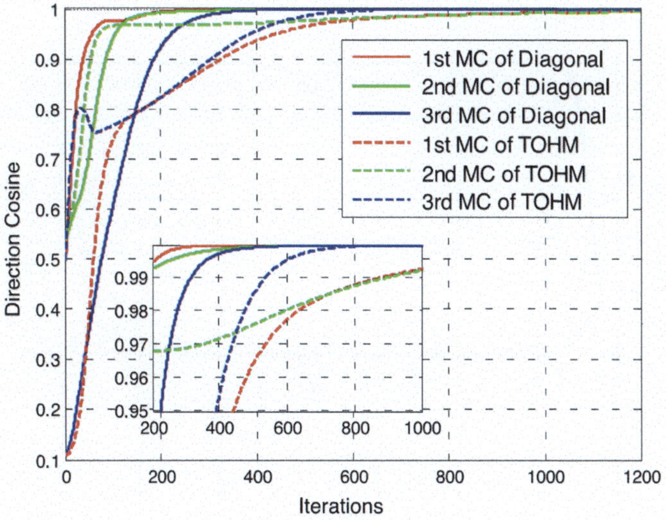

Fig. 6.1 DC curves of the two algorithms

6.3.4.2 Experiment 2

Figure 6.2 depicts the element values of $\mathbf{W}^T\mathbf{W}$ during the iterations, which are obtained by adding a random variation on the state matrix at the 600th step, where $\mathbf{W}^T\mathbf{W}(ij)$ is the ith row and the jth column of the matrix $\mathbf{W}^T\mathbf{W}$. Since $\mathbf{W}^T\mathbf{W}$ is a symmetric matrix, we only provide the behaviors of the elements on lower triangular matrix. From Fig. 6.2, it is obvious that after some iterations, the state matrix is stable on the manifold of $\mathbf{W}^T\mathbf{W} = \mathbf{A} = \text{diag}\{[1, 0.8, 0.6]\}$. Although there is some volatility on the manifold due to the adding perturbation, the volatility quickly disappears after several iterations. This further confirms the above analytical results.

6.3.4.3 Conclusion

In this section, a novel algorithm for multiple MCs extraction by adding a diagonal matrix to the Douglas MS tracking algorithm, is presented. Through analyzing all of the fixed points, it is proven that the state matrix in the proposed algorithm must converge to the desired multiple MCs after some iterations. Simulation experiments further verify the efficiency of the proposed Diagonal-Douglas algorithm.

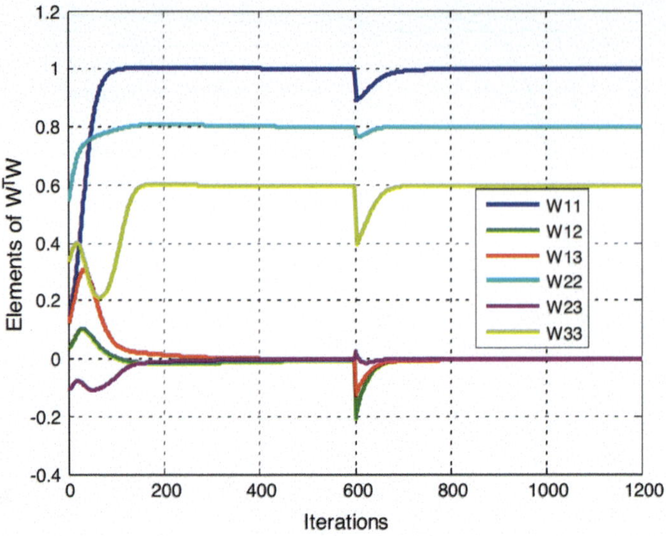

Fig. 6.2 Element values of $W^T W$

6.4 A Weighted Information Criterion and Multiple MC Algorithm

In this section, based on the concepts of weighted subspace and optimum theory, a weighted information criterion is proposed for searching the optimum solution of a linear neural network. This information criterion exhibits a unique global minimum attained if and only if the state matrix is composed of the desired MCs of an autocorrelation matrix of an input signal. A gradient algorithm and a RLS algorithm are developed for multiple MCs extraction. The proposed algorithms can extract the multiple MCs in parallel and has advantage in dealing with high dimension matrices. Since the weighted matrix does not require an accurate value, it facilitates the system design of the proposed algorithms for practical applications.

6.4.1 Preliminaries

Suppose that the input vector sequence $\{\mathbf{x}(k)|\mathbf{x}(k) \in \Re^{n\times 1}, k = 1, 2, \ldots\}$ is a stationary stochastic process with zeromean and the autocorrelation matrix $\mathbf{R} = E[\mathbf{x}(k)\mathbf{x}^T(k)]$. Let λ_i and $\mathbf{v}_i (i = 1, 2, \ldots, n)$ denote the eigenvalue and the corresponding orthonormal eigenvector of $\mathbf{R}$, respectively. From matrix theories, we know that $\mathbf{R}$ is a symmetric non-negative definite matrix and all the eigenvalues of $\mathbf{R}$ are non-negative. Suppose that the orthonormal eigenvectors $\mathbf{v}_1, \mathbf{v}_2, \ldots, \mathbf{v}_n$ are arranged in such a way that the corresponding eigenvalues are in a non-decreasing order: $\lambda_1 \geq \lambda_2 \cdots \geq \lambda_n > 0$. According to the definition of MCs, the eigenvectors $\mathbf{v}_{n-r+1}, \mathbf{v}_{n-r+2}, \ldots, \mathbf{v}_n$ are called the first r MCs of the input vector $\mathbf{x}(k)$.

For convenience, the autocorrelation matrix $\mathbf{R}$ is decomposed as follows:

$$\mathbf{R} = \mathbf{U}_1\mathbf{\Lambda}_1\mathbf{U}_1^T + \mathbf{U}_n\mathbf{\Lambda}_n\mathbf{U}_n^T, \tag{6.43}$$

where $\mathbf{U}_1 = [\mathbf{v}_1, \mathbf{v}_2, \ldots, \mathbf{v}_{n-r}]$ is composed of the eigenvectors corresponding to the first n-r eigenvalues of $\mathbf{R}$, the matrix $\mathbf{U}_n = [\mathbf{v}_{n-r+1}, \mathbf{v}_{n-r+2}, \ldots, \mathbf{v}_n]$ contains the rest of the r eigenvectors, $\mathbf{\Lambda}_1 = \text{diag}(\lambda_1, \lambda_2, \ldots, \lambda_{n-r})$ and are two diagonal matrices. Next, the neural network is used for extracting the first r MCs of the input vector. Consider the multiple-input-multiple-output (MIMO) linear neural network with the following relationship: $\mathbf{y}(k) = \mathbf{W}^T\mathbf{x}(k), k = 1, 2, \ldots$ where $\mathbf{W} \in \Re^{n\times r}$ is the state matrix of the neural network, $\mathbf{y}(k)$ is the output of the neural network, and $\mathbf{x}(k)$ is the input signal. The objective of developing neural-based algorithms is to construct a suitable learning rule so that the state matrix can converge to the first r MCs of the input vector.

6.4.2 A Weighted Information Criterion and Its Landscape

6.4.2.1 A Weighted Information Criterion

In [9], an information criterion (called OJAm) for MSA was proposed, which is given by the following:

$$\begin{aligned} \mathbf{W}^* &= \arg\min_{\mathbf{W}\in\Omega} J_{OJAm}(\mathbf{W}) \\ J_{OJAm}(\mathbf{W}) &= \frac{1}{2}\text{tr}\{\ln(\mathbf{W}^T\mathbf{R}\mathbf{W})\} + \frac{1}{2}\{(\mathbf{W}^T\mathbf{W})^{-1}\}, \end{aligned} \tag{6.44}$$

where tr(·) denotes a matrix trace. In [9], OJAm criterion is a subspace criterion, and the OJAm algorithm can only yield an orthonormal basis of the MS. If the MCs are required, one commonly used method is spatial rotation, in which the state matrix $\mathbf{W}$ is orthogonalized. Obviously, this procedure will increase the cost of hardware and produce a long delay. In order to solve this problem, a weighed information criterion (called WOJAm) is proposed by introducing a weight matrix into the OJAm criterion, which is given below

$$\begin{aligned} \mathbf{W}^* &= \arg\min_{\mathbf{W}\in\Omega} J_{WOJAm}(\mathbf{W}) \\ J_{WOJAm}(\mathbf{W}) &= \frac{1}{2}\text{tr}\{\ln(\mathbf{W}^T\mathbf{R}\mathbf{W})\} + \frac{1}{2}\{(\mathbf{W}^T\mathbf{W}\mathbf{A})^{-1}\}, \end{aligned} \tag{6.45}$$

where $\mathbf{A}$ is a $r \times r$ diagonal matrix with the positive entries $a_r > a_{r-1} > \cdots > a_1 > 0$. It is worth noting that when $\mathbf{A} = \mathbf{I}_r$, where $\mathbf{I}_r$ is an identity matrix, the information criterion (6.45) is reduced to the OJAm criterion (6.44).

From (6.45), we can see that $J_{\mathrm{WOJAm}}(\mathbf{W})$ is actually an optimization function. Herein we are interested in the minimum of $J_{\mathrm{WOJAm}}(\mathbf{W})$. In particular, we want to know:

(1) Is there a global minimum of $J_{\mathrm{WOJAm}}(\mathbf{W})$?
(2) What is the relation between this minimum and the first r MCs of $\mathbf{R}$?
(3) Are there any other local minima of $J_{\mathrm{WOJAm}}(\mathbf{W})$?

These questions are answered in the following section.

6.4.2.2 *Landscape of the W_{OJAm} Criterion*

The landscape of the proposed criterion is depicted by the following two theorems.

Theorem 6.4 *Given* $\mathbf{W} \in \Re^{n \times r}$ *in the domain* $\Omega = \{\mathbf{W} | 0 < \mathbf{W}^T\mathbf{R}\mathbf{W} < \infty, \mathbf{W}^T\mathbf{W}\mathbf{A} \neq 0\}$, $\mathbf{W}$ *is a stationary point of* $J_{WOJAm}(\mathbf{W})$ *in this domain if and only if* $\mathbf{W} = \mathbf{U}'_r\mathbf{A}^{-1/2}\mathbf{Q}$, *where* $\mathbf{U}'_r$ *is a* $n \times r$ *matrix and is composed of any* r *distinct orthonormal eigenvectors of* $\mathbf{R}$, *and* $\mathbf{Q}$ *is an arbitrary orthogonal matrix.*

Proof
Since $\mathbf{W}^T\mathbf{R}\mathbf{W}$ is a positive definite matrix and $\mathbf{W}^T\mathbf{W}\mathbf{A}$ is non-singular, it is easy to compute the first order differential of $J_{WOJAm}(\mathbf{W})$, which is given by the following:

$$\begin{aligned} dJ_{WOJAm}(\mathbf{W}) &= \frac{1}{2} tr[(\mathbf{W}^T\mathbf{R}\mathbf{W})^{-1} d(\mathbf{W}^T\mathbf{R}\mathbf{W})] - \frac{1}{2} tr[(\mathbf{W}^T\mathbf{W}\mathbf{A})^{-2} d(\mathbf{W}^T\mathbf{W}\mathbf{A})] \\ &= tr[(\mathbf{W}^T\mathbf{R}\mathbf{W})^{-1}\mathbf{W}^T\mathbf{R} d\mathbf{W}] - tr[(\mathbf{A}\mathbf{W}^T\mathbf{W})^{-2}\mathbf{A}\mathbf{W}^T d\mathbf{W}], \end{aligned} \tag{6.46}$$

where d denotes the differential. By using (6.46), the gradient of $J_{WOJAm}(\mathbf{W})$ with respect to $\mathbf{W}$ can be obtained as follows:

$$\nabla J_{WOJAm}(\mathbf{W}) = \mathbf{R}\mathbf{W}(\mathbf{W}^T\mathbf{R}\mathbf{W})^{-1} - \mathbf{W}\mathbf{A}(\mathbf{W}^T\mathbf{W}\mathbf{A})^{-2} \tag{6.47}$$

Clearly, if $\mathbf{W} = \mathbf{U}_r\mathbf{Q}\mathbf{A}^{-1/2}$, where $\mathbf{U}_r \in \Re^{n \times r}$ contains any r distinct eigenvectors of $\mathbf{R}$ and $\mathbf{Q}$ is an arbitrary orthogonal matrix, it is easy to verify that $\nabla J_{WOJAm}(\mathbf{W}) = \mathbf{0}$. Conversely, by definition, at the stationary point of $J_{WOJAm}(\mathbf{W})$, $\mathbf{W}$ satisfies $\nabla J_{WOJAm}(\mathbf{W}) = \mathbf{0}$, which yields

$$\mathbf{R}\mathbf{W}(\mathbf{W}^T\mathbf{R}\mathbf{W})^{-1} = \mathbf{W}\mathbf{A}(\mathbf{W}^T\mathbf{W}\mathbf{A})^{-2} \tag{6.48}$$

Premultiplying both sides of (6.48) with $\mathbf{W}^T$, it holds that $\mathbf{W}^T\mathbf{W} = \mathbf{A}^{-1}$.

Let us denote $\mathbf{P} = \mathbf{U}_r^T \mathbf{W} \mathbf{A}^{1/2}$ and substitute it into (6.48), we have $\mathbf{P}^T\mathbf{P} = \mathbf{I}$. Clearly, at any stationary point of $J_{WOJAm}(\mathbf{W})$, the columns of $\mathbf{P}$ are orthonormal, which implies that $\mathbf{P}$ is a orthogonal matrix. Hence, $\mathbf{W} = \mathbf{U}_r\mathbf{Q}\mathbf{A}^{-1/2}$ is a stationary point of $J_{WOJAm}(\mathbf{W})$, where $\mathbf{Q}$ is an arbitrary orthogonal matrix.

This completes the proof of Theorem 6.4.

Theorem 6.4 establishes the property of all the stationary points of $J_{WOJAm}(\mathbf{W})$. The latter theorem will distinguish the global minimum point set attained by $\mathbf{W}$ consisting of the first r MCs from the other stationary points which are saddle points.

Theorem 6.5 *In the domain* $\Omega = \{\mathbf{W}|0 < \mathbf{W}^T\mathbf{R}\mathbf{W} < \infty, \mathbf{W}^T\mathbf{W}\mathbf{A} \neq 0\}$,$J_{WOJAm}(\mathbf{W})$ *has a global minimum point that is attained if and only if* $\mathbf{W} = \mathbf{U}_n\mathbf{A}^{-1/2}$, *where* $\mathbf{U}_n$ *is composed of the r eigenvectors corresponding to the last r eigenvalues of the autocorrelation matrix* $\mathbf{R}$. *All the other stationary points are saddle (unstable) points of* $J_{WOJAm}(\mathbf{W})$. *At this global minimum,*

$$J_{WOJAm}(\mathbf{W}) = \frac{1}{2}\left(\sum_{i=1}^{r} \ln(\lambda_{n-r+i}/a_i) + r\right) \tag{6.49}$$

Proof

Let $\mathbf{H}J_{WOJAm}(\mathbf{W})$ be the $nr \times nr$ Hessian matrix of $J_{WOJAm}(\mathbf{W})$ with respect to the nr dimensional vector $\mathrm{vec}(\mathbf{W}) = [\mathbf{w}_1^T, \mathbf{w}_2^T, ..., \mathbf{w}_r^T]^T$, which is defined by the following:

$$HJ_{WOJAm}(\mathbf{W}) = \frac{\partial}{\partial(\mathrm{vec}(\mathbf{W}))^T}\left(\frac{\partial J(\mathbf{W})}{\partial(vec(\mathbf{W}))^T}\right)^T \tag{6.50}$$

By (6.46) and (6.50), the second order of $J_{WOJAm}(\mathbf{W})$ can be calculated as follows:

$$\begin{aligned}
&\mathbf{H}J_{WOJAm}(\mathbf{W}) = d\{tr[(\mathbf{W}^T\mathbf{R}\mathbf{W})^{-1}\mathbf{W}^T\mathbf{R}d\mathbf{W}] - tr[(\mathbf{A}\mathbf{W}^T\mathbf{W})^{-2}\mathbf{A}\mathbf{W}^T d(\mathbf{W})]\}\\
&= -tr[(\mathbf{W}^T\mathbf{R}\mathbf{W})^{-1}d(\mathbf{W}^T\mathbf{R}\mathbf{W})(\mathbf{W}^T\mathbf{R}\mathbf{W})^{-1}\mathbf{W}^T\mathbf{R}d\mathbf{W}]\\
&\quad +2tr[(\mathbf{A}\mathbf{W}^T\mathbf{W})^{-2}d(\mathbf{A}\mathbf{W}^T\mathbf{W})(\mathbf{A}\mathbf{W}^T\mathbf{W})^{-1}\mathbf{A}\mathbf{W}^T d(\mathbf{W})]\\
&\quad +tr\,[(\mathbf{W}^T\mathbf{R}\mathbf{W})^{-1}(d\mathbf{W}^T)\mathbf{R}d\mathbf{W}] - tr[(\mathbf{A}\mathbf{W}^T\mathbf{W})^{-2}\mathbf{A}(\mathbf{W}^T)d(\mathbf{W})]\\
&= -tr[(\mathbf{W}^T\mathbf{R}\mathbf{W})^{-1}d(\mathbf{W}^T)\mathbf{R}\mathbf{W}(\mathbf{W}^T\mathbf{R}\mathbf{W})^{-1}\mathbf{W}^T\mathbf{R}d\mathbf{W}]\\
&\quad -tr[(\mathbf{W}^T\mathbf{R}\mathbf{W})^{-1}\mathbf{W}^T\mathbf{R}(d\mathbf{W})(\mathbf{W}^T\mathbf{R}\mathbf{W})^{-1}\mathbf{W}^T\mathbf{R}d\mathbf{W}]\\
&\quad +2tr[(\mathbf{A}\mathbf{W}^T\mathbf{W})^{-2}\mathbf{A}\mathbf{W}^T d(\mathbf{W})(\mathbf{A}\mathbf{W}^T\mathbf{W})^{-1}\mathbf{A}\mathbf{W}^T d(\mathbf{W})]\\
&\quad +2tr[(\mathbf{A}\mathbf{W}^T\mathbf{W})^{-2}\mathbf{A}d(\mathbf{W}^T)\mathbf{W}(\mathbf{A}\mathbf{W}^T\mathbf{W})^{-1}\mathbf{A}\mathbf{W}^T d(\mathbf{W})]\\
&\quad +tr[(\mathbf{W}^T\mathbf{R}\mathbf{W})^{-1}(d\mathbf{W}^T)\mathbf{R}d\mathbf{W}]\\
&\quad +tr[(\mathbf{A}\mathbf{W}^T\mathbf{W})^{-2}\mathbf{A}d(\mathbf{W}^T)d(\mathbf{W})]
\end{aligned} \tag{6.51}$$

By using (6.51), we have the Hessian matrix of $J_{WOJAm}(\mathbf{W})J_{WOJAm}(\mathbf{W})$

$$
\begin{aligned}
&\mathbf{H}J_{\text{WOJAm}}(\mathbf{W}) = -\mathbf{K}_{rn}[\mathbf{RW}(\mathbf{W}^T\mathbf{RW})^{-1}] \otimes [(\mathbf{W}^T\mathbf{RW})^{-1}\mathbf{W}^T\mathbf{R}] \\
&+ (\mathbf{W}^T\mathbf{RW})^{-1} \otimes \mathbf{R} + \mathbf{A}(\mathbf{W}^T\mathbf{WA})^{-2} \otimes \mathbf{W}(\mathbf{AW}^T\mathbf{W})^{-1}\mathbf{AW}^T \\
&-\tfrac{1}{2}[\mathbf{A}(\mathbf{W}^T\mathbf{WA})^{-2} \otimes \mathbf{I}_n + (\mathbf{AW}^T\mathbf{W})^{-2}\mathbf{A} \otimes \mathbf{I}_n] \\
&+ (\mathbf{AW}^T\mathbf{W})^{-2}\mathbf{A} \otimes \mathbf{WA}(\mathbf{W}^T\mathbf{WA})^{-1}\mathbf{W}^T - (\mathbf{W}^T\mathbf{RW})^{-1} \otimes \mathbf{RW}(\mathbf{W}^T\mathbf{RW})^{-1}\mathbf{W}^T\mathbf{R} \\
&+ \mathbf{K}_{rn}[\mathbf{WA}(\mathbf{W}^T\mathbf{WA})^{-2} \otimes (\mathbf{AW}^T\mathbf{W})^{-1}\mathbf{AW}^T + \mathbf{WA}(\mathbf{W}^T\mathbf{WA})^{-1} \otimes (\mathbf{AW}^T\mathbf{W})^{-2}\mathbf{AW}^T],
\end{aligned}
\tag{6.52}
$$

where $\otimes$ is the Kronecker product and $\mathbf{K}_{rn}$ is a $rn \times rn$ commutation matrix such that $\mathbf{K}_{rn}vec(\mathbf{M}) = vec(\mathbf{M}^T)$ for any matrix $\mathbf{M}$. The commutation matrix $\mathbf{K}_{rn}$ has some good properties, for any matrices $\mathbf{M}_1 \in \Re^{m\times n}$ and $\mathbf{M}_2 \in \Re^{p\times q}$, there is

$$
\mathbf{K}_{pm}(\mathbf{M}_1 \otimes \mathbf{M}_2) = (\mathbf{M}_2 \otimes \mathbf{M}_1)\mathbf{K}_{qn} \tag{6.53}
$$

Substituting $\boldsymbol{I}_n = \mathbf{U}_1\mathbf{U}_1^T + \mathbf{U}_n\mathbf{U}_n^T$ and the EVD of $\mathbf{R} = \mathbf{U}_1\boldsymbol{\Lambda}_1\mathbf{U}_1^T + \mathbf{U}_n\boldsymbol{\Lambda}_n\mathbf{U}_n^T$ into (6.52) and calculating $\mathbf{H}J_{WOJAm}(\mathbf{W})$ at the stationary point $\mathbf{W} = \mathbf{U}_n\mathbf{A}^{-1/2}$, we have

$$
\begin{aligned}
\hat{\mathbf{H}} &= \mathbf{H}J_{WOJAm}(\mathbf{W})\big|_{\mathbf{W}=\mathbf{U}_n\mathbf{A}^{-1/2}} \\
&= -\boldsymbol{\Lambda}_n^{-1}\mathbf{A} \otimes \mathbf{U}_n\boldsymbol{\Lambda}_n\mathbf{U}_n^T - \mathbf{K}_{rn}\left[\mathbf{U}_n\mathbf{A}^{1/2} \otimes \mathbf{A}^{1/2}\mathbf{U}_n^T\right] + \boldsymbol{\Lambda}_n^{-1}\mathbf{A} \otimes \mathbf{R} \\
&\quad +2\mathbf{A} \otimes \mathbf{U}_n\mathbf{A}\mathbf{U}_n^T + 2\mathbf{K}_{rn}\left(\mathbf{U}_n\mathbf{A}^{1/2} \otimes \mathbf{A}^{1/2}\mathbf{U}_n^T\right) \\
&\quad -\mathbf{A} \otimes (\mathbf{U}_1\mathbf{U}_1^T + \mathbf{U}_n\mathbf{U}_n^T) \\
&= (\mathbf{I}_r \otimes \mathbf{U}_n)\left[\mathbf{K}_{rr}(\mathbf{A}^{1/2} \otimes \mathbf{A}^{1/2}) + 2(\mathbf{A} \otimes \mathbf{A}) - \mathbf{A} \otimes \mathbf{I}_r\right](\mathbf{I}_r \otimes \mathbf{U}_n^T) \\
&\quad +(\mathbf{I}_r \otimes \mathbf{U}_1)(\boldsymbol{\Lambda}_n^{-1}\mathbf{A} \otimes \boldsymbol{\Lambda}_1 - \mathbf{A} \otimes \mathbf{I}_{n-r})(\mathbf{I}_r \otimes \mathbf{U}_1^T)
\end{aligned}
\tag{6.54}
$$

Denote

$$
\mathbf{G} = \mathbf{K}_{rr}\left(\mathbf{A}^{1/2} \otimes \mathbf{A}^{1/2}\right) + 2(\mathbf{A} \otimes \mathbf{A}) - \mathbf{A} \otimes \mathbf{I}_r \tag{6.55}
$$

It is easy to verify that $\mathbf{G}$ is symmetrical. Thus, there is the following EVD:

$$
\mathbf{G} = \mathbf{VSV}^T, \tag{6.56}
$$

where $\mathbf{S} = \text{diag}(s_1, s_2..., s_{r^2})$ is a diagonal matrix whose diagonal elements are composed of the r^2 eigenvalues of $\mathbf{G}$, and $\boldsymbol{V}$ is an $r^2 \times r^2$ orthogonal matrix consisting of corresponding eigenvectors.

Inserting (6.55) and (6.56) into (6.54) and after some rearrangements, we have the EVD of $\hat{\mathbf{H}}$ as

$$
\hat{\mathbf{H}} = \begin{bmatrix} \mathbf{V}^T(\mathbf{I}_r \otimes \mathbf{U}_n^T) \\ \mathbf{I}_r \otimes \mathbf{U}_1^T \end{bmatrix}^T \begin{bmatrix} \mathbf{S} & \\ & \boldsymbol{\Theta} \end{bmatrix} \begin{bmatrix} \mathbf{V}^T(\mathbf{I}_r \otimes \mathbf{U}_n^T) \\ \mathbf{I}_r \otimes \mathbf{U}_1^T \end{bmatrix} \tag{6.57}
$$

where $\boldsymbol{\Theta} = \boldsymbol{\Lambda}_n^{-1}\mathbf{A} \otimes \boldsymbol{\Lambda}_1 - \mathbf{A} \otimes \mathbf{I}_{n-r}$. Clearly, $\boldsymbol{\Theta}$ is a diagonal matrix. It is easy to verify that the eigenvalues of $\hat{\mathbf{H}}$ are the diagonal elements of both $\boldsymbol{S}$ and $\boldsymbol{\Theta}$.

Clearly, all diagonal elements of $\boldsymbol{\Theta}$ must be positive as long as $\max\{\lambda_{n-r+1}, \lambda_{n-r+2}, ..., \lambda_n\}$.

$< \min\{\lambda_1, \lambda_2, ..., \lambda_{n-r}\}$. So we only need to determine whether the eigenvalues of $\boldsymbol{G}$ are positive. In order to illustrate this problem, we follow the proof procedures mentioned in [31]. Let us consider one case for $r = 2$, then, we have

$$\mathbf{G} = \begin{bmatrix} 2a_1^2 & 0 & 0 & 0 \\ 0 & 2a_1a_2 - a_1 & \sqrt{a_1a_2} & 0 \\ 0 & \sqrt{a_1a_2} & 2a_1a_2 - a_2 & 0 \\ 0 & 0 & 0 & 2a_2^2 \end{bmatrix}, \tag{6.58}$$

where a_1, a_2 are the diagonal elements of weighted matrix $\mathbf{A}$. We can calculate that the eigenvalues of $\mathbf{G}$ are $s_1 = 2a_1$, $s_4 = 2a_2$, $s_2 = 2a_1a_2$, and $s_3 = 2a_1a_2 - a_1 - a_2$. It is obvious that we only need to determine the sign of s_3. In other words, we should calculate the conditions where the inequality $2a_1a_2 - a_1 - a_2 > 0$ holds. After some manipulations, we have

$$a_1 > a_2 > a1/(2a_1 - 1). \tag{6.59}$$

From (6.59), we can conclude that $a_1/(2a_1 - 1)$ will tend to ∞ when $a_1 = 0.5$, and converges to 0.5 as $s_4 = 2a_2$,. For illustration, Fig. 6.3 shows the feasible domain of a_2 versus a_1, which is highlighted by the shadow.

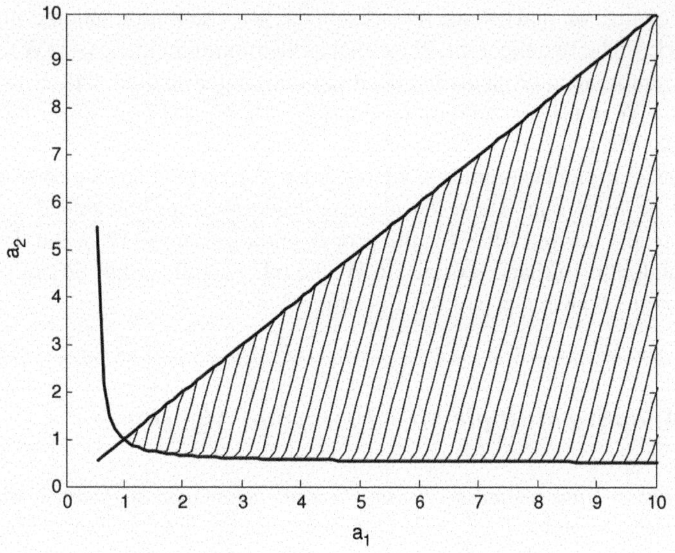

Fig. 6.3 Feasible domain of a_2 versus a_1

From Fig. 6.3, it holds that there always exists the properly chosen a_2 so that (6.59) holds for different a_1 as long as $a_1 > 0.5$. For example, if we set $a_1 = 2$, the allowable a_2 is from 1/3 to 2. From Fig. 6.3, it is obvious that the larger the a_1, the wider the allowable a_2. So it is very easy to choose a suitable a_2. Since (6.55) has no relationship with the eigenvalues of the autocorrelation matrix **R**, it is easy to verify the above conclusions holds for $r > 2$. Thus, it is concluded that all eigenvalues of $\hat{\mathbf{H}}$ are positive as long as the following inequality holds

$$\max\{\lambda_{n-r+1}, \lambda_{n-r+2}, ..., \lambda_n\} < \min\{\lambda_1, \lambda_2, ..., \lambda_{n-r}\}. \tag{6.60}$$

Since $\hat{\mathbf{H}}$ is positive definite if and only if (6.60) holds, and when $J_{WOJAm}(\mathbf{W})$ is unbounded as **W** tends to infinity, this local minimum is also the global one. Except for $\mathbf{W} = \mathbf{U}_n\mathbf{A}^{-1/2}$, it is easy to verify that all other stationary points result in $\hat{\mathbf{H}}$ being indefinite (having both positive and negative eigenvalues) and, thus, are saddle points. Furthermore, we can obtain the minimum value of $J_{WOJAm}(\mathbf{W})$ given by the following:

$$J_{WOJAm}(\mathbf{W} = \mathbf{U}_n\mathbf{A}^{-1/2}) = 1/2\left(\sum_{i=1}^{r} \ln(\lambda_{n-r+i}/a_i) + r\right) \tag{6.61}$$

This completes the proof of Theorem 6.5.

6.4.2.3 Remarks

In the following, we make some remarks on the proposed information criterion.

Remark 1 The two theorems show that at the minimum of $J_{WOJAm}(\mathbf{W})$, the state matrix **W** automatically yields the desired minor eigenvectors of **R**. However, the state matrix **W** in OJAm can only produce an orthonormal basis of the MS but not the multiple MCs.

Remark 2 From Theorem 6.5, we have that $J_{WOJAm}(\mathbf{W})$ has a global minimum point and no local one. Thus, the iterative algorithms are guaranteed to globally converge to the desired MCs for the proper initialization of **W** in the domain Ω. Even if random disturbances occur on **W**, the other saddle points do not cause any problem on the convergence of the algorithms.

6.4.3 Adaptive Multiple MCs Extraction Algorithms

Assume that at time instant k, the input vector sequences $\mathbf{x}(i)$ are available and $\mathbf{W}(i-1)$ are known for $1 \le i < k$. Our objective here is to develop some adaptive algorithms that can calculate an estimate of the MCs at time instant k from the estimate at k–1 and the newly observed sample $\mathbf{x}(k)$.

6.4.3.1 Gradient Descent Algorithm

The gradient of $J_{\mathrm{WOJAm}}(\mathbf{W})$ respect to $\mathbf{W}$ is given by (6.47). By using gradient descent search method, the following algorithm can be obtained for updating $\mathbf{W}$:

$$\begin{aligned}\mathbf{W}(k) &= \mathbf{W}(k-1) \\ &-[\hat{\mathbf{R}}(k)\mathbf{W}(k-1)(\mathbf{W}^T(k-1)\mathbf{R}(k)\mathbf{W}(k-1))^{-1} \\ &-\mathbf{W}(k-1)\mathbf{A}(\mathbf{W}^T(k-1)\mathbf{W}(k-1)\mathbf{A})^{-2}],\end{aligned} \tag{6.62}$$

where μ is the learning rate and satisfies $0 < \mu \leq 1$. $\hat{\mathbf{R}}(k)$ is the estimate of the autocorrelation matrix $\mathbf{R}$ at the time instant k and is given by

$$\hat{\mathbf{R}}(k) = \frac{1}{k}\sum_{i=1}^{k}\alpha^{k-i}\mathbf{x}(i)\mathbf{x}^T(i) = \frac{k-1}{k}\alpha\hat{\mathbf{R}}(k-1) + \frac{1}{k}\mathbf{x}(k)\mathbf{x}^T(k), \tag{6.63}$$

where α is the forgetting factor. For slow time-varying situation, α is chosen near one, whereas α should be chosen near zero for fast time-varying inputs.

6.4.3.2 Recursive Least Squares (RLS) Algorithm

For online learning algorithm, it is expected that the algorithm should extract the desired MCs directly from the input vector sequence. Such an algorithm is derived below.

Firstly, suppose that the projection of the input vector $\boldsymbol{x}(i)$ on the state matrix $\mathbf{W}(k-1)$ can be obtained by the following approximation: $\mathbf{y}(i) \triangleq \mathbf{W}^T(k-1)\mathbf{x}(i)$, and this approximation can be generally satisfied for many situations [32]. By using this approximation and the state matrix estimate $\mathbf{W}(i-1)$, it is easy to obtain the estimation of $\mathbf{y}(i)$ at any time instant $i(i = 1, 2, ..., k)$. Substituting (6.63) into (6.62), we can obtain

$$\begin{aligned}\mathbf{W}(k) = \mathbf{W}(k-1) - \mu\Bigg[\left(\sum_{i=1}^{k}\alpha^{k-i}\mathbf{x}(i)\mathbf{y}^T(i)\right) \times \left(\sum_{i=1}^{k}\alpha^{k-i}\mathbf{y}(i)\mathbf{y}^T(i)\right)^{-1} \\ -\mathbf{W}(k-1)\mathbf{A}\left(\mathbf{W}^T(k-1)\mathbf{W}(k-1)\mathbf{A}\right)^{-2}\Bigg].\end{aligned} \tag{6.64}$$

Denote $\mathbf{P}(k) = [\sum_{i=1}^{k}\alpha^{k-i}\mathbf{y}(i)\mathbf{y}^T(i)]^{-1}$ and $\tilde{\mathbf{W}}_k = [\sum_{i=1}^{k}\alpha^{k-i}\mathbf{x}(i)\mathbf{y}^T(i)]\mathbf{P}(k)$. By using the matrix inversion lemma to $\mathbf{P}(k)$, we have

$$\mathbf{P}(k)=\left(\sum_{i=1}^{k}\alpha^{k-i}\mathbf{y}(i)\mathbf{y}^T(i)\right)^{-1}=\left(\alpha\sum_{i=1}^{k-1}\alpha^{k-i-1}\mathbf{y}(i)\mathbf{y}^T(i)+\mathbf{y}(k)\mathbf{y}^T(k)\right)^{-1}$$
$$=\left(\alpha\mathbf{P}(k-1)+\mathbf{y}(k)\mathbf{y}^T(k)\right)^{-1}=\alpha^{-1}\left(\mathbf{P}(k-1)-\mathbf{g}(k)\mathbf{y}^T(k)\mathbf{P}(k-1)\right), \tag{6.65}$$

where $\mathbf{g}(k)=\mathbf{P}(k-1)\mathbf{y}(k)/(\alpha+\mathbf{y}^T(k)\mathbf{P}(k-1)\mathbf{y}(k))$ is the gain vector. By using (6.65), it is easy to obtain that $\mathbf{g}(k)=\mathbf{P}(k-1)\mathbf{y}(k)$. Then, we have

$$\begin{aligned}\tilde{\mathbf{W}}(k)&=\left(\sum_{i=1}^{k}\alpha^{k-i}\mathbf{x}(i)\mathbf{y}^T(i)\right)\mathbf{P}(k)\\&=\left(\sum_{i=1}^{k-1}\alpha^{k-i-1}\mathbf{x}(i)\mathbf{y}^T(i)\right)\mathbf{P}(k-1)+\mathbf{x}(k)\mathbf{y}_k^T\mathbf{P}(k)\\&-\left(\sum_{i=1}^{k-1}\alpha^{k-i-1}\mathbf{x}(i)\mathbf{y}^T(i)\right)\mathbf{P}(k-1)\mathbf{y}(k)\mathbf{g}^T(k)\\&=\tilde{\mathbf{W}}(k-1)+\left(\mathbf{x}(k)-\tilde{\mathbf{W}}(k-1)\mathbf{y}(k)\right)\mathbf{g}^T(k)\end{aligned} \tag{6.66}$$

During the above derivation procedures, the following conclusions are used:

$$\begin{aligned}\mathbf{g}(k)\mathbf{y}^T(k-1)\mathbf{P}(k-1)&=\left(\mathbf{g}(k)\mathbf{y}^T(k-1)\mathbf{P}(k-1)\right)^T\\&=\mathbf{P}(k-1)\mathbf{y}(k-1)\ \mathbf{g}^T(k)\end{aligned} \tag{6.67}$$

Summarizing the above derivations, we can obtain the RLS implementation of the WOJAm algorithm, which is given by Table 6.1.

Remark First of all, for convenience, the gradient algorithm is denoted as WOJAm algorithm, and its RLS implementation is denoted as RLS-WOJAm algorithm. The

Table 6.1 The RLS algorithm for multiple MCs extraction

Initialization: Set $k=0$, and set the learning rate μ and forgetting factor α. The other initial parameters are set as follows:
$\mathbf{P}(0)=\delta\mathbf{I}_r(\delta$ is a small positive number)
$\tilde{\mathbf{W}}(0)=\mathbf{0}$
$\mathbf{W}(0)$ ($\mathbf{W}(0)$ is randomly generated)
Iteration: Set $k=k+1$, and update the state matrix by the following equations:
$\mathbf{y}(k)=\mathbf{W}^T(k-1)\mathbf{x}(k)$
$\mathbf{g}(k)=\mathbf{P}(k-1)\mathbf{y}(k)/(\alpha+\mathbf{y}^T(k)\mathbf{P}(k-1)\mathbf{y}(k))$
$\mathbf{P}(k)=\alpha^{-1}(\mathbf{P}(k-1)-\mathbf{g}(k)\mathbf{y}^T(k)\mathbf{p}(k-1))$
$\tilde{\mathbf{W}}(k)=\tilde{\mathbf{W}}(k-1)+\left(\mathbf{x}(k)-\tilde{\mathbf{W}}(k-1)\mathbf{y}(k)\right)\mathbf{g}^T(k)$
$\mathbf{W}(k)=\mathbf{W}(k-1)-\mu[\tilde{\mathbf{W}}(k)-\mathbf{W}(k-1)\mathbf{A}(\mathbf{W}^T(k-1)\mathbf{W}(k-1)\mathbf{A})^{-2}]$

two algorithms (WOJAM and RLS-WOJAm) can extract the multiple MCs of $\boldsymbol{R}$ in parallel, while the OJAm algorithm in [9] can only provide an orthonormal basis of the minor space.

6.4.4 Convergence Analysis

Here, the convergence of the gradient algorithm is analyzed by exploring the Eq. (6.62). Suppose that the input vector $\mathbf{x}(k)$ is sampled from a stationary stochastic process. The ODE for the WOJAm algorithm is given below.

$$\begin{aligned}\frac{\mathrm{d}\mathbf{W}(t)}{\mathrm{d}t} &= \mathbf{RW}(t)(\mathbf{W}^T(t)\mathbf{RW}(t))^{-1} \\ &- \mathbf{W}(t)\mathbf{A}(\mathbf{W}^T(t)\mathbf{W}(t)\mathbf{A})^{-2},\end{aligned} \tag{6.68}$$

where $t = \mu k$. By analyzing the convergence of (6.68), we can establish the condition for the global convergence of (6.62). In particular, we will answer the following questions.

(1) Is the dynamical system described by (6.68) able to globally converge to the desired MCs of the autocorrelation matrix $\boldsymbol{R}$?
(2) What is the domain of attraction around the equilibrium attained at the MCs, or equivalently, what is the initial condition to ensure the global convergence?

Since the Lyapunov function will be used later, herein we provide it as Theorem 5.5 [33].

Theorem 6.5 *Suppose that there exists a function $L(\mathbf{u})$. Define a region Ω' consisting of all points of $\mathbf{u}$ such as $L(\mathbf{u}) < c$ for some constant, with the boundary of Ω' given by the all points $L(\mathbf{u}) = c$, such that.*
(a) *$dL(\mathbf{u})/dt < 0$ for all $\mathbf{u} \neq \mathbf{u}^*$ in Ω', and.*
(b) *$dL(\mathbf{u}^*)/dt = 0$.*
Then, the equilibrium point $\mathbf{u} = \mathbf{u}^*$ is asymptotically stable with a domain of attraction Ω'.

To utilize Theorem 6.5 to analyze the global convergence of the WOJAm algorithm, the following function is constructed:

$$L(\mathbf{W}) = -\frac{1}{2}\mathrm{tr}\{\mathrm{In}[\mathbf{W}^T\mathbf{R}_r\mathbf{W}] + (\mathbf{W}^T\mathbf{WA})^{-1}\}, \tag{6.69}$$

where $\mathbf{R}_r = \mathbf{U}_n\mathbf{\Lambda}_n\mathbf{U}_n^T$. Next, we will prove that (6.69) is a Lyapunov function for the ODE (6.68).

Theorem 6.6 *$L(\mathbf{W})$ is a Lyapunov function for the* ODE (6.68), *which defines the domain of attraction for convergence of $\mathbf{W}$ to the equilibrium stable point $\mathbf{W} = \mathbf{U}_n\mathbf{A}^{-1/2}$ as.*

$$\Omega = \{\mathbf{W} | 0 < \mathbf{W}^T\mathbf{R}_r\mathbf{W} < \infty,\ \mathbf{W}^T\mathbf{W}\mathbf{A} \neq \mathbf{0}\} \tag{6.70}$$

Proof

From (6.69), we have that $L(\mathbf{W})$ is bounded for any $\mathbf{W}$ in the domain Ω, and $L(\mathbf{W}) = \infty$ if and only if $\mathbf{W}^T\mathbf{R}_r\mathbf{W}$ is singular. Since $L(\mathbf{W})$ is continuous and has the following continuous derivative:

$$\nabla L(\mathbf{W}) = -[\mathbf{R}_r\mathbf{W}(\mathbf{W}^T\mathbf{R}_r\mathbf{W})^{-1} - \mathbf{W}\mathbf{A}(\mathbf{W}^T\mathbf{W}\mathbf{A})^{-2}] \tag{6.71}$$

By using the chain rule, we can obtain the derivation of $L(\mathbf{W})$ with respect to t

$$\begin{aligned} \frac{dL(\mathbf{W}(t))}{dt} &= tr\left[\nabla L(\mathbf{W}) \cdot \frac{d\mathbf{W}^T}{dt}\right] \\ &= tr\{[\mathbf{R}_r\mathbf{W}(\mathbf{W}^T\mathbf{R}_r\mathbf{W})^{-1} - \mathbf{W}\mathbf{A}(\mathbf{W}^T\mathbf{W}\mathbf{A})^{-2}]^T \\ &\cdot [\mathbf{R}\mathbf{W}(\mathbf{W}^T\mathbf{R}\mathbf{W})^{-1} - \mathbf{W}\mathbf{A}(\mathbf{W}^T\mathbf{W}\mathbf{A})^{-2}]\} \end{aligned} \tag{6.72}$$

Notice that we have dropped the dependence on t in the above formula. Following Theorem 6.5, next we will prove that $(dL(\mathbf{W})/dt) < 0$ for any $\mathbf{W} \in \Omega^*$, where $\Omega^* = \Omega - \{\mathbf{W}|\mathbf{W} = \mathbf{U}_n\mathbf{A}^{-1/2}\}$, and $(dL(\mathbf{W})/dt) = 0$ for $\mathbf{W} = \mathbf{U}_n\mathbf{A}^{-1/2}$. Substituting (6.43) into (6.72), we have

$$\begin{aligned} \frac{dL(\mathbf{W})}{dt} &= -tr\left\{\begin{bmatrix} -\mathbf{Z}_1\mathbf{A}(\mathbf{Z}_2^T\mathbf{Z}_2\mathbf{A})^{-2} \\ \Lambda_2\mathbf{Z}_2(\mathbf{Z}_2^T\Lambda_2\mathbf{Z}_2)^{-1} - \mathbf{Z}_2\mathbf{A}(\mathbf{Z}_2^T\mathbf{Z}_2\mathbf{A})^{-2} \end{bmatrix}^T\right. \\ &\times \left.\begin{bmatrix} \Lambda_1\mathbf{Z}_1(\mathbf{Z}_1^T\Lambda_1\mathbf{Z}_1 + \mathbf{Z}_2^T\Lambda_2\mathbf{Z}_2)^{-1} - \mathbf{Z}_1\mathbf{A}(\mathbf{Z}_2^T\mathbf{Z}_2\mathbf{A})^{-2} \\ \Lambda_2\mathbf{Z}_2(\mathbf{Z}_1^T\Lambda_1\mathbf{Z}_1 + \mathbf{Z}_2^T\Lambda_2\mathbf{Z}_2)^{-1} - \mathbf{Z}_2\mathbf{A}(\mathbf{Z}_2^T\mathbf{Z}_2\mathbf{A})^{-2} \end{bmatrix}\right\} \end{aligned} \tag{6.73}$$

where $\mathbf{Z}_1 = \mathbf{U}_1^T\mathbf{W}$ and $\mathbf{Z}_2 = \mathbf{U}_n^T\mathbf{W}$. Since $\mathbf{W}^T\mathbf{R}_r\mathbf{W} \neq \mathbf{0}$, $\mathbf{Z}_2$ must be invertible. Therefore, there must exist a matrix $\mathbf{B} \in \Re^{(n-r)\times r}$ such that $\mathbf{Z}_1 = \mathbf{B}\mathbf{Z}_2$. Substituting this equation into (6.73), we can obtain

$$\begin{aligned} \frac{dL(\mathbf{W})}{dt} &= -tr\{-2\mathbf{C} + \mathbf{C}\mathbf{Z}_2^T(\mathbf{I} + \mathbf{B}^T\mathbf{B})\mathbf{Z}_2\mathbf{C}^T \\ &+ \mathbf{Z}_2^{-1}(\mathbf{I} + \mathbf{B}^T\Lambda_1\mathbf{B}\Lambda_2^{-1})^{-1}\mathbf{Z}_2^{-1}\} \end{aligned} \tag{6.74}$$

where $\mathbf{C} = (\mathbf{Z}_2^T\mathbf{Z}_2)^{-1}\mathbf{A}^{-1}(\mathbf{Z}_2^T\mathbf{Z}_2)^{-1}$. When $\mathbf{W}$ is near the stable point, the matrix $\mathbf{C}$ can be approximated as a diagonal matrix: $\mathbf{C} \approx \mathbf{C}' = diag(c_1, c_2, ..., c_r)$. Then, we write (6.74) as

$$\begin{aligned} \frac{dL(\mathbf{W})}{dt} &\approx -tr\{-2\mathbf{C}' + \mathbf{C}'\mathbf{Z}_2^T(\mathbf{I} + \mathbf{B}^T\mathbf{B})\mathbf{Z}_2\mathbf{C}' \\ &+ \mathbf{Z}_2^{-1}(\mathbf{I} + \mathbf{B}^T\Lambda_1\mathbf{B}\Lambda_2^{-1})^{-1}\mathbf{Z}_2^{-1}\} \end{aligned} \tag{6.75}$$

Firstly, we consider the situation where $\mathbf{Z}_1 \neq \mathbf{0}$. From matrix theory, we have

$$tr\{\mathbf{Z}_2^T(\mathbf{I}+\mathbf{B}^T\mathbf{B})\mathbf{Z}_2\} < tr\{\mathbf{Z}_2^T(\mathbf{I}+\mathbf{B}^T\mathbf{\Lambda}_1\mathbf{B}\Lambda_2^{-1})\mathbf{Z}_2\} \tag{6.76}$$

Denote $\mathbf{D} = \mathbf{Z}_2^T(\mathbf{I}+\mathbf{B}^T\mathbf{B})\mathbf{Z}_2$, and the eigen decomposition of matrix $\mathbf{D}$ is given by:

$$\mathbf{D} = \Phi\Psi\Phi^T, \tag{6.77}$$

where $\boldsymbol{\Psi} = diag(\mu_1, \mu_2, ..., \mu_r)$, μ_i are the eigenvalues of $\mathbf{D}$ for $i = 1, 2, ..., r$, and $\boldsymbol{\Phi}$ is composed of the eigenvectors of $\mathbf{D}$. Substituting (6.77) into (6.76), we have

$$\begin{aligned} \frac{\mathrm{d}L(\mathbf{W})}{\mathrm{d}t} &< -tr\{-2\mathbf{C}' + \mathbf{C}'\mathbf{D}\mathbf{C}' + \mathbf{D}^{-1}\} \\ &= 2tr(\mathbf{C}') - tr(\mathbf{C}'\Psi\mathbf{C}') - tr[(\boldsymbol{\Psi})^{-1}] \\ &= \sum_{i=1}^{r}\left(2c_i - c_i^2\mu_i - 1/\mu_i\right) \le 0. \end{aligned} \tag{6.78}$$

Next, when $\mathbf{Z}_1 = \mathbf{0}$, (6.74) becomes

$$\frac{\mathrm{d}L(\mathbf{Z})}{\mathrm{d}t} = -tr\{\mathbf{E}^T\mathbf{E}\} \le 0, \tag{6.79}$$

where $\mathbf{E} = \mathbf{\Lambda}_2\mathbf{Z}_2(\mathbf{Z}_2^T\mathbf{\Lambda}_2\mathbf{Z}_2)^{-1} - \mathbf{Z}_2\mathbf{A}(\mathbf{Z}_2^T\mathbf{Z}_2\mathbf{A})^{-2}$.

From (6.79), it is obvious that $\mathrm{d}L(\mathbf{Z})/\mathrm{d}t = 0$ if and only if $\mathbf{E} = \mathbf{0}$. After some simplification, we have $\mathbf{Z}_2 = \mathbf{A}^{-1/2}$. It turns out that $\mathbf{W} = \mathbf{U}_n\mathbf{A}^{-1/2}$. From (6.78) and (6.79), it can be seen that $\mathrm{d}(L(\mathbf{Z}))/\mathrm{d}t < 0$ all any $\mathbf{W} \in \Omega^*$. Thus, the point $\mathbf{W} = \mathbf{U}_n\mathbf{A}^{-1/2}$ is asymptotically stable with the domain of attraction Ω. In other words, starting from any $\mathbf{W}$ in Ω, the solution of (6.68) will globally asymptotically converge to the first r MCs of the autocorrelation matrix $\mathbf{R}$.

This completes the proof of Theorem 6.6.

6.4.5 *Simulations Experiments and Conclusions*

Here, four simulations are provided to demonstrate the behavior of the proposed algorithms. In the first simulation, the proposed algorithm is used for multiple MCs extraction. In the second simulation, the numerical stability of the proposed algorithm is shown. In the third simulation, some performances of the proposed algorithm are contrasted with other algorithms. The fourth simulation demonstrates the efficiency of the proposed algorithms with high dimension data.

Two evaluation functions are adopted for evaluating the convergence speed and the estimation accuracy of the algorithms. The first one is the Direction Cosine (DC for short), which is defined as follows:

$$DC(k) = \frac{\left|\mathbf{w}_i^T(k)\mathbf{v}_i\right|}{\|\mathbf{w}_i(k)\|\,\|\mathbf{v}_i\|}, \tag{6.80}$$

where $\mathbf{w}_i(k)$ is the estimated vector of the ith MC at time instant k, and $\mathbf{v}_i$ denotes the real ith MC. The second one is the norm of the state matrix

$$\rho_i(k) = \left|\mathbf{w}_i^T(k)\mathbf{w}_i(k)\right| \; i = 1, 2, ..., r, \tag{6.81}$$

where $\mathbf{w}_i(k)$ is the ith column of the state matrix, i.e., the estimated vector of the ith MC at time instant k. The DC is used to examine the ability of algorithms to extract the MCs. If the DC curves converge to one, this means that the state matrix has converged to the direction of the MCs. The norm of the state matrix $\rho_i(k)$ is used to evaluate the transient behavior of the state matrix during the iterations.

6.4.5.1 Convergence of the Proposed Algorithm

Here, we use the same way to generate the input signal as in [9, 34]. Let $\mathbf{T} = \text{randn}(10, 10)/10$ and $\mathbf{z}_k \in \Re^{10\times 1}$ be Gaussian, spatially–temporally white and randomly generated, where "randn" is an order in MATLAB, then the input data sequence is generated by $\mathbf{x}_k = \mathbf{T}\mathbf{z}_k$. Next, the RLS-WOJAm algorithm is used to extract the first 3 MCs of this data sequence. Some initial parameters of the RLS-WOJAm algorithm are set below. The weighted matrix is set as $\mathbf{A} = \text{diag}[3, 2, 1][3, 2, 1]$, and the learning rate η is equal to 0.1. Figures 6.4 and 6.5 show the transient behavior of the RLS-WOJAm algorithm for extracting in parallel the first three MCs.

From Fig. 6.4, we can see that the state matrix in the proposed algorithm converge to the directions of the MCs. From Fig. 6.5, it is obvious that the norms of the state matrix converge to a constant value. From the two figures, it is concluded that the proposed algorithm can extract multiple MCs in parallel.

6.4.5.2 Numerical Stability

In the above experiment, the Direction Cosine and the norm of the state matrix are evaluated over 100 independent runs. In order to measure the numerical stability of the proposed algorithm, the confidence intervals (CI) is calculated at each iteration time. Assume that the evaluation function DC(k) obeys a Gaussian distribution at time $k(k = 1, 2, \ldots, 600)$, i.e., $DC(k) \sim N(\eta_1, \sigma_1^2)$, where η_1 and σ_1 are the mean and standard deviation, respectively. According to the mathematical theory of statistics, the CI of DC(k) can be calculated by the following two equations

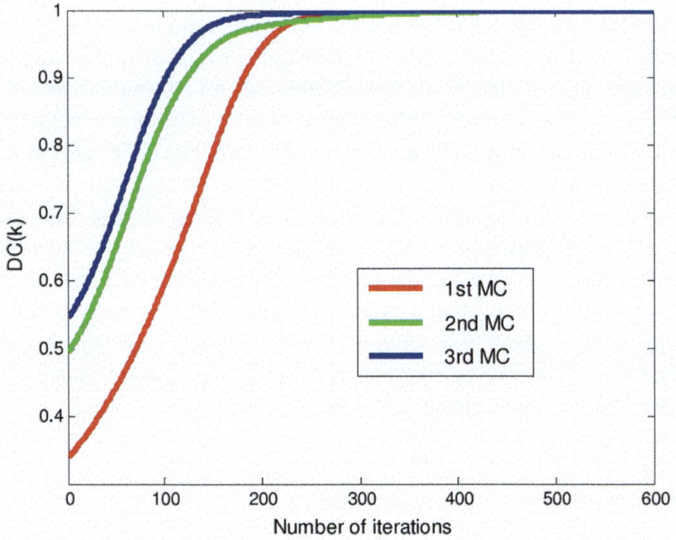

Fig. 6.4 DC curves of RLS-WOJAm algorithm

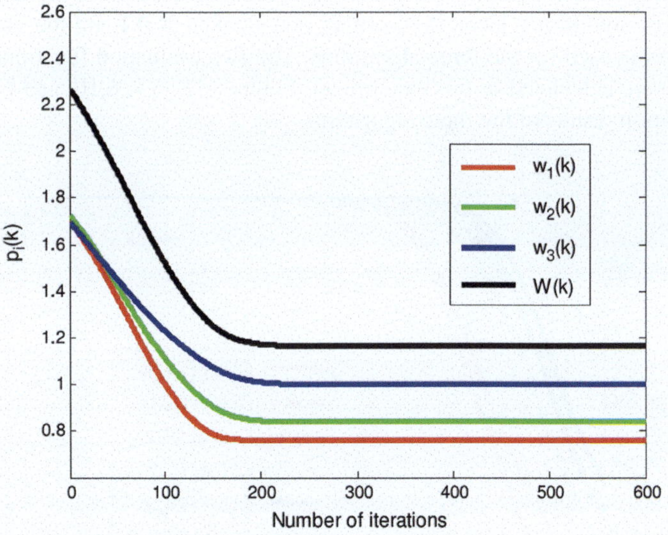

Fig. 6.5 Norm curves of RLS-WOJAm algorithm

$$
\begin{cases} \mathrm{CUK}(k) = \overline{DC}(k) + \frac{S(k)}{\sqrt{n}} t_{\frac{\beta}{2}}(n-1) \\ \mathrm{CLL}(k) = \overline{DC}(k) - \frac{S(k)}{\sqrt{n}} t_{\frac{\beta}{2}}(n-1) \end{cases} \tag{6.82}
$$

where CUL means confidence upper limit, CLL means confidence lower limit, $\overline{DC}(k)$ is the mean value of the DC(k) over 100 independent runs at time instant k, S(k) is the sample variance, t means Student's distribution, β is the significance level, and n is equal to 100 here. Besides, if the other evaluation function $\rho_i(k)$ also obeys a Gaussian distribution, we can obtain the method to calculate CI of $\rho_i(k)$ in the similar way.

Figures 6.6 and 6.7 provide the CIs of the two evaluation functions at the 0.05 significance level, respectively. From the two figures, it is obvious that at the beginning of the iteration, the two evaluation functions have wide confidence intervals. However, the length of confidence interval becomes more and more shorter as the iterations go on. After about 300 iteration steps for DC(k) (220 for $\rho_i(k)$), the length of CI converges to zero. From this experiment, it is concluded that the proposed algorithm has good numerical stability.

6.4.5.3 Contrasts with Other Algorithms

Here, the RLS-WOJAm algorithm is compared with the AMMD algorithm [35] and the MCA-OFRANS algorithm [36]. The simulation data is produced using the method mentioned in Sect. 6.4.5.1. The initial parameters are set as the same values for the three algorithms. Here, the learning rate is taken as 0.1 and the same initial state matrix is used for the three algorithms. The two evaluation functions (DC(k) and $p_i(k)$) are calculated in this experiment. Figures 6.8, 6.9, 6.10, and 6.11 show the simulation results of the three algorithms.

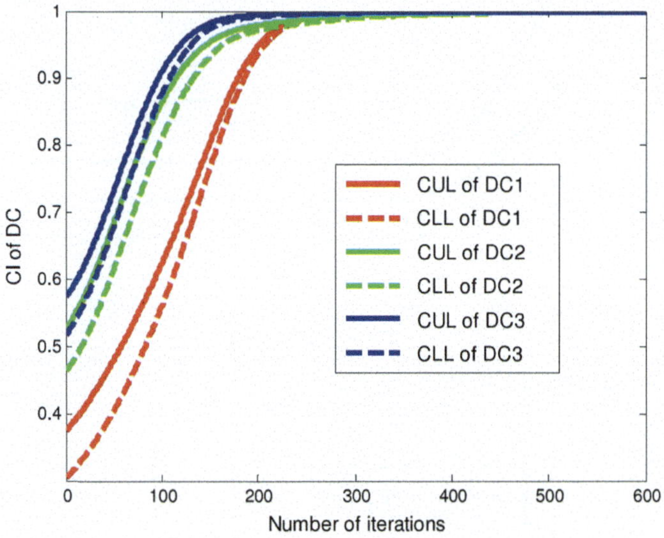

Fig. 6.6 CI curves of DC

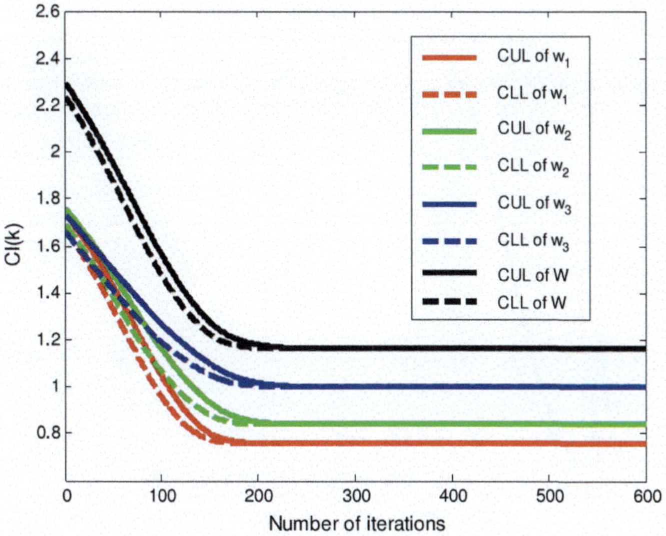

Fig. 6.7 CI curves of norm

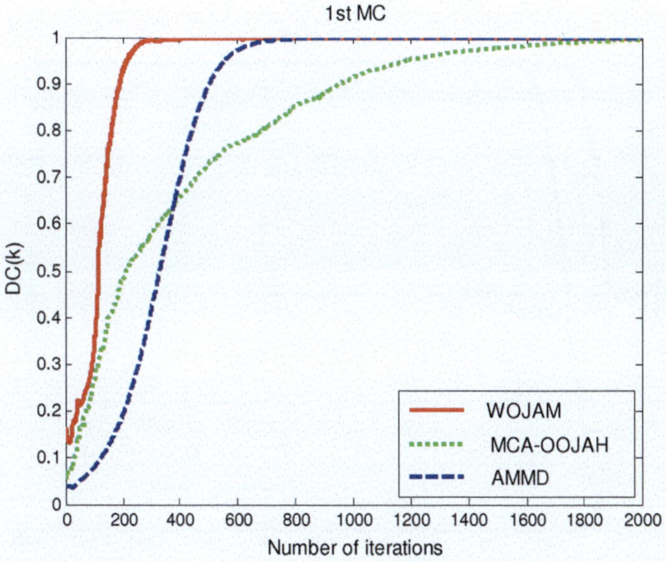

Fig. 6.8 DC curves of 1st MC

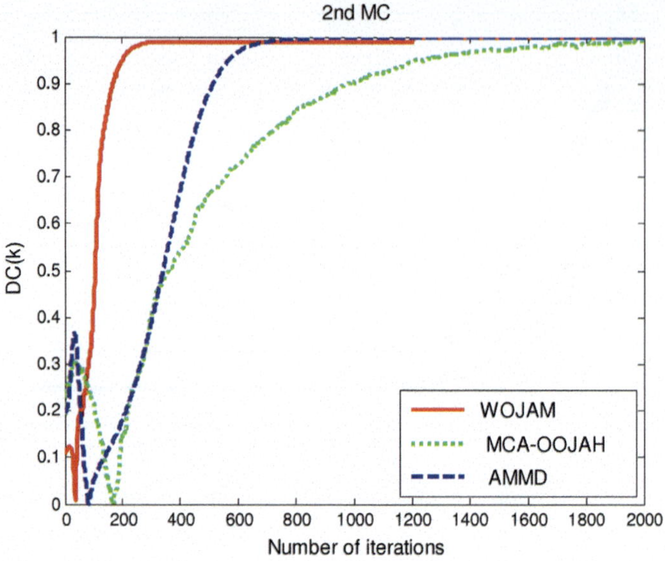

Fig. 6.9 DC curves of 2nd MC

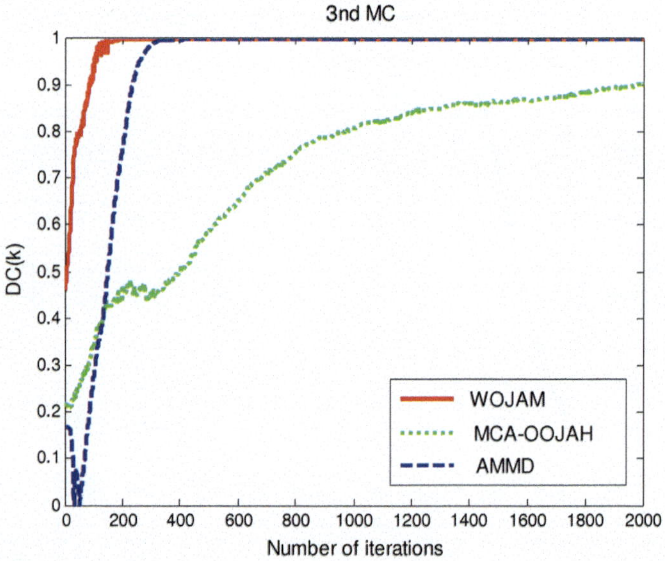

Fig. 6.10 DC curves of 2nd MC

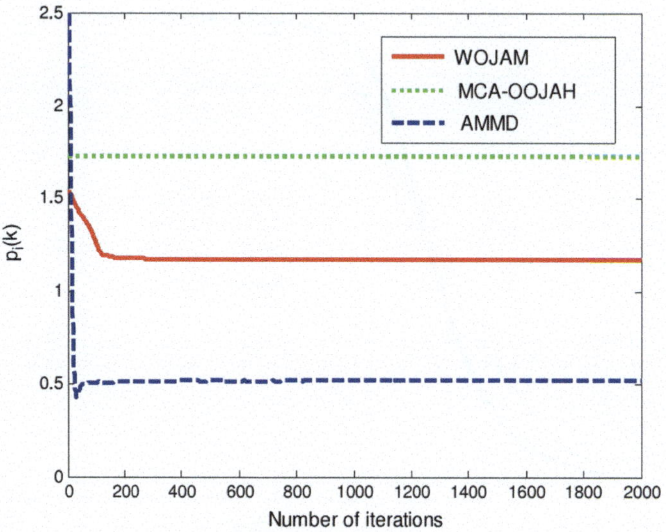

Fig. 6.11 Norm curves of three algorithms

6.4.5.4 Simulation for High Dimension Data

Here, some simulations are provided to illustrate the superiority of the proposed algorithm over traditional algebraic approaches in dealing with high dimension data. Consider the following method to generate symmetric positive definite matrix as in [37]

$$\mathbf{R}_1 = \mathbf{Q}_1 \Lambda \mathbf{Q}_1^T, \tag{6.83}$$

where $\mathbf{Q}_1$ is an orthogonal matrix and satisfies $\mathbf{Q}_1\mathbf{Q}_1^T = \mathbf{I}$, $\mathbf{I}$ stands for the identity matrix, and $\mathbf{\Lambda}$ is a diagonal matrix, whose diagonal elements compose the eigenvalues of $\mathbf{R}_1$. Next, the WOJAm algorithm and the EVD method are used to extract the first 3 minor eigenvector of the autocorrelation matrix, which is produced by using (6.83). Here, the dimensions of the autocorrelation matrix are set as 100, 300, 500, 800, 1000, respectively. Table 6.2 provides the computation times of the two methods in dealing with different dimension matrices. Figures 6.12 and 6.13 show the evaluation results (DC(k) and $\rho_i(k)$) of the WOJAm algorithm in dealing with 1000 dimension matrix.

From two figures, it is obvious that although the dimension of the autocorrelation matrix is very high, the WOJAm algorithm can effectively extract multiple MCs. Compared the traditional EVD, the WOJAm algorithm has less computation time, especially for high dimension matrix. From this experiment, it is concluded that the WOJAm can deal with high dimension matrix and has fast convergence speed.

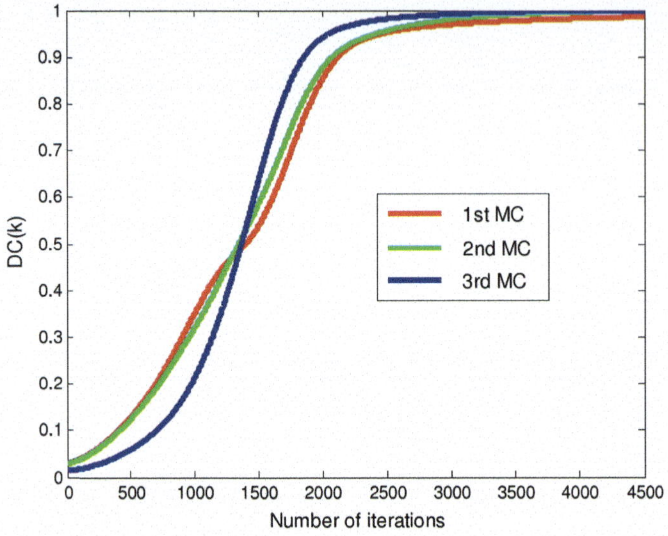

Fig. 6.12 DC curves of WOJAm algorithm

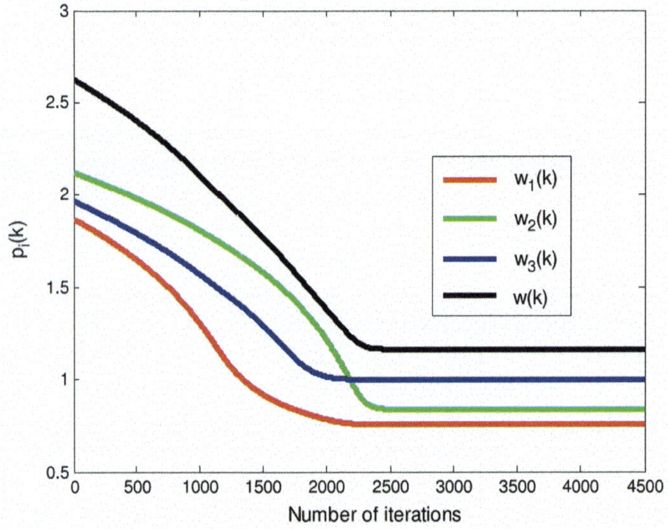

Fig. 6.13 Norm curves of WOJAm algorithm

Table 6.2 Computation times of the two algorithms

	100	300	500	800	1000
WOJAm(s)	0.078	0.1073	0.4347	1.5670	2.772
EVD(s)	0.0103	0.9070	3.2830	26.9100	64.7800

6.4.5.5 Conclusions

In this section, a novel information criterion is proposed by introducing a weighted matrix into an existing information criterion. A dependent relation of choosing the weighted matrix on statistics of the input signal is established. Two algorithms (WOJAm and RLS-WOJAm) are derived, which can effectively extract the desired MCs in parallel. Besides, the proposed algorithms have no restriction on the eigenvalues and do not need other orthogonalization operations on the state matrix. Simulation results show that the proposed algorithms have good numerical stability and are advantageous for high-dimensional data. The proposed algorithms can provide a fast and flexible method for many potential real applications.

6.5 A Coupled Minor Component Analysis Algorithm

In this section, based on a novel information criterion, we propose a self-stabilizing couple MCA algorithms.

6.5.1 *Coupled Dynamical System*

To derive neural-based algorithms often starts with an information criterion, e.g., by maximization of the variance of the projected data or by minimization of the reconstruction error [38]. However, as stated in [39], the freedom of choosing an information criterion is greater if Newton's method is applied because the criterion just has to have stationary points in the desired solutions.

Here, we use the same information criterion as Hou's [16], which is

$$p = \mathbf{w}^T \mathbf{C} \mathbf{w} - \mathbf{w}^T \mathbf{w} \lambda + \lambda, \tag{6.84}$$

where $\mathbf{C} = E\{\mathbf{x}\mathbf{x}^T\} \in \Re^{n \times n}$, and $\mathbf{w} \in \Re^{n \times 1}$ and $\lambda \in \Re$ denote the estimation of eigenvector (weight vector) and eigenvalue of $\boldsymbol{C}$, respectively.

From (6.84), we can obtain that

$$\frac{\partial p}{\partial \mathbf{w}} = 2\mathbf{C}\mathbf{w} - 2\lambda \mathbf{w} \tag{6.85}$$

$$\frac{\partial p}{\partial \lambda} = -\mathbf{w}^T \mathbf{w} + 1 \tag{6.86}$$

thus the stationary points $(\overline{\mathbf{w}}, \overline{\lambda})$ of (6.84) are defined by the following:

$$\mathbf{C}\overline{\mathbf{w}} = \overline{\lambda}\overline{\mathbf{w}} \tag{6.87}$$

$$\overline{\mathbf{w}}^T\overline{\mathbf{w}} = 1 \tag{6.88}$$

from which we have $\overline{\mathbf{w}}^T\mathbf{C}\overline{\mathbf{w}} = \overline{\lambda}$. Thus, the criterion (6.84) fulfills the aforementioned requirement: the stationary points include all associated eigenvectors and eigenvalues of $\boldsymbol{C}$. The Hessian of the criterion is given by the following:

$$\mathbf{H}(\mathbf{w},\lambda) = \left(\begin{array}{c|c} \frac{\partial^2 p}{\partial \mathbf{w}^2} & \frac{\partial^2 p}{\partial \mathbf{w}\partial\lambda} \\ \frac{\partial^2 p}{\partial\lambda\partial\mathbf{w}} & \frac{\partial^2 p}{\partial\lambda^2} \end{array}\right) = 2\left(\begin{array}{c|c} \mathbf{C}-\lambda\mathbf{I} & -\mathbf{w} \\ -\mathbf{w}^T & 0 \end{array}\right) \tag{6.89}$$

Based on the Newton's method, the equation used by Möller and Hou to derive the differential equations can be written as follows:

$$\begin{pmatrix} \dot{\mathbf{w}} \\ \dot{\lambda} \end{pmatrix} = -\mathbf{H}^{-1}(\mathbf{w},\lambda)\begin{pmatrix} \frac{\partial p}{\partial \mathbf{w}} \\ \frac{\partial p}{\partial\lambda} \end{pmatrix} \tag{6.90}$$

Using different information criterion, both Möller and Hou tried to find the inverse of their Hessian $\mathbf{H}^{-1}(\mathbf{w},\lambda)$. Although the inverse Hessian of Möller's and Hou's are different, they finally obtained the same CPCA and CMCA algorithms [16]. However, using another method, we derive the differential equation as follows:

$$\mathbf{H}(\mathbf{w},\lambda)\begin{pmatrix} \dot{\mathbf{w}} \\ \dot{\lambda} \end{pmatrix} = -\begin{pmatrix} \frac{\partial p}{\partial \mathbf{w}} \\ \frac{\partial p}{\partial\lambda} \end{pmatrix} \tag{6.91}$$

In this case, we do not need to calculate the inverse Hessian. Substituting (6.85), (6.86), and (6.88) into (6.91), we have

$$2\left(\begin{array}{c|c} \mathbf{C}-\lambda\mathbf{I} & -\mathbf{w} \\ -\mathbf{w}^T & 0 \end{array}\right)\begin{pmatrix} \dot{\mathbf{w}} \\ \dot{\lambda} \end{pmatrix} = -\begin{pmatrix} 2\mathbf{C}\mathbf{w} - 2\lambda\mathbf{w} \\ -\mathbf{w}^T\mathbf{w}+1 \end{pmatrix} \tag{6.92}$$

Then, it holds that

$$(\mathbf{C}-\lambda\mathbf{I})\dot{\mathbf{w}} - \mathbf{w}\dot{\lambda} = -(\mathbf{C}-\lambda\mathbf{I})\mathbf{w} \tag{6.93}$$

$$-2\mathbf{w}^T\dot{\mathbf{w}} = \mathbf{w}^T\mathbf{w} - 1 \tag{6.94}$$

In the vicinity of the stationary point $(\mathbf{w}_1, \lambda_1)$, by approximating $\mathbf{w} = \mathbf{w}_1, \lambda = \lambda_1 \ll \lambda_j (2 \le j \le n)$, and after some manipulations (see Appendix 1 [17]), we obtain the following coupled dynamical system

$$\dot{\mathbf{w}} = \frac{(\mathbf{w}^T\mathbf{w}+1)\mathbf{C}^{-1}\mathbf{w}}{2\mathbf{w}^T\mathbf{C}^{-1}\mathbf{w}} - \mathbf{w} \tag{6.95}$$

$$\dot{\lambda} = \frac{\mathbf{w}^T\mathbf{w} + 1}{2}\left(\frac{1}{\mathbf{w}^T\mathbf{C}^{-1}\mathbf{w}} - \lambda\right) \tag{6.96}$$

The approximation error analysis of (6.95, 6.96) can be found in Appendix 1 in [17].

6.5.2 *Coupled MCA Learning Algorithms*

6.5.2.1 Single MC Extraction Algorithm

The online form of an algorithm is usually used in practical application. The differential equations can be turned into the online form by informally approximating $\mathbf{C} = \mathbf{x}(k)\mathbf{x}^T(k)$. That is, the expression of the algorithms in online form can be approximated by slowly following $\mathbf{w}(k+1) = f(\mathbf{x}(k)\mathbf{x}^T(k), \mathbf{w}(k))$ for subsequent observations of $\mathbf{x}$. In [15], Möller has pointed out that this transition is infeasible if the equation contains $\mathbf{C}^{-1}$, because it is hard to replace the inverse matrix $\mathbf{C}^{-1}$ by an expression containing the input vector $\boldsymbol{x}$. However, this problem can be solved in another way [32, 40], in which $\mathbf{C}^{-1}$ can be updated as follows:

$$\hat{\mathbf{C}}^{-1}(k+1) = \frac{k+1}{k}\left[\hat{\mathbf{C}}^{-1}(k) - \frac{\hat{\mathbf{C}}^{-1}(k)\mathbf{x}(k+1)\mathbf{x}^T(k+1)\hat{\mathbf{C}}^{-1}(k)}{k + \mathbf{x}^T(k+1)\hat{\mathbf{C}}^{-1}(k)\mathbf{x}(k+1)}\right], \tag{6.97}$$

where $\hat{\mathbf{C}}^{-1}(k)$ starts with $\hat{\mathbf{C}}^{-1}(0) = \mathbf{I}$ and converges to $\mathbf{C}^{-1}$ as $k \to \infty$. Then, the CMCA system (6.95, 6.96) has the online form as

$$\mathbf{w}(k+1) = \mathbf{w}(k) + \gamma(k)\left\{\frac{[\mathbf{w}^T(k)\mathbf{w}(k) + 1]\mathbf{Q}(k)\mathbf{w}(k)}{2\mathbf{w}^T(k)\mathbf{Q}(k)\mathbf{w}(k)} - \mathbf{w}(k)\right\} \tag{6.98}$$

$$\lambda(k+1) = \lambda(k) + \gamma(k)\frac{\mathbf{w}^T(k)\mathbf{w}(k) + 1}{2}\left[\frac{1}{\mathbf{w}^T(k)\mathbf{Q}(k)\mathbf{w}(k)} - \lambda(k)\right] \tag{6.99}$$

$$\mathbf{Q}(k+1) = \frac{k+1}{\alpha k}\left[\mathbf{Q}(k) - \frac{\mathbf{Q}(k)\mathbf{x}(k+1)\mathbf{x}^T(k+1)\mathbf{Q}(k)}{\alpha k + \mathbf{x}^T(k+1)\mathbf{Q}(k)\mathbf{x}(k+1)}\right], \tag{6.100}$$

where $0 < \alpha \leq 1$ denotes the forgetting factor and $\gamma(k)$ is the learning rate. If all training samples come from a stationary process, we choose $\alpha = 1$. $\mathbf{Q}(k) = \mathbf{C}^{-1}(k)$ starts with $\mathbf{Q}(0) = \mathbf{I}$. Here, we refer to the Eqs. (6.95, 6.96) and its online form (6.98–6.100) as "fMCA", where f means fast.

If we further approximate $\mathbf{w}^T\mathbf{w} \approx 1$ (which fulfills in the vicinity of the stationary points) in (6.95, 6.96), we can obtain a simplified CMCA system

$$\dot{\mathbf{w}} = \frac{\mathbf{C}^{-1}\mathbf{w}}{\mathbf{w}^T\mathbf{C}^{-1}\mathbf{w}} - \mathbf{w} \tag{6.101}$$

$$\dot{\lambda} = \left(\frac{1}{\mathbf{w}^T\mathbf{C}^{-1}\mathbf{w}} - \lambda\right) \tag{6.102}$$

and the online algorithm is

$$\mathbf{w}(k+1) = \mathbf{w}(k) + \gamma(k)\left\{\frac{\mathbf{Q}(k)\mathbf{w}(k)}{\mathbf{w}^T(k)\mathbf{Q}(k)\mathbf{w}(k)} - \mathbf{w}(k)\right\} \tag{6.103}$$

$$\lambda(k+1) = \lambda(k) + \gamma(k)\left[\frac{1}{\mathbf{w}^T(k)\mathbf{Q}(k)\mathbf{w}(k)} - \lambda(k)\right], \tag{6.104}$$

where $\mathbf{Q}(k)$ is updated by (6.6.100). In the following, we will refer to this algorithm as "aMCA", where *a* means adaptive.

6.5.2.2 Multiple Minor Eigen Pairs Estimation

The following shows how to estimate the remaining p-1 minor eigen pairs.

For the CMCA case, considering the following equations

$$\hat{\mathbf{C}}_j = \hat{\mathbf{C}}_{j-1} + \eta\lambda_{j-1}\mathbf{w}_{j-1}\mathbf{w}_{j-1}^T,\ j = 2, \ldots, p, \tag{6.105}$$

where $\hat{\mathbf{C}}_1 = \hat{\mathbf{C}}$ and η is larger than the largest eigenvalue of $\hat{\mathbf{C}}$, and $(\mathbf{w}_{j-1}, \lambda_{j-1})$ is the $(j-1)$th minor eigen pair of $\hat{\mathbf{C}}$ that has been extracted. It is found that

$$\begin{aligned}\hat{\mathbf{C}}_j\mathbf{w}_q &= (\hat{\mathbf{C}}_{j-1} + \eta\lambda_{j-1}\mathbf{w}_{j-1}\mathbf{w}_{j-1}^T)\mathbf{w}_q = (\hat{\mathbf{C}}_1 + \eta\sum_{r=1}^{j-1}\lambda_r\mathbf{w}_r\mathbf{w}_r^T)\mathbf{w}_q \\ &= \hat{\mathbf{C}}_1\mathbf{w}_q + \eta\sum_{r=1}^{j-1}\lambda_r\mathbf{w}_r\mathbf{w}_r^T\mathbf{w}_q \\ &= \begin{cases}\hat{\mathbf{C}}_1\mathbf{w}_q + \eta\lambda_q\mathbf{w}_q = (1+\eta)\lambda_q\mathbf{w}_q & \text{for } q = 1, 2, \ldots, j-1 \\ \hat{\mathbf{C}}_1\mathbf{w}_q = \lambda_q\mathbf{w}_q & \text{for } q = j, \ldots, p\end{cases}\end{aligned} \tag{6.106}$$

Suppose that matrix $\hat{\mathbf{C}}_1$ has eigenvectors $\mathbf{w}_1, \mathbf{w}_2, \ldots, \mathbf{w}_n$ corresponding to eigenvalues $(0 <)\sigma_1 < \sigma_2 < \cdots < \sigma_n$, then matrix $\boldsymbol{C}_j$ has eigenvectors $\mathbf{w}_j, \ldots, \mathbf{w}_n, \mathbf{w}_1, \ldots, \mathbf{w}_{j-1}$ corresponding to eigenvalues $(0 <)\sigma_j < \cdots\sigma_n < (1+\eta)\sigma_1 < \cdots < (1+\eta)\sigma_{j-1}$. In this case, σ_j is the smallest eigenvalue of $\mathbf{C}_j$. Based on the SM-formula, we have

$$
\begin{aligned}
\hat{\mathbf{Q}}_j = \hat{\mathbf{C}}_j^{-1} &= (\hat{\mathbf{C}}_{j-1} + \eta\lambda_{j-1}\mathbf{w}_{j-1}\mathbf{w}_{j-1}^T)^{-1} \\
&= \hat{\mathbf{C}}_{j-1}^{-1} - \frac{\eta\lambda_{j-1}\hat{\mathbf{C}}_{j-1}^{-1}\mathbf{w}_{j-1}\mathbf{w}_{j-1}^T\hat{\mathbf{C}}_{j-1}^{-1}}{1+\eta\lambda_{j-1}\mathbf{w}_{j-1}^T\hat{\mathbf{C}}_{j-1}^{-1}\mathbf{w}_{j-1}} \\
&= \hat{\mathbf{Q}}_{j-1} - \frac{\eta\lambda_{j-1}\hat{\mathbf{Q}}_{j-1}\mathbf{w}_{j-1}\mathbf{w}_{j-1}^T\hat{\mathbf{Q}}_{j-1}}{1+\eta\lambda_{j-1}\mathbf{w}_{j-1}^T\hat{\mathbf{Q}}_{j-1}\mathbf{w}_{j-1}},\ j = 2, \ldots, p.
\end{aligned}
\tag{6.107}
$$

Thus, by replacing $\hat{\mathbf{Q}}$ with $\hat{\mathbf{Q}}_j$ in (6.98, 6.99) or (6.103, 6.104), they can be used to estimate the jth minor eigen pair $(\mathbf{w}_j, \lambda_j)$ of $\hat{\mathbf{C}}$.

6.5.3 Analysis of Convergence and Self-stabilizing Property

In analyzing the convergence of coupled algorithms, the major work is to find the eigenvalues of the Jacobian

$$
\mathbf{J}(\mathbf{w}_1, \lambda_1) = \left(\begin{array}{c|c} \frac{\partial\dot{\mathbf{w}}}{\partial\mathbf{w}^T} & \frac{\partial\dot{\mathbf{w}}}{\partial\lambda} \\ \frac{\partial\dot{\lambda}}{\partial\mathbf{w}^T} & \frac{\partial\dot{\lambda}}{\partial\lambda} \end{array}\right)
\tag{6.108}
$$

of the differential equations for a stationary point $(\mathbf{w}_1, \lambda_1)$. For fMCA rule, after some manipulations (see Appendix 2 in [17]), we have

$$
\mathbf{J}(\mathbf{w}_1, \lambda_1) = \left(\begin{array}{c|c} \mathbf{C}^{-1}\lambda_1 - \mathbf{I} - \mathbf{w}_1\mathbf{w}_1^T & \mathbf{0} \\ -2\lambda_1\mathbf{w}_1^T & -1 \end{array}\right).
\tag{6.109}
$$

The Jacobian can be simplified by an orthogonal transformation with

$$
\mathbf{U} = \left(\begin{array}{c|c} \overline{\mathbf{W}} & \mathbf{0} \\ \mathbf{0}^T & 1 \end{array}\right).
\tag{6.110}
$$

where $\overline{\mathbf{W}}$ contains all eigenvectors of $\boldsymbol{C}$ in its columns $(\mathbf{w}_1, \ldots, \mathbf{w}_n)$, coincides with eigenvalues $\lambda_1 < \cdots < \lambda_n$.

The transformed Jacobian $\mathbf{J}^* = \mathbf{U}^T\mathbf{J}\mathbf{U}$ has the same eigenvalues as $\mathbf{J}$. In the vicinity of a stationary point $(\mathbf{w}_1, \lambda_1)$, we approximate $\overline{\mathbf{W}}^T\mathbf{w} \approx \mathbf{e}_1$ and obtain

$$
\mathbf{J}^*(\mathbf{w}_1, \lambda_1) = \left(\begin{array}{c|c} \overline{\Lambda}^{-1}\lambda_1 - \mathbf{I} - \mathbf{e}_1\mathbf{e}_1^T & \mathbf{0} \\ -2\lambda_1\mathbf{e}_1^T & -1 \end{array}\right).
\tag{6.111}
$$

where $\overline{\Lambda}$ is a diagonal matrix containing all eigenvalues of $\boldsymbol{C}$, which is, $\overline{\Lambda} = \mathrm{diag}(\lambda_1, \ldots, \lambda_n)$. The eigenvalues α of $\mathbf{J}^*$ are determined as $\det(\mathbf{J}^* - \alpha\mathbf{I}) = 0$, which are

$$
\alpha_1 = \alpha_{n+1} = -1,\ \alpha_j = \tfrac{\lambda_1}{\lambda_j} - 1,\ j = 2, \ldots, n.
\tag{6.112}
$$

Since stability requires $\alpha < 0$ and thus $\lambda_1 < \lambda_j, j = 2, \ldots, n$, it is easy to find that only minor eigen pairs are stable stationary points, while all others are saddles or repellers. What's more, if we further assume $\lambda_1 \leq \lambda_j$, all eigenvalues are $\alpha \approx -1$. Hence, the system converges with approximately equal speed in all its eigendirections, and this speed is widely independent of the eigenvalues λ_j of the covariance matrix [15]. That is to say, the speed-stability problem does not exist in fMCA algorithm.

Similarly, for aMCA rule, we analyze the stability by finding the eigenvalues of

$$\mathbf{J}^*(\mathbf{w}_1, \lambda_1) = \left(\begin{array}{c|c} \overline{\Lambda}^{-1}\lambda_1 - \mathbf{I} - \mathbf{e}_1\mathbf{e}_1^T & \mathbf{0} \\ -2\lambda_1\mathbf{e}_1^T & -1 \end{array}\right). \tag{6.113}$$

which are

$$\alpha_1 = -2,\ \alpha_{n+1} = -1,\ \alpha_j = \tfrac{\lambda_1}{\lambda_j} - 1,\ j = 2, \ldots, n. \tag{6.114}$$

The situation of aMCA is similar to that of fMCA, and the only difference is that the first eigenvalue of Jacobian is $\alpha_1 = -1$ for fMCA and $\alpha_1 = -2$ for aCMA. Thus, the convergence speed of fMCA and aMCA are almost the same.

Next we analyze the self-stabilizing property of the proposed algorithms. For fMCA rules, we have from (6.98) that

$$\begin{aligned} \Xi &= \frac{\|\mathbf{w}(k+1)\|^2}{\|\mathbf{w}(k)\|^2} = \frac{\mathbf{w}^T(k+1)\mathbf{w}(k+1)}{\mathbf{w}^T(k)\mathbf{w}(k)} \\ &= 1 + \gamma(k)\left[1 - \mathbf{w}^T(k)\mathbf{w}(k)\right] \end{aligned} \tag{6.115}$$

where the second-order terms associated with the learning rate have been neglected. Similarly, for aMCA rules, we have that

$$\Xi = 1 + \gamma(k)\left[1 - \mathbf{w}^T(k)\mathbf{w}(k)\right] \tag{6.116}$$

That is to say, for fMCA, aMCA, it holds that

$$\Xi \begin{cases} > 1 & \text{for } \mathbf{w}^T(k)\mathbf{w}(k) < 1 \\ = 1 & \text{for } \mathbf{w}^T(k)\mathbf{w}(k) = 1 \\ < 1 & \text{for } \mathbf{w}^T(k)\mathbf{w}(k) > 1 \end{cases} \tag{6.117}$$

This shows the self-stabilizing property of the weight vector: if $\mathbf{w}^T(k)\mathbf{w}(k) < 1$, the weight vector length increases in the next iteration, and if $\mathbf{w}^T(k)\mathbf{w}(k) > 1$, it decreases. So at each step, the weight vector moves toward unity.

Then, we analyze the self-stabilizing property of the eigenvalue estimate for the proposed algorithms. While $\lambda(k) > 0$ for all $k \geq 0$ when $0 < \gamma < 1$ and $\lambda(0) > 0$ (see Appendix 3 in [17]), and based on the result of convergence analysis and the self- stabilizing property of the weight vector, which is, $\mathbf{w}^T(k)\mathbf{w}(k)$ converges to 1 and $\mathbf{w}^T(k)\mathbf{Q}(k)\mathbf{w}(k)$ converges to λ_1^{-1} and $\mathbf{w}^T(k)\mathbf{C}(k)\mathbf{w}(k)$ converges to λ_1 fastly and stably. Then, for some $K > 0$, we have $\mathbf{w}^T(k)\mathbf{Q}(k)\mathbf{w}(k) \approx \lambda_1^{-1}$ and $\mathbf{w}^T(k)\mathbf{C}(k)\mathbf{w}(k) \approx \lambda_1$ for all $k \geq K$. Then, take fMCA as an example, we have

$$\lambda(k+1) \approx \tilde{\lambda}(k+1) = \lambda(k) + \gamma(k)[\lambda_1 - \lambda(k)] \ \ (k \geq K) \tag{6.118}$$

and the approximation error is

$$\begin{aligned} &\lambda(k+1) - \tilde{\lambda}(k+1) \\ &= \gamma(k)\left\{\left[\frac{\mathbf{w}^T(k)\mathbf{w}(k)+1}{2} - 1\right]\left[\frac{1}{\mathbf{w}^T(k)\mathbf{Q}\mathbf{w}(k)} - \lambda\right] - [\lambda_1 - \lambda(k)]\right\} \end{aligned} \tag{6.119}$$

Since $\mathbf{w}^T(k)\mathbf{w}(k) \to 1$ and $\mathbf{w}^T(k)\mathbf{Q}(k)\mathbf{w}(k) \to \lambda_1^{-1}$ as $k \to \infty$, we deduce that.

$$\lim_{k\to\infty} \left|\lambda(k+1) \approx \tilde{\lambda}(k+1)\right| = 0. \tag{6.120}$$

This means that approximation error of $\lambda(k+1)$ approaches to 0 as $k \to \infty$. Then, we have that.

$$\frac{\lambda(k+1)}{\lambda(k)} = 1 + \gamma(k)\left(\frac{\lambda_1}{\lambda(k)} - 1\right)\begin{cases} > 1 \ \text{for} \ \lambda(k) < \lambda_1 \\ = 1 \ \text{for} \ \lambda(k) = \lambda_1 \\ < 1 \ \text{for} \ \lambda(k) > \lambda_1 \end{cases} \tag{6.121}$$

This shows the self-stabilizing property of the eigenvalue estimate of fMCA: if $\lambda(k) < \lambda_1$, the value of $\lambda(k+1)$ increases in the next iteration, and if $\lambda(k) > \lambda_1$, it decreases. Thus, in each step, the value of $\lambda(k)$ moves toward the true eigenvalue λ_1. Similarly, It is concluded that the eigenvalue estimate of aMCA are all self-stabilized, and the approximation error of $\lambda(k+1)$ approaches to 0 as $k \to \infty$.

To sum up, fMCA and aMCA are self-stabilizing learning algorithms.

6.5.4 *Simulation and Conclusions*

Here, several experiments are provided to illustrate the performance of the proposed algorithms. In experiments 1, the stability of proposed CMCA algorithms is shown in comparison with existing CMCA algorithm. The experiment 2 shows the self-stabilizing property of the proposed algorithm. Experiments 3 illustrates an example of practical application.

In experiments 1 and 2, all algorithms are used to extract the MC from a high-dimensional input data sequence, which is generated by.

$$\mathbf{x} = \mathbf{B} \cdot \mathbf{y}(k), \tag{6.122}$$

where each column of $\mathbf{B} \in \Re^{30\times 30}$ is Gaussian with variance 1/30, and $\mathbf{y}(k) \in \Re^{30\times 1}$ is Gaussian and randomly produced.

In all experiments, to measure the estimation accuracy, we compute the norm of eigenvector estimation (weight vector) $\|\mathbf{w}(k)\|$ and the projection $(\psi(k))$ of the weight vector onto the true eigenvector at the each step:

$$\psi(k) = \left|\mathbf{w}^T(k)\mathbf{w}_1\right| / \|\mathbf{w}(k)\|, \tag{6.123}$$

where $\boldsymbol{w}_1$ is the true minor (for MCA) eigenvector with unit length.

Note that in all experiments, the convergence speed of all algorithms is judged by the terms of number of iterations.

6.5.4.1 Multiple Minor Eigen Pairs Estimation

In experiments 1 and 2, the initial conditions are set as follows:

(1) The weight vector is initialized with a random vector (unit length).
(2) The learning rate $\gamma(k)$ starts at $\gamma(0) = 10^{-2}$ and decays exponentially toward zero with a final value $\gamma(k_{\max}) = 10^{-4}$.
(3) We set $\alpha = 1$ (if used), and $\lambda(0) = 0.001$ for all CMCA algorithms.

In experiments 1, $k_{\max} = 20000$ training steps are executed for all algorithms. In order to test the stability of algorithms, after 10,000 training steps, we drastically change the input signals, thus the eigen information changed suddenly. The new signals are also generated by similar method as shown in (6.122). All algorithms start to extract the new eigen pair since $k = 10{,}001$. The learning rate for nMCA is 10 times smaller than that for the others. Then, 20 times of Monte Carlo simulation are executed for all experiments.

Figure 6.14 shows the time course of the projection of minor weight vector. Obviously, except mMCA, in all algorithms the projection converges toward unity, thus these weight vector aligns with the true eigenvector. The convergence speed of mMCA is lower than that of the others. We can also find that the convergence speed of fMCA and aMCA rules are similar, and higher than that of the others. At time step $k = 10{,}001$, where the input signals changed suddenly, all algorithms start to extract the new eigen pair.

Figure 6.15 shows the time course of weight vector length. It is found that the vector length of nMCA converges to a non-unit length. The convergence speed and the stability of fMCA and aMCA are higher and better than that of the others. The convergence speed of aMCA is a bit higher than that of fMCA.

Figure 6.16 shows the time course of the minor eigenvalue estimation. Obviously, mMCA cannot extract the minor eigenvalue as effective as the other algorithms after the input signals changed. From Figures 6.14, 6.15, and 6.16, it is concluded that the performance of fMCA and aMCA is better that that of the other CMCA algorithms.

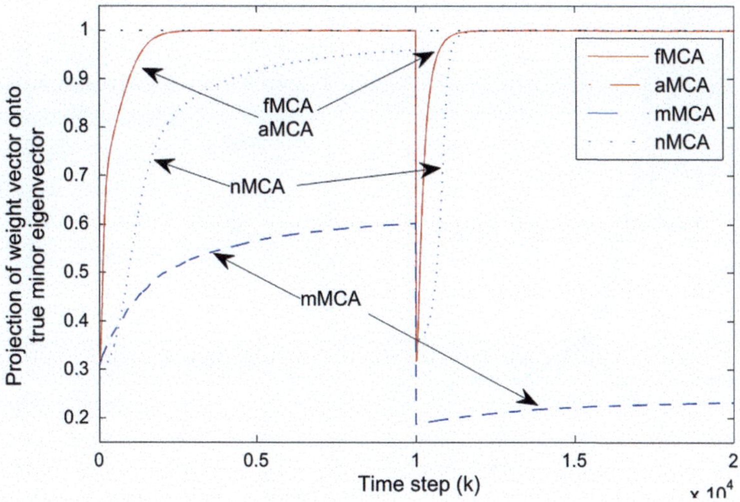

Fig. 6.14 Experiment 1: the projection of weight

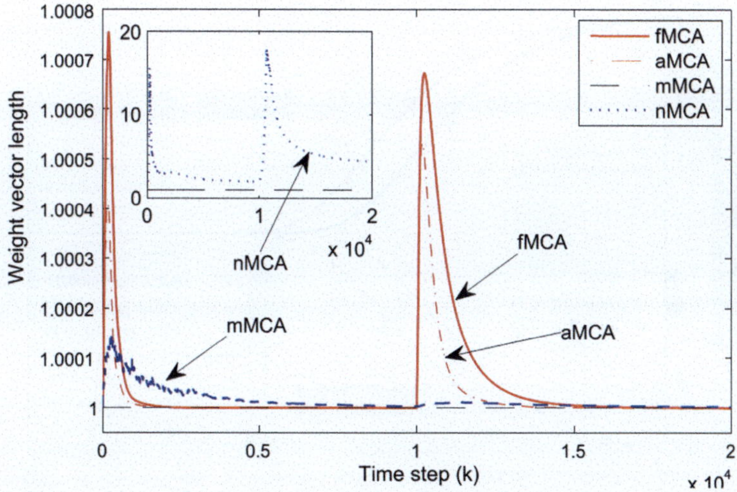

Fig. 6.15 Experiment 1: the weight vector length vector onto the true minor eigenvector

It is noted that nMCA contains $\boldsymbol{C}$ and $\boldsymbol{C}^{-1}$ simultaneously in the equations, we can prove that mMCA also has the "speed-stability" problem though it is a coupled rules. These may be the reason why our algorithms perform better than nMCA and mMCA.

The result of experiment 2 is drawn in Fig. 6.17, which shows the time course of weight vector length estimation of fMCA and aMCA. It is found that these two algorithms converge to unit length rapidly, which shows the self-stabilizing property.

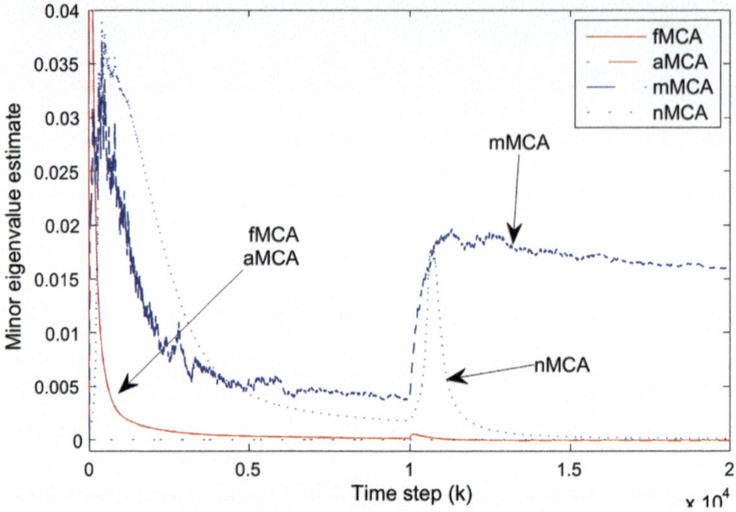

Fig. 6.16 Experiment 1: the minor eigenvalue estimation

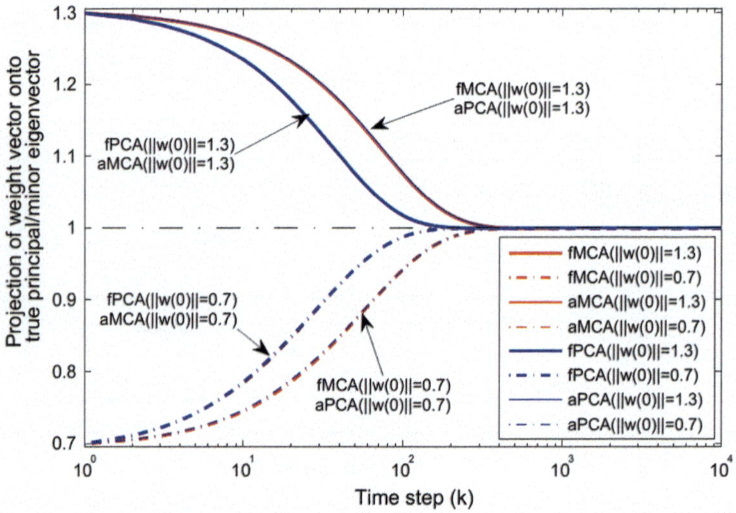

Fig. 6.17 Experiment 2: weight vector length

6.5.4.2 Examples: Practical Application of the Proposed Algorithms

An important application of MCA is to solve the TLS problem. In experiment 3, we use fMCA, aMCA, mMCA and nMCA, where the dimension is taken as one, to conduct the line fitting under the TLS criterion. By adding Gaussian noises to the 400 sample points on the line $x_2 = -0.5x_1$, we obtain a set of two-dimensional data

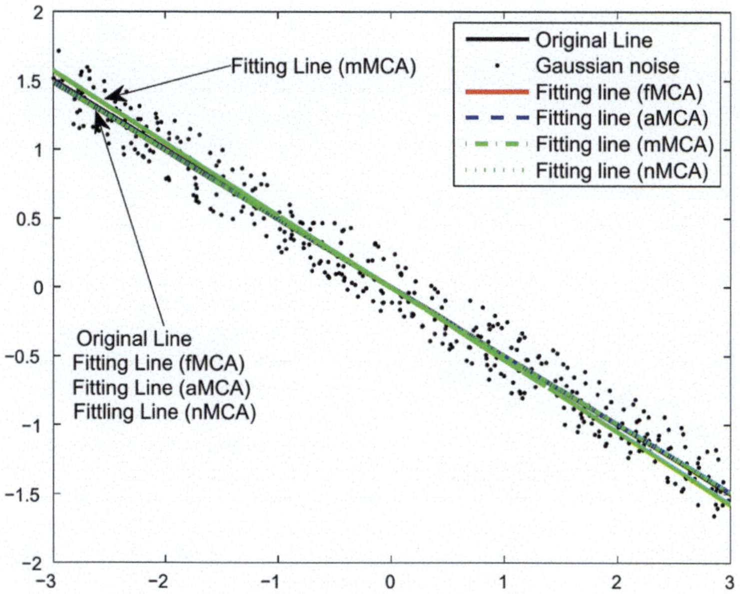

Fig. 6.18 Experiment 3: the line fitting problem

points $\mathbf{T} = \{[x_1(k), x_2(k)]^T, k = 1, 2, ..., 400\}$ as shown in Fig. 6.18. The problem of line fitting is to find a parameterized line model (e.g.,$\omega_1 x_1 + \omega_2 x_2 = 0$) to fit the point set, such that the sum of the squared perpendicular distances of the line from these data points is minimized. This line fitting problem can be reduced to a problem of estimating minor component of this data set $\mathbf{T}$. After the mean removal and normalization, randomly select vectors from the data set $\mathbf{T}$ as the inputs for all aforementioned algorithms. Figure 6.19 shows the convergence of the weight vector with initial length $\mathbf{w}(0) = 0.2$ under the learning rate $\gamma = 0.05$ and the projection of the weight vector onto the true minor eigenvector for all algorithms. After 200 iterations, the weight vector $\boldsymbol{w}(k)$ of all algorithms converges to a vector $\mathbf{w}^* = [0.4362, 0.7925]^T$. Figure 6.18 gives the fitting results. From Fig. 6.18, we can see that all algorithms have similar performance.

6.5.4.3 Conclusion

In this section, we propose a novel method to derive neural-based algorithms based on a special information criterion. Based on the modified Newton's method, we derive a coupled dynamical system and obtain several coupled MCA algorithms. The proposed MCA algorithms can mitigate the speed-stability problem that plagues most non-coupled algorithms. The proposed MCA algorithms are self-stabilized and converge faster than existing algorithms.

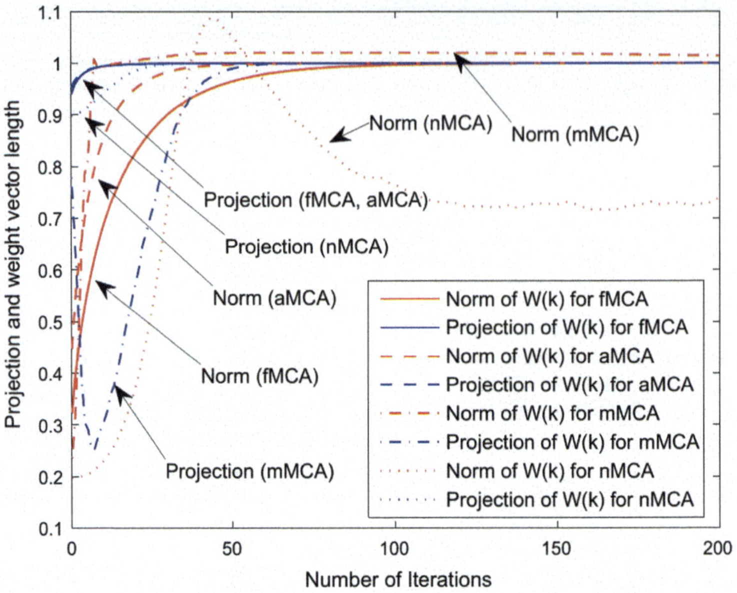

Fig. 6.19 Experiment 3: projection and weight vector length and fitting results

6.6 Summary

In this chapter, four types of neural-based MCA algorithms including the Oja's MCA algorithms, the self-stabilizing MCA algorithms, the orthogonal Oja's algorithms and the coupled MCA algorithms, have been reviewed. Three MCA algorithms we proposed, i.e., a novel parallel multiple MCs extraction algorithm by diagonal matrix, a weighted information criterion for multiple MC and two adaptive MCA, and a self-stabilizing coupled MCA algorithm, are presented, and their performance, e.g., convergence, stability, self-stabilizing, etc. are mathematically proved and numerically verified. The proposed MCA algorithms can provide a fast and flexible method for many potential real TLS applications.

References

1. L. Xu, E. Oja, C. Suen, Modified Hebbian learning for curve and surface fitting. Neural Netw. **5**(3), 441–457 (1992)
2. E. Oja, Principal components, minor components and linear neural networks. Neural Netw. **5**(6), 927–935 (1992)
3. F. Luo, R. Unbehauen, A. Cichocki, A minor component analysis algorithm. Neural Netw. **10**(2), 291–297 (1997)
4. G. Cirrincione, M. Cirrincione, The MCA EXIN neuron for the minor component analysis. IEEE Trans. Neural Netw. **13**(1), 160–187 (2002)

5. D.Z. Feng, Z. Bao, L.C. Jiao, Total least mean squares algorithm. IEEE Trans. Signal Process. **46**(8), 2122–2130 (1998)
6. D. Peng, Y. Zhang, Y. Xiang, H. Zhang, A globally convergent MC algorithm with an adaptive learning rate. IEEE Trans. Neural Netw. Learn. Syst. **23**, 359–365 (2012)
7. F. Luo, Y. Li, C. He, Neural network approach to the TLS linear prediction frequency estimation problem. Neurocomputing **11**(1), 31–42 (1996)
8. S.C. Douglas, S.Y. Kung, S. Amari, A self-stabilized minor subspace rule. IEEE Signal Process. Lett. **5**(12), 328–330 (1998)
9. D.Z. Feng, W.X. Zheng, Y. Jia, Neural network Learning algorithms for tracking minor subspace in high-dimensional data stream. IEEE Trans. Neural Netw. **16**(3), 513–521 (2005)
10. R. Möller, A self-stabilizing learning rule for minor component analysis. Int. J. Neural Syst. **14**(1), 1–8 (2004)
11. T. Chen, S. Amari, Unified stabilization approach to principal and minor components extraction algorithms. Neural Netw. **14**(10), 1377–1387 (2001)
12. G. Mathew, V.U. Reddy, Orthogonal eigensubspace estimation using neural networks. IEEE Trans. Signal Process. **42**(7), 1803–1811 (1994)
13. G. Mathew, V.U. Reddy, S. Dasgupta, Adaptive estimation of eigensubspace. IEEE Trans. Signal Process. **43**(2), 401–411 (1995)
14. X.Y. Kong, C.H. Hu, C.Z. Han, On the discrete-time dynamics of a class of self-stabilizing MCA extraction algorithms. IEEE Trans. Neural Netw. **21**(1), 175–181 (2010)
15. R. Moller, A. Konies, Coupled principal component analysis. IEEE Trans. Neural Netw. **15**(1), 214–222 (2004)
16. L. Hou, T.P. Chen, Online algorithm of coupled principal (minor) component analysis. J. Fudan Univ. **45**(2), 158–169 (2006)
17. X.W. Feng, X.Y. Kong, H.G. Ma, H.M. Liu, Unified and coupled self-stabilizing algorithm for minor and principal eigen-pair extraction. Neural. Process. Lett. (2016). https://doi.org/10.1007/s11063-016-9520-3
18. S. Ouyang, Z. Bao, G.S. Liao, P.C. Ching, Adaptive minor component extraction with modular structure. IEEE Trans. Signal Process. **49**(9), 2127–2137 (2001)
19. S. Attallah, K. Abed-Meraim, Low-cost adaptive algorithm for noise subspace estimation. Electron. Lett. **38**(12), 609–611 (2002)
20. X.G. Doukopoulos, G.V. Moustakides, Fast and stable subspace tracking. IEEE Trans. Signal Process. **56**(4), 1452–1465 (2008)
21. X.G. Doukopoulos, G.V. Moustakides, Blind adaptive channel estimation in OFDM systems. IEEE Trans. Wireless Commun. **5**(7), 1716–1725 (2006)
22. S. Attallah, The generalized Rayleigh's quotient adaptive noise subspace algorithm: a householder transformation-based implementation. IEEE Trans. Circ. Syst. II **53**(1), 3–7 (2006)
23. S. Bartelmaos, K. Abed-Meraim, Principal and minor subspace tracking: algorithms and stability analysis, in *Proceeding of IEEE International Conference on Acoustic, Speech, and Signal Processing*, vol. 3, pp 560–563
24. R. Badeau, G. Richard, B. David, Fast and stable YAST algorithm for principal and minor subspace tracking. IEEE Trans. Signal Process. **56**(8), 3437–3446 (2008)
25. J.F. Yang, M. Kaveh, Adaptive eigensubspace algorithms for direction or frequency estimation and tracking. IEEE Trans. Acoustic Speech Signal Process. **36**(2), 241–251 (1988)
26. Y.B. Gao, X.Y. Kong, H.D. Dong, Novel parallel multiple minor components extraction algorithm by diagonal matrix method. Neural Process. Lett. https://doi.org/10.1007/s11063-023-11186-4
27. Y.B. Gao, X.Y. Kong, H.H. Zhang, L.A. Hou, A weighted information criterion for multiple minor components and its adaptive extraction algorithms. Neural Netw. **89**, 1–10 (2017)
28. K. Abed-Meraim, S. Attallah, A. Chkeif, Y. Hua, Orthogonal Oja algorithm. IEEE Signal Process. Lett. **7**(5), 116–119 (2000)
29. S. Attallah, K. Abed-Meraim, Fast algorithms for subspace tracking. IEEE Signal Process. Lett. **8**(7), 3–206 (2001)

30. M.V. Jankovic, B. Reljin, A new minor component analysis method based on Douglas-Kung-Amari minor subspace analysis method. IEEE Signal Process. Lett. **12**(2), 859–862 (2005)
31. S. Ouyang, Z. Bao, Fast principal component extraction by a weighted information criterion. IEEE Trans. Signal Process. **50**, 1994–2002 (2002)
32. Y. Miao, Y. Hua, Fast subspace tracking and neural network learning by a novel information criterion. IEEE Trans. Signal Process. **46**, 1967–1979 (1998)
33. M.D. Plumbley, Lyapunov function for convergence of principal component algorithms. Neural Netw. **8**, 11–23 (1995)
34. X. Kong, C. Hu, C. Han, A dual purpose principal and minor subspace gradient flow. IEEE Trans. Signal Process. **60**, 197–210 (2012)
35. K.K. Tan, J. Lv, Y. Zhang, S. Huang, Adaptive multiple minor directions extraction in parallel using a PCA neural network. Theoret. Comput. Sci. **411**, 4200–4215 (2010)
36. S. Bartelmaos, K. Abed-Meraim, Fast adaptive algorithms for minor component analysis using Householder transformation. Digit. Signal Process. **21**, 667–678 (2011)
37. T. Duong Nguyen, Y. Isao, Necessary and sufficient conditions for convergence of the DDT systems of the normalized PAST algorithms. Signal Process. **94**, 288–299 (2014)
38. K.I. Diamantaras, S.Y. Kung, *Principal component neural networks: theory and applications* (Wiley, New York, 1996)
39. A. Hyvärinen, J. Karhunen, E. Oja, *Independent component analysis* (Wiley, New York, 2004)
40. S. Ouyang, P. Ching, T. Lee, Quasi-Newton algorithm for adaptive minor component extraction. Electron. Lett. **38**(19), 1142–1144 (2002)

Chapter 7
Neural-Based SVD Algorithms for Adaptive TLS

7.1 Introduction

The TLS method approximately solves a set of m linear equations in $n \times 1$ unknowns $\boldsymbol{x}$ represented in matrix form by $\mathbf{Ax} = \mathbf{b}$, where $\mathbf{A} \in \Re^{m\times n}$ is the data matrix, $\mathbf{b} \in \Re^{m\times 1}$ is the observation vector and $m > n$. The TLS solution $\boldsymbol{x}_{\text{TLS}}$ can be obtained by $[\mathbf{x}_{\text{TLS}}, -1]^T = -\mathbf{v}_{n+1}/v_{n+1,n+1}$, where $\mathbf{v}_{n+1}$ is the right singular vector associated with the smallest singular value of $\boldsymbol{C} = [\boldsymbol{A};\boldsymbol{b}]$, and $v_{n+1,n+1}$ is the last component of $\mathbf{v}_{n+1}$. Thus, an adaptive algorithm for extracting the right singular vector associated with the smallest singular value of the augmented input data, in which a subsequent normalization (the last component of weight $\boldsymbol{x}_{n+1} = -1$) is needed, can be used to solve the TLS problem. Compared with neural-based MCA algorithms, the field of neural-based algorithms that perform the SVD received relatively little attention [1].

In an earlier time, some methods for SVD were developed by using purely matrix algebra [6–9], but these methods are batching methods and thus cannot be used for real-time processing. To extract feature information online, some SVD dynamical systems based on gradient flows are developed [10–12], but these algorithms converge slowly. To overcome this problem, Diamantaras and Kung [5] extended the online correlation Hebbian rule to the cross-correlation Hebbian rule to perform the SVD of a cross-correlation matrix. However, the models are sometimes divergent for some initial states [4]. In [26], Sanger proposed double generalized Hebbian algorithm (DGHA) for SVD, which was derived from a two-fold optimization problem. It adapts the left singular vector estimate by the generalized Hebbian algorithm. In [27–29], the so-called "trace"-algorithm and "Riccati"-algorithm were proposed. For one-unit case, the "trace" and "Riccati" algorithms coincide with the cross-coupled Hebbian rule [1]. The algorithm proposed by Helmke-Moore [30] resembles the "trace"-algorithm, and if the weights are chosen mutually different, the system converges to the exact SVD of cross-correlation matrix up to permutations of the principal singular vectors [1]. In [13], Feng et al. provided a theoretical foundation for a novel model based on an extension of the Hebbian rule. In order to improve the

X. Kong and D. Feng, *Efficient Online Learning Algorithms for Total Least Square Problems*, Engineering Applications of Computational Methods 21,
https://doi.org/10.1007/978-981-97-1765-1_7

convergence speed and stability, Feng proposed a cross-correlation neural network (CNN) learning rule for SVD [4] in which the learning rate is independent of the singular value distribution of a non-squared data or cross-correlation matrix. It is found that all efforts of aforementioned work are only focused on single singular component analysis. Actually, there are also some principal singular subspace (PSS) extraction algorithms that have been proposed. In [14], a novel CNN model for finding the PSS of a cross-correlation matrix was proposed. Hasan [15] proposed an algorithm to track PSS in which the equilibrium point relates to the principal singular values. Later, Kong et al. [16] proposed an effective neural network algorithm for cross-correlation feature extraction, which is also a parallel algorithm. As known, neural network algorithm for extracting the principal singular vectors/ subspace, which is always derived based on Hebbian rule or gradient method, is a special case of optimization problem. Actually, solving the optimization problems by using the recurrent neural network has also been well studied in [17–19].

In [1], neural-based algorithms that only consider the singular vectors extraction are coined as non-coupled algorithms. In most of non-coupled algorithms, the convergence speed depends on the singular values of the matrix, but it is hard to select an appropriate learning rate. This problem was coined as the speed-stability problem [1]. In order to resolve the speed-stability problem that plagues most non-coupled learning algorithms, Kaiser et al. proposed a coupled online learning algorithms for the SVD of a cross-covariance matrix [1], which is called coupled SVD (CSVD) algorithms. In the CSVD rules, the singular value is estimated alongside the singular vectors, and the effective learning rates for the singular vector rules are influenced by the singular value estimates. But unfortunately, there is only one CSVD algorithm that has been proposed and analyzed so far.

The aim of this chapter is to introduce and develop some new neural-based SVD algorithms to applied into the TLS parameter estimation. In this chapter, a neural network for SVD of non-squared data matrix is firstly introduced in order to improve the convergence speed in [4]. Then, a fast and effective neural-based algorithm to perform SVD of non-squared data matrix is presented in [24], which is derived from a newly proposed information criterion. Finally, a novel information criterion (NIC) is proposed, the learning rule was derived by applying Newton's method to this NIC, and an algorithm is obtained based on the gradient and the inverse Hessian in [25], which solves the speed-stability problem.

In this chapter, the existing neural-based SVD algorithms will be reviewed and discussed. Three algorithms proposed by us, including a neural network for SVD of non-squared data matrix, a fast and effective neural network algorithm to perform SVD of non-squared data matrix and a coupled neural-based algorithm for PST extraction of a non-squared data matrix, will be analyzed in detail. The remainder of this chapter is organized as follows. An overview of the existing neural-based SVD algorithms is presented in Sect. 7.2. a neural network for SVD of non-squared data matrix is presented in in Sect. 7.3. Section 7.4 is devoted to a fast and effective neural-based algorithm to perform SVD of non-squared data matrix. A coupled neural-based algorithm for PST extraction of a non-squared data matrix is proposed in in Sect. 7.5, followed by summary in Sect. 7.6.

7.2 Review of Neural-Based SVD Algorithms

7.2.1 Parallel SVD Learning Algorithms on Double Stiefel Manifold

Several parallel SVD learning algorithms simultaneously compute the SVD vectors. The considered neural algorithms were developed in [28, 30] and have been analyzed in [27]. These algorithms are utilized to train in an unsupervised way a three-layer neural network with the classical "buttery" topology [27–29]. The first layer has connection matrix $\boldsymbol{A}$, the second one has connection matrix $\boldsymbol{B}$, and the middle (hidden) layer provides network's output. It has been shown that when proper initial conditions are chosen, the associated learning trajectories lie on the double Stiefel manifold [27].

The following matrix set is used for analysis purpose in this section: $St(m, n, \mathbf{K}) \overset{def}{=} \{\mathbf{X} \in \mathbf{K}^{m\times n} | \mathbf{X}^{*}\mathbf{X} = \mathbf{I}_n\}$ with $m-1, n-1 \in N; K \in \Re$ or C. When $m = n$, the manifold coincides with the orthogonal group $O(m, \mathbf{K}) \overset{def}{=} \{\mathbf{X} \in \mathbf{K}^{m\times m} | \mathbf{X}^{*}\mathbf{X} = \mathbf{I}_m\}$. Here, the product $O(m, \mathbf{K}) \times O(n, \mathbf{K})$ is referred to as double orthogonal group and the product $St(m, p, \mathbf{K}) \times St(n, p, \mathbf{K})$ as double Stiefel manifold.

Denote as $\mathbf{Z} \in \Re^{m\times n}$ the matrix whose SVD is to be computed and as $r \leq \min\{m, n\}$ the rank of $\mathbf{Z}$, the SVD can be written as $\mathbf{Z} = \mathbf{UDV}^{*}$, where $\mathbf{U} \in \Re^{m\times m}$ and $\mathbf{V} \in \Re^{n\times n}$ are orthogonal matrices and $\boldsymbol{D}$ is a pseudo-diagonal matrix whose elements are all zeros except for the first r diagonal entries, termed as singular values.

Denote as $\mathbf{A}(t) \in \Re^{m\times p}$ the network connection matrix stream that should learn p left singular vectors and as $\mathbf{B}(t) \in \Re^{n\times p}$ the estimator for p right singular vectors of the SVD of matrix $\mathbf{Z} \in \Re^{m\times n}$, with $p \leq r \leq \min\{m, n\}$. For simplicity, parallel SVD learning algorithms are described in terms of their ordinary differential equations.

The algorithm WH2 [25] can be written as: (also called as the "trace"-algorithm):

$$\dot{\mathbf{A}} = \mathbf{ZB} - \mathbf{AB}^{T}\mathbf{Z}^{T}\mathbf{A}, \quad \mathbf{A}(0) = \mathbf{A}_0, \tag{7.1}$$

$$\dot{\mathbf{B}} = \mathbf{Z}^{T}\mathbf{A} - \mathbf{BA}^{T}\mathbf{ZB}, \quad \mathbf{B}(0) = \mathbf{B}_0. \tag{7.2}$$

Equations (7.1) and (7.2) have been derived by extending Brockett's work on isospectral flow systems [31] from single to double orthogonal group. The initial state $\boldsymbol{A}_0$, $\boldsymbol{B}_0$ may be freely chosen. In particular, one can consider the following choice: $\mathbf{A}_0 \in St(m, p, \Re)$ and $\mathbf{B}_0 \in St(n, p, \Re)$, as for instance $\mathbf{A}_0 = \mathbf{I}_{m,p}$ and $\mathbf{B}_0 = \mathbf{I}_{n,p}$.

Then, for the algorithm WH2, it has been proved that: (1) If the initial states of the WH2 system belong to the Stiefel manifold, then the whole dynamics is double Stiefel. (2) The steady states of WH2 learning system can be written as $\mathbf{A} = \mathbf{U}_p\mathbf{K}$ and $\mathbf{B} = \mathbf{V}_p\mathbf{K}$, where $\boldsymbol{K}$ is arbitrary in $O(p, \Re)$ and $\mathbf{U}_p$ and $\mathbf{V}_p$ denote the submatrices whose columns are p right and left singular vectors of matrix $\mathbf{Z}$, respectively.

Obviously, the WH2 algorithm does not actually compute the true SVD, but a SVD subspace of dimension p.

The WH3 learning system was derived as an extension of the well-known Oja's subspace rule. The algorithm WH3 can be written as [27]:

$$\dot{\mathbf{A}} = \mathbf{ZB} - \mathbf{A}(\mathbf{A}^T\mathbf{ZB} + \mathbf{B}^T\mathbf{Z}^T\mathbf{A}), \quad \mathbf{A}(0) = \mathbf{A}_0, \tag{7.3}$$

$$\dot{\mathbf{B}} = \mathbf{Z}^T\mathbf{A} - \mathbf{B}(\mathbf{A}^T\mathbf{ZB} + \mathbf{B}^T\mathbf{Z}^T\mathbf{A}), \quad \mathbf{B}(0) = \mathbf{B}_0. \tag{7.4}$$

For the algorithm WH3, it has been proved that under the hypotheses $\mathbf{A}_0/\sqrt{2} \in St(m, p, \Re)$ and $\mathbf{B}_0/\sqrt{2} \in St(n, p, \Re)$, the learning equation WH3 can keep $\mathbf{A}(t)/\sqrt{2}$ and $\mathbf{B}(t)/\sqrt{2}$ within the Stiefel manifold.

The WH4 learning system can be written as: (also called as the "Riccati"-algorithm)

$$\dot{\mathbf{A}} = \mathbf{ZB} - \frac{1}{2}\mathbf{A}(\mathbf{A}^T\mathbf{ZB} + \mathbf{B}^T\mathbf{Z}^T\mathbf{A}), \quad \mathbf{A}(0) = \mathbf{A}_0, \tag{7.5}$$

$$\dot{\mathbf{B}} = \mathbf{Z}^T\mathbf{A} - \frac{1}{2}\mathbf{B}(\mathbf{A}^T\mathbf{ZB} + \mathbf{B}^T\mathbf{Z}^T\mathbf{A}), \quad \mathbf{B}(0) = \mathbf{B}_0. \tag{7.6}$$

For the algorithm WH4, it has been proved that under the hypotheses $\mathbf{A}_0 \in St(m, p, \Re)$ and $\mathbf{B}_0 \in St(n, p, \Re)$, the learning equations WH4 can keep $\mathbf{A}(t)$ and $\mathbf{B}(t)$ within the Stiefel manifold.

The structure of the stationary points of the WH3-4 algorithms is similar to the structure of the equilibria of the WH2 system. It has been proved that the steady states of WH3 and WH4 learning systems can be written as $\mathbf{A} = \mathbf{U}_p\mathbf{K}$ and $\mathbf{B} = \mathbf{V}_p\mathbf{K}$, where K is arbitrary in $O(p, \Re)$ and $\mathbf{U}_p$ and $\mathbf{V}_p$ denote the submatrices whose columns are p right and left singular vectors of the matrix $\mathbf{Z}$, respectively [27].

The HM dynamics [30] arises from the maximization of a specific metric (criterion) $\Phi_W : O(m, C) \times O(n, C) \to \Re$ defined as follows:

$$\Phi_{\mathbf{W}}(\mathbf{A}, \mathbf{B}) \stackrel{def}{=} 2\mathrm{Re}\,tr(\mathbf{WA}^*\mathbf{ZB}), \tag{7.7}$$

where $\mathbf{W} \in \Re^{n\times m}$ is a weighting matrix, $\mathbf{Z} \in C^{m\times n}$ and $m \geq n$. The dynamic system, derived as a Riemannian gradient flow on $O(m, C) \times O(n, C)$, can be written as [27] follows:

$$\dot{\mathbf{A}} = \mathbf{A}(\mathbf{W}^*\mathbf{B}^*\mathbf{Z}^*\mathbf{A} - \mathbf{A}^*\mathbf{ZBW}), \quad \mathbf{A}(0) = \mathbf{A}_0, \tag{7.8}$$

$$\dot{\mathbf{B}} = \mathbf{B}(\mathbf{WA}^*\mathbf{ZB} - \mathbf{B}^*\mathbf{Z}^*\mathbf{AW}^*), \quad \mathbf{B}(0) = \mathbf{B}_0. \tag{7.9}$$

It has been proved that $\mathbf{A}(t) \in O(m, C)$ and $\mathbf{B}(t) \in O(n, C)$. In the particular case that $\mathbf{W} = -\mathbf{I}_{n,m}$ and the involved quantities are real-valued, the above system is equivalent to WH2 when $p = n$. Obviously, the Weingessel–Hornik SVD learning equations can be regarded as special cases of the Helmke-Moore system.

7.2.2 Cross-Associative Neural Network for SVD (CANN)

In [14], Feng et al. proposed a novel CNN model to improve the convergence speed of the CNN models, in which the learning rate is independent of the singular value distribution of the cross-correlation matrix and its state matrix maintains orthonormality if the initial state matrix is orthonormal. Later, based on [14], Feng et al. also proposed a novel CNN model for finding the PSS of the cross-correlation matrix between two high-dimensional data streams, and introduced a novel non-quadratic criterion (NQC) for searching the optimum weights of two linear neural networks (LNN). An adaptive algorithm based on the NQC for tracking the PSS of the cross-correlation matrix between two high-dimensional vector sequences were developed, and the NQC algorithm provides fast online learning of the optimum weights for two LNN.

In order to improve the cost surface for the PSS and the convergence of gradient searching, a novel NQC for the PSS was presented in [14]. Given $\mathbf{U} \in \Re^{M\times r}$ and $\mathbf{V} \in \Re^{N\times r}$ in the domain $\{(\mathbf{U},\mathbf{V})|\mathbf{U}^T\mathbf{C}\mathbf{V} > 0\}$, the following framework for PSS was proposed:

$$\begin{aligned} &\min_{\mathbf{U},\mathbf{V}} J_{NQC}(\mathbf{U},\mathbf{V}) \\ &J_{NQC}(\mathbf{U},\mathbf{V}) = -tr[\ln(\mathbf{U}^T\mathbf{C}\mathbf{V})] + \frac{1}{2}tr[\mathbf{U}^T\mathbf{U} + \mathbf{V}^T\mathbf{V}]. \end{aligned} \tag{7.10}$$

The landscape of NQC is depicted by the following two theorems. *Theorem* (1): $(\mathbf{U},\mathbf{V})$ is a stationary point of $J_{NQC}(\mathbf{U},\mathbf{V})$ in the domain $\{(\mathbf{U},\mathbf{V})|\mathbf{U}^T\mathbf{C}\mathbf{V} > 0\}$ if and only if $\mathbf{U}\doteqdot\mathbf{L}_r$ and $\mathbf{V}\doteqdot\mathbf{R}_r$, where $\mathbf{L}_r \in \Re^{M\times r}$ and $\mathbf{R}_r \in \Re^{N\times r}$ consist of the left and right singular vectors of C, respectively. Note that $(\mathbf{U},\mathbf{V})\doteqdot(\mathbf{L}_r,\mathbf{R}_r)$ shows a stationary set of $J_{NQC}(\mathbf{U},\mathbf{V})$. And for the definition of "$\doteqdot$", see [14]. *Theorem* (2): In the domain $\{(\mathbf{U},\mathbf{V})|\mathbf{U}^T\mathbf{C}\mathbf{V} > 0\}$, $J_{NQC}(\mathbf{U},\mathbf{V})$ has a global minimum that is achieved if and only if $\mathbf{U}\doteqdot\mathbf{L}_s$ and $\mathbf{V}\doteqdot\mathbf{R}_s$. And the global minimum is $J_{NQC}(\mathbf{U},\mathbf{V}) = \sum_{i=1}^{r}\ln\sigma_i - r$. All the other stationary points $(\mathbf{U},\mathbf{V})(\not\doteqdot(\mathbf{L}_s,\mathbf{R}_s))$ are saddle (unstable) points of $J_{NQC}(\mathbf{U},\mathbf{V})$. In fact $(\mathbf{U},\mathbf{V})\doteqdot(\mathbf{L}_s,\mathbf{R}_s)$ shows a global minimizer set of $J_{NQC}(\mathbf{U},\mathbf{V})$, where $\mathbf{L}_s$ and $\mathbf{R}_s$ are the left and right singular vector matrix associated with signal, respectively.

One can apply the gradient descent searching to the unconstrained minimization of $J_{NQC}(\mathbf{U},\mathbf{V})$. In [14], a batch algorithm and a recursive algorithm were derived.

Given the gradient of $J_{NQC}(\mathbf{U},\mathbf{V})$ with respect to $\mathbf{U}$ and $\mathbf{V}$, the following gradient descent rule can be used for updating $\boldsymbol{U}(k)$ and $\boldsymbol{V}(k)$:

$$\mathbf{U}(k) = (1-\eta)\mathbf{U}(k-1) + \eta\mathbf{C}(k)\mathbf{V}(k)\times(\mathbf{U}^T(k)\mathbf{C}(k)\mathbf{V}(k))^{-1}, \tag{7.11}$$

$$\mathbf{V}(k) = (1 - \eta)\mathbf{V}(k - 1) + \eta\mathbf{C}^T(k)\mathbf{U}(k) \times (\mathbf{V}^T(k)\mathbf{C}^T(k)\mathbf{U}(k))^{-1}, \tag{7.12}$$

$$\mathbf{C}(k) = \alpha\mathbf{C}(k - 1) + \mathbf{x}(k)\mathbf{y}^T(k), \tag{7.13}$$

where $0 < \eta < 1$ denotes the learning rate, $0 < \alpha \leq 1$ is the forgetting factor. This batch implementation, however, is mainly suitable for the adaptive singular subspace estimation and tracking, where the cross-correlation matrix $\boldsymbol{C}(k)$ is explicitly involved in computations [14]. For online learning of neural networks, it is expected that the network should learn the PSS directly from the input data sequences $\boldsymbol{x}(k)$ and $\boldsymbol{y}(k)$ [14].

Following the projection approximation method of [32], the recursive implementation of the NQC algorithm was derived in [14], and the algorithm can be summarized as follows:

Initializations:

$\mathbf{P}(0) = \varepsilon\mathbf{I}_r$ (ε is a very large positive number)

$\hat{\mathbf{U}}(0) = \mathbf{0}$ and $\hat{\mathbf{V}}(0) = \mathbf{0}$.

$\mathbf{U}(0) =$ a random $M \times r$ matrix with very small Frobenius norm.

$\mathbf{V}(0) =$ a random $N \times r$ matrix with very small Frobenius norm.

Update:

$$\begin{cases} \mathbf{g}(k) = \mathbf{U}^T(k - 1)\mathbf{x}(k) \\ \mathbf{h}(k) = \mathbf{V}^T(k - 1)\mathbf{y}(k) \\ \hat{\mathbf{g}}(k) = \mathbf{P}(k - 1)\mathbf{g}(k) \\ \hat{\mathbf{h}}(k) = \mathbf{P}^T(k - 1)\mathbf{h}(k) \\ \gamma(k) = \frac{1}{(\alpha + \mathbf{h}^T(k)\hat{\mathbf{g}}(k))} \\ \mathbf{P}(k) = \alpha^{-1}\left[\mathbf{P}(k - 1) - \gamma(k)\hat{\mathbf{g}}(k)\mathbf{h}^T(k)\right] \\ \hat{\mathbf{U}}(k) = \hat{\mathbf{U}}(k - 1) + \gamma(k)\left[\mathbf{x}(k) - \hat{\mathbf{U}}(k)\mathbf{g}(k)\right]\hat{\mathbf{h}}^T(k) \\ \hat{\mathbf{V}}(k) = \hat{\mathbf{V}}(k - 1) + \gamma(k)\left[\mathbf{y}(k) - \hat{\mathbf{V}}(k)\mathbf{h}(k)\right]\hat{\mathbf{g}}^T(k) \\ \mathbf{U}(k) = (1 - \eta)\mathbf{U}(k - 1) + \eta\hat{\mathbf{U}}(k) \\ \mathbf{V}(k) = (1 - \eta)\mathbf{V}(k - 1) + \eta\hat{\mathbf{V}}(k). \end{cases} \tag{7.14}$$

The above update equation yields an online learning algorithm for two linear neural networks. $\mathbf{U}^T(k)$ and $\mathbf{V}^T(k)$ denote the weight matrices of these linear neural networks. In [14], the convergence of batch algorithm was proved via the Lyapunov theory.

In [14], it has been proved that the iterative algorithm for computing $\mathbf{U}^T(k)\mathbf{U}(k)$ and $\mathbf{V}^T(k)\mathbf{V}(k)$ can be written as follows:

$$\begin{cases} \mathbf{U}^T(k)\mathbf{U}(k) = (1 - 2\eta)\mathbf{U}^T(k - 1)\mathbf{U}(k - 1) + 2\eta\mathbf{I}_r \\ \mathbf{V}^T(k)\mathbf{V}(k) = (1 - 2\eta)\mathbf{V}^T(k - 1)\mathbf{V}(k - 1) + 2\eta\mathbf{I}_r \end{cases}, \tag{7.15}$$

where an appropriate learning rate is $0 < \eta < 0.5$. Since (7.15) is linear, the learning rate can be selected as a fixed constant close to 0.5. For example, $\eta = 0.49$. That is, the learning rate can be selected as a constant independent of the singular value distribution of the underlying matrix, which evidently increases the convergence speed of the CNN.

7.2.3 *Coupled SVD of a Cross-Covariance Matrix*

It is known that there exists the speed-stability problem in most non-coupled learning algorithms. In order to resolve the speed-stability problem that plagues most non-coupled learning algorithms, Kaiser et al. proposed a coupled online coupled learning algorithms for the SVD of a cross-covariance matrix [1]. In this algorithm, the singular value is estimated alongside the singular vectors, and the effective learning rates for the singular vector rules are influenced by the singular value estimates.

7.2.3.1 Single Component Learning Rules

In [1], Kaiser et al. proposed an information criterion, which is designed in a way that its stationary points coincide with the singular triplets of the general rectangular matrix $\boldsymbol{A}$. However, since the stationary points correspond to the saddle points, Newton's method was applied to derive a learning rule to turn even saddle points into attractors which guarantees the stability of the resulting learning rules.

The information criterion is as follows:

$$\rho = \sigma^{-1}\mathbf{u}^T\mathbf{A}\mathbf{v} - \frac{1}{2}\mathbf{u}^T\mathbf{u} - \frac{1}{2}\mathbf{v}^T\mathbf{v} + \ln\sigma. \tag{7.16}$$

From the gradient of (7.16), it is clear that the stationary points of (7.16) are the singular triplets of A, and $\|\mathbf{u}\| = \|\mathbf{v}\| = 1$ for these points. Since the Hessian of (7.16) is indefinite for all stationary points, the first principal singular triplet is only a saddle point of (7.16), and hence gradient-based methods would not converge to this solution. Here, Newton's method was applied to derive a learning rule which converges toward the principal singular triplet. In [1], the derived individual component equations are as follows:

$$\begin{cases} \dot{\mathbf{u}} = \sigma^{-1}(\mathbf{A}\mathbf{v} - \mathbf{u}\mathbf{u}^T\mathbf{A}\mathbf{v}) + \frac{1}{2}(\mathbf{u}^T\mathbf{u} - 1)\mathbf{u} \\ \dot{\mathbf{v}} = \sigma^{-1}(\mathbf{A}^T\mathbf{u} - \mathbf{v}\mathbf{v}^T\mathbf{A}^T\mathbf{u}) + \frac{1}{2}(\mathbf{v}^T\mathbf{v} - 1)\mathbf{v} \\ \dot{\sigma} = \mathbf{u}^T\mathbf{A}\mathbf{v} - \frac{1}{2}\sigma(\mathbf{u}^T\mathbf{u} + \mathbf{v}^T\mathbf{v}). \end{cases} \tag{7.17}$$

By using a linear stability analysis which is based on the ED of the stability matrix, i.e. the system's Jacobian, evaluated at an equilibrium point, i.e., the k-th principal singular triplet $(\overline{\mathbf{u}}_k, \overline{\mathbf{v}}_k, \overline{\sigma}_k)$, it has been proved that only the first principal singular triplet is stable [1]. Moreover, it has been proved that the system converges with approximately equal speed in all its eigendirections and is widely independent of σ.

If one further approximates $\mathbf{u}^T\mathbf{u} \approx 1$ and $\mathbf{v}^T\mathbf{v} \approx 1$ in (7.17), then the approximated system can be obtained

$$\begin{cases} \dot{\mathbf{u}} = \sigma^{-1}(\mathbf{Av} - \mathbf{uu}^T\mathbf{Av}) \\ \dot{\mathbf{v}} = \sigma^{-1}(\mathbf{A}^T\mathbf{u} - \mathbf{vv}^T\mathbf{A}^T\mathbf{u}) \\ \dot{\sigma} = \mathbf{u}^T\mathbf{Av} - \sigma. \end{cases} \tag{7.18}$$

It can be easily shown that the approximated system has the same equilibria as the original system. The convergence properties are widely unchanged, i.e., the first principal singular triplet is the only stable equilibrium, all eigenvalues are widely independent of the singular values, and the convergence speed is approximately the same in all its eigendirections [1].

Ignoring the factor σ^{-1} in (7.18) yields the cross-coupled Hebbian rule in its averaged form

$$\begin{cases} \dot{\mathbf{u}} = (\mathbf{Av} - \mathbf{uu}^T\mathbf{Av}) \\ \dot{\mathbf{v}} = (\mathbf{A}^T\mathbf{u} - \mathbf{vv}^T\mathbf{A}^T\mathbf{u}). \end{cases} \tag{7.19}$$

By analyzing the stability of these ODEs, it has been shown that their convergence speed strongly depends on the singular values of $\boldsymbol{A}$.

It is worth noting that for the online learning rule, the learning rates have to be selected. In the cross-coupled Hebbian rule, the convergence speed depends on the principal singular value of $\boldsymbol{A}$, which leads to a problem in the selection of the learning rate. Thus, there exists a typical speed-stability dilemma. Obviously, the influence of the singular value estimate (7.18) on the equations of the singular vector estimates potentially improves the convergence properties.

7.2.3.2 Multiple Component Learning Rules

For the estimation of multiple principal singular triplets, the above single singular component analyzer has to be combined using some decorrelation method. These methods can be interpreted as descendents of the Gram–Schmidt orthonormalization method. In [1], four methods were introduced, i.e., full Gram–Schmidt orthonormalization method, first-order Gram–Schmidt approximation, the deflation and double deflation. Applying the Gram–Schmidt method would lead to perfectly orthonormal left and right singular vectors. However, the complexity of this method is in the order of mM^2 for the left and nM^2 for the right singular vectors. First-order approximation of Gram–Schmidt orthonormalization reduces the computational effort to the order of mM and nM. Deflation methods have the same order of complexity but approximately halving the number of computation steps.

Here, we take multiple component learning of the coupled SVD rule using first-order approximation of Gram–Schmidt orthonormalization as an example. The algorithms can be summarized as follows:

$$\hat{\mathbf{u}}_k = \mathbf{u}_k + \gamma \left[\sigma_k^{-1} \left(\mathbf{y} - \sum_{j=1}^{k} \eta_j \mathbf{u}_j \right) \xi_k - \eta_k \sum_{j=1}^{k-1} \sigma_j^{-1} \xi_j \mathbf{u}_j \right], \tag{7.20}$$

$$\hat{\mathbf{v}}_k = \mathbf{v}_k + \gamma \left[\sigma_k^{-1} \left(\mathbf{x} - \sum_{j=1}^{k} \eta_j \mathbf{v}_j \right) \xi_k - \eta_k \sum_{j=1}^{k-1} \sigma_j^{-1} \xi_j \mathbf{v}_j \right], \tag{7.21}$$

$$\sigma_k'' = \sigma_k + \gamma (\eta_k \xi_k - \sigma_k), \tag{7.22}$$

where $\xi_k = \mathbf{v}_k^T \mathbf{x}$, $\eta_k = \mathbf{u}_k^T \mathbf{y}$. The weight vector of the next time step is either the same as this intermediate vector, i.e.,

$$\mathbf{u''}_k = \hat{\mathbf{u}}_k, \quad \mathbf{v''}_k = \hat{\mathbf{v}}_k, \tag{7.23}$$

or a normalized version:

$$\mathbf{u}_k'' = \frac{\hat{\mathbf{u}}_k}{\|\hat{\mathbf{u}}_k\|}, \quad \mathbf{v}_k'' = \frac{\hat{\mathbf{v}}_k}{\|\hat{\mathbf{v}}_k\|}. \tag{7.24}$$

For $k = 1$, these rules coincide with the coupled SVD rule. For other estimation rules of multiple principal singular triplets, see [1].

7.3 A Neural Network for SVD of Non-squared Data Matrix

In [4], a neural network and learning algorithm for SVD of a non-squared data matrix was proposed, in order to improve the convergence speed and avoid the potential instability of the deterministic networks associated with the cross-correlation neural-network models. Compared with the DFCNN in [5], the proposed algorithm has two remarkable advantages: (1) its learning rate can be a fixed constant independent of the singular value distribution of the underlying matrix, which evidently increases the convergent speed of the algorithm; (2) its state vectors have the unit-norm conservation.

7.3.1 A Novel Recurrent Neural Network

Here, we consider a neural network for performing SVD of a data matrix. Given a non-squared matrix $\mathbf{A} \in \Re^{n \times m}$, without loss of generality, let $n \geq m$. Let σ_i, $\mathbf{l}_i$, and $\mathbf{r}_i$ be the th singular value, left singular vector, and right singular vector, respectively, and $\sigma_1 \geq \sigma_2 \geq \cdots \geq \sigma_m > 0$. All the σ_i,$\mathbf{l}_i$, and $\mathbf{r}_i$ is called the th singular component, then the standard SVD of A is $\mathbf{A} = \sum_{i=1}^{m} \sigma_i l_i r_i^T = \mathbf{L}\Lambda\mathbf{R}^T$ and its equivalent form $\mathbf{A} = \sum_{i=1}^{m} \sigma_i(-l_i)(-r_i^T) = (-\mathbf{L})\Lambda(-\mathbf{R}^T)$ in which all the singular values are positive, where both $\mathbf{L} = [\mathbf{l}_1, \ldots, \mathbf{l}_n] \in \Re^{n\times n}$ and $\mathbf{R} = [\mathbf{r}_1, \ldots, \mathbf{r}_m] \in \Re^{m\times m}$ are unitary, and Λ denotes the $n \times m$ diagonal matrix made up of all the singular values [33]. Noticeably, in algebra, the SVD of has some non-standard forms that are described by $\mathbf{A} = \sum_{i=1}^{m} (\pm\sigma_i)(\pm l_i) r_i^T$ or their equivalent form $\mathbf{A} = \sum_{i=1}^{m} (\pm\sigma_i) l_i (\pm r_i^T)$ in which any nonzero singular value may be positive or negative. Our objective is to perform the (standard or non-standard) SVD of A. The most direct methods are the matrix-algebra-based methods [6–9], while the SVD can also be obtained by using the recurrent neural network such as the DFCNN [5].

We propose the following recurrent network for SVD of A:

$$\frac{d\mathbf{u}_i(t)}{dt} = \frac{\mathbf{A}_i(t)\mathbf{v}_i(t)}{\mathbf{u}_i^T(t)\mathbf{A}_i(t)\mathbf{v}_i(t)} - \mathbf{u}_i(t) \tag{7.25a}$$

$$\frac{d\mathbf{v}_i(t)}{dt} = \frac{\mathbf{A}_i^T(t)\mathbf{u}_i(t)}{\mathbf{u}_i^T(t)\mathbf{A}_i(t)\mathbf{v}_i(t)} - \mathbf{v}_i(t) \tag{7.25b}$$

for $i = 1, 2, \ldots, m$, where $\mathbf{u}_i(t) \in \Re^n$, $\mathbf{v}_i(t) \in \Re^m$, and

$$\mathbf{A}_1(t) = \mathbf{A} \tag{7.26a}$$

$$\begin{aligned}\mathbf{A}_i(t) &= \mathbf{A} - \sum_{j=1}^{i-1} \mathbf{u}_j(t)\mathbf{u}_j^T(t)\mathbf{A}\mathbf{v}_j(t)\mathbf{v}_j^T(t) \\ &= \mathbf{A}_{i-1}(t) - \mathbf{u}_{i-1}(t)\mathbf{u}_{i-1}^T(t)\mathbf{A}\mathbf{v}_{i-1}(t)\mathbf{v}_{i-1}^T(t), \quad (i = 2, 3, \ldots, m)\end{aligned} \tag{7.26b}$$

The design objective of (7.25a) is to make $\mathbf{u}_j(t) \to \pm\mathbf{l}_j$ and $\mathbf{v}_j(t) \to \pm\mathbf{r}_j (j = 1, 2, \ldots, m)$ as $t \to \infty$. $\mathbf{A}_i(t)$ in (7.26b) is called the deflation transformation [5]. The design objective of (7.26a–7.26b) is to make $\mathbf{A}_i(t) \to \sum_{j=i}^{m} \sigma_j \mathbf{l}_j \mathbf{r}_j^T$ in order to let $\mathbf{u}_i(t) \to \pm\mathbf{l}_i$ and $\mathbf{v}_i(t) \to \pm\mathbf{r}_i$ as $t \to \infty$. In fact, if $\mathbf{u}_j(t) \to \pm\mathbf{l}_j$ and $\mathbf{v}_j(t) \to \pm\mathbf{r}_j (j = 1, 2, \ldots, i-1)$ as $t \to \infty$, we directly verify that (7.26a–7.26b) can achieve its design objective.

The neural network for finding the first singular component is shown in Fig. 7.1. Its adaptability is indicated by change of connection weights with data matrix. Thus, the entire neural network has complex topologies.

The iterative algorithm corresponding to (7.25a–7.25b) is as follows:

$$\mathbf{u}_i(k+1) = \mathbf{u}_i(k) + \alpha[\mathbf{A}_i\mathbf{v}_i(k)/\mathbf{u}_i^T(k)\mathbf{A}_i(k)\mathbf{v}_i(k) - \mathbf{u}_i(k)] \tag{7.27a}$$

$$\mathbf{v}_i(k+1) = \mathbf{v}_i(k) + \alpha[\mathbf{A}_i^T\mathbf{u}_i(k)/\mathbf{u}_i^T(k)\mathbf{A}_i(k)\mathbf{v}_i(k) - \mathbf{v}_i(k)] \tag{7.27b}$$

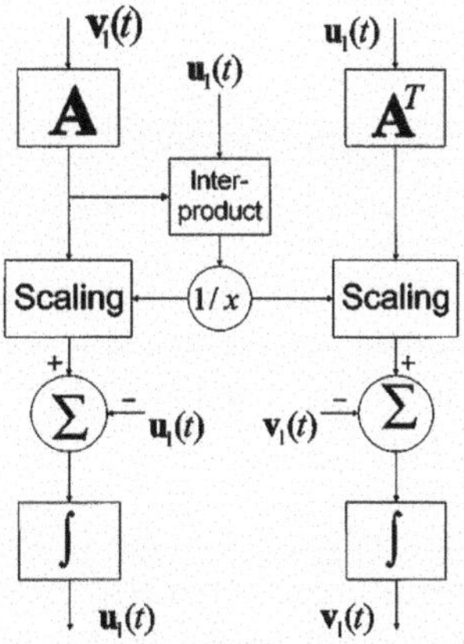

Fig. 7.1 Block diagram of the neural network for tracking the first singular component

for $i = 1, 2, \ldots, m$, α is called the time-step length or learning rate. It is easily known that convergence of (7.27a–7.27b) greatly replies on the learning rate. An important problem is how to choose a good learning rate.

It is worth noting that in the neural-network literature, many algorithms can be extended to solve the SVD problem, but their learning rate are difficult to be determined in advance. Hence, the main reason to adopt (7.25a–7.25b) is that the learning rate in (7.27a–7.27b) can be selected in advance. Moreover, in order to avoid dividing by zero, the initial state vectors should appropriately been selected so that. Once obtaining all the singular components associated with all the nonzero singular values, we can also obtain those associated with the zero singular values by Gram–Schmidt orthogonalization.

At this point, we establish an essential result.

Lemma 7.1 *Given* (7.25a–7.25b) *and arbitrary initial values* $\|\mathbf{u}_i(0)\|^2$ *and* $\|\mathbf{v}_i(0)\|^2$, *then* $\|\mathbf{u}_i(t)\|^2$ *and* $\|\mathbf{v}_i(t)\|^2$ *converge exponentially to 1 as* $t \to \infty$, *and their convergence is independent of the matrix* $\mathbf{A}$.

Proof Differentiating $\|\mathbf{u}_i(t)\|^2$ and $\|\mathbf{v}_i(t)\|^2$ with respect to and considering (7.25a–7.25b), we can obtain

$$\frac{d\|\mathbf{u}_i(t)\|^2}{dt} = 2\left(1 - \|\mathbf{u}_i(t)\|^2\right) \tag{7.28a}$$

$$\frac{d\|\mathbf{v}_i(t)\|^2}{dt} = 2\left(1 - \|\mathbf{v}_i(t)\|^2\right) \tag{7.28b}$$

The analytical solutions of (7.28a) and (7.28b), respectively, are given by the following:

$$\|\mathbf{u}_i(t)\|^2 = 1 + (\|\mathbf{u}_i(0)\|^2 - 1)\exp(-2t) \tag{7.29a}$$

$$\|\mathbf{v}_i(t)\|^2 = 1 + (\|\mathbf{v}_i(0)\|^2 - 1)\exp(-2t) \tag{7.29b}$$

Clearly, if $\|\mathbf{u}_i(0)\|^2 = 1$ and $\|\mathbf{v}_i(0)\|^2 = 1$, then $\|\mathbf{u}_i(t)\|^2 \equiv 1$ and $\|\mathbf{v}_i(t)\|^2 \equiv 1$. This completes the proof of Lemma 7.1.

Remark 7.1 We refer to $\|\mathbf{u}_i(t)\|^2 \equiv 1$ and $\|\mathbf{v}_i(t)\|^2 \equiv 1$ if $\|\mathbf{u}_i(0)\|^2 = 1$ and $\|\mathbf{v}_i(0)\|^2 = 1$ as the unit-norm conservation. As seen from (7.29a) and (7.29b), if $\|\mathbf{u}_i(0)\|^2 > 1$ and $\|\mathbf{v}_i(0)\|^2 > 1$, then there are always $\|\mathbf{u}_i(t)\|^2 > 1$ and $\|\mathbf{v}_i(t)\|^2 > 1$ for any finite positive t; similarly, if $\|\mathbf{u}_i(0)\|^2 < 1$ and $\|\mathbf{v}_i(0)\|^2 < 1$, then there are always $\|\mathbf{u}_i(t)\|^2 < 1$ and $\|\mathbf{v}_i(t)\|^2 < 1$ for any finite positive t. Moreover, since the analytical solution (7.29a) and (7.29b) is independent of the data matrix, the convergence of $\|\mathbf{u}_i(t)\|^2$ and $\|\mathbf{v}_i(t)\|^2$ is uniform with respect to the data matrix. Importantly, since the decay rate of linear Eq. (7.28a) and (7.28b) is 2, the suitable learning rate α in (7.27a–7.27b) can be taken as $0 < \alpha < 0.5$. In order to guarantee that iteration algorithm (7.27a–7.27b) has the reliable, fast and global convergence, we take the fixed learning rate $\alpha = 0.49$ that is confirmed to be suitable through many simulation tests.

From Lemma 7.1, we directly deduce the following Corollary.

Corollary 7.1 *(Boundedness): For any bounded initial values* $\mathbf{u}_i(0)$ *and* $\mathbf{v}_i(0)(i =$ *1,2,...,m), the state vectors of nonlinear system* (7.25a–7.25b) *are bounded.*

7.3.2 *Stability Analysis of Algorithm*

First, we consider stability of the first component in (7.25a–7.25b) and assume that $\boldsymbol{x}(t)$ is stationary. We will prove that the stable equilibrium point set of the first component case in (7.25a–7.25b) is given by the following:

$$\begin{aligned} P_1 = \{(\mathbf{u}_1, \mathbf{v}_1) | \mathbf{A}\mathbf{v}_1 = (\mathbf{u}_1^T\mathbf{A}\mathbf{v}_1)\mathbf{u}_1, \\ \mathbf{A}^T\mathbf{u}_1 = (\mathbf{u}_1^T\mathbf{A}\mathbf{v}_1)\mathbf{v}_1, \mathbf{u}_1^T\mathbf{A}\mathbf{v}_1 = \pm\tilde{\sigma}_1\} \end{aligned} \tag{7.30a}$$

$$\begin{aligned} P_1 = \Bigg\{ (\mathbf{u}_1, \mathbf{v}_1) | \mathbf{u}_1 = \sum_{i=1}^{m_1} \alpha_i \mathbf{l}_i, \mathbf{v}_1 = \sum_{i=1}^{m_1} \beta_i \mathbf{r}_i, \sum_{i=1}^{m_1} \alpha_i^2 = 1 \\ \sum_{i=1}^{m_1} \beta_i^2 = 1, \quad \text{and} \quad \sum_{i=1}^{m_1} \alpha_i\beta_i = \pm 1 \Bigg\}, \end{aligned} \tag{7.30b}$$

where $\boldsymbol{A}$ has the L distinct nonzero singular values $\tilde{\sigma}_1 > \tilde{\sigma}_2 > \cdots > \tilde{\sigma}_L > 0$ with multiplicity $m_1, \cdots, m_L (L \le m)$. It is easily shown that (7.30b) is equivalent to (7.30a). It is easily shown that for $m_1 = 1$, P_1 is a set made up of the finite points, and the direction of the first singular vectors $\boldsymbol{l}_1$ and $\boldsymbol{r}_1$ can uniquely be determined, while for $m_1 > 1$, P_1 is continuous, and the choice of the first singular vectors is not unique. Any point in P_1 can be regarded as the first singular component. Hence, once such a point in P_1 is obtained, we can get the first singular component.

Two distinct subsets in P_1 are defined as follows:

$$\hat{P}_1 = \left\{ (\mathbf{u}_1, \mathbf{v}_1) | \mathbf{u}_1 = \sum_{i=1}^{m_1} \alpha_i \mathbf{l}_i, \mathbf{v}_1 = \sum_{i=1}^{m_1} \beta_i \mathbf{r}_i, \sum_{i=1}^{m_1} \alpha_i^2 = 1, \right.$$

$$\left. \sum_{i=1}^{m_1} \beta_i^2 = 1, \text{ and } \sum_{i=1}^{m_1} \alpha_i \beta_i = 1 \right\} \tag{7.31a}$$

$$\tilde{P}_1 = \left\{ (\mathbf{u}_1, \mathbf{v}_1) | \mathbf{u}_1 = \sum_{i=1}^{m_1} \alpha_i \mathbf{l}_i, \mathbf{v}_1 = \sum_{i=1}^{m_1} \beta_i \mathbf{r}_i, \sum_{i=1}^{m_1} \alpha_i^2 = 1 \right.$$

$$\left. \sum_{i=1}^{m_1} \beta_i^2 = 1, \text{ and } \sum_{i=1}^{m_1} \alpha_i \beta_i = -1 \right\} \tag{7.31b}$$

There is obviously $P_1 = \hat{P}_1 \cup \tilde{P}_1$, and $\hat{P}_1$ corresponds to the standard SVD of $\mathbf{A}$.

Lemma 7.2 (*Globally Asymptotic Stability of the First Component*): *Let m nonzero singular values of* $\mathbf{A}$ *be* $\sigma_1 \geq \sigma_2 \geq \cdots > \sigma_m > 0$, *with the corresponding normalized left singular vectors* $\boldsymbol{l}_1, \boldsymbol{l}_2, \ldots, \boldsymbol{l}_m$ *and right singular vectors* $\boldsymbol{r}_1, \mathbf{r}_2, \ldots, \boldsymbol{r}_m$, *respectively. Furthermore, let* $\mathbf{A}$ *have the L distinct singular values* $\tilde{\sigma}_1 > \tilde{\sigma}_2 > \cdots > \tilde{\sigma}_L > 0$ *with multiplicity* $m_1, \cdots, m_L$. *Assume* $\mathbf{l}_i^T \mathbf{u}_1(0) \neq 0$ *or* $\mathbf{r}_j^T \mathbf{v}_1(0) \neq 0$ *for any* $i(\leq m_1)$ *and* $j(\leq m_1)$. *Then* $(\mathbf{u}_1(t), \mathbf{v}_1(t))$ *in* (7.25a–7.25b) *globally asymptotically converges to a point in* P_1 *as* $t \to \infty$.

Proof See Appendix B in [4].

Remark 7.2 From Lemma 7.2, we conclude that when $\sum_{i=1}^{m_1} [\mathbf{l}_i^T \mathbf{u}_1(0)][\mathbf{r}_i^T \mathbf{v}_1(0)] > 0$, $(\mathbf{u}_1(t), \mathbf{v}_1(t))$ in (6.26) globally asymptotically converges to a point in $\hat{P}_1$, while $\sum_{i=1}^{m_1} [\mathbf{l}_i^T \mathbf{u}_1(0)][\mathbf{r}_i^T \mathbf{v}_1(0)] < 0$, $(\mathbf{u}_1(t), \mathbf{v}_1(t))$ in (7.25a–7.25b) globally asymptotically converges to a point in $\tilde{P}_1$, as $t \to \infty$.

Theorem 7.1 (*Globally Asymptotic Stability*): *Let m nonzero singular values of* $\mathbf{A}$ *be* $\sigma_1 \geq \sigma_2 \geq \cdots > \sigma_m > 0$, *with the corresponding normalized left singular vectors* $\boldsymbol{l}_1, \boldsymbol{l}_2, \ldots, \boldsymbol{l}_m$ *and right singular vectors* $\boldsymbol{r}_1, \boldsymbol{r}_2, \ldots, \boldsymbol{r}_m$, *respectively. Assume* $\mathbf{l}_i^T \mathbf{u}_1(0) \neq 0$ *or* $\mathbf{r}_j^T \mathbf{v}_1(0) \neq 0$ *for* $i = 1, 2, \ldots, m$. *Then* $(\mathbf{u}_i(t), \mathbf{v}_i(t))$ *in* (7.25a–7.25b) *globally asymptotically converges to* $(\mathbf{l}_i, \pm\mathbf{r}_i)$ *as* $t \to \infty$.

Proof From (7.25a–7.25b) and (7.26a–7.26b), it is known that the state vectors $\boldsymbol{u}_i(t)$ and $\boldsymbol{v}_i(t)$ $(i > 1)$ are governed by $\boldsymbol{u}_j(t)$ and $\boldsymbol{v}_j(t)$ $(j = 1, 2, \ldots, i-1)$ through $\boldsymbol{A}_i(t)$, while $\boldsymbol{u}_j(t)$ and $\boldsymbol{v}_j(t)$ $(j = 1, 2, \ldots, i-1)$ are not affected by $\boldsymbol{u}_i(t)$ and $\boldsymbol{v}_i(t)$. This feature provides the convenience for analysis.

By Lemma 7.2, it is known that $(\mathbf{u}_1(t), \mathbf{v}_1(t)) \to (\mathbf{l}_1, \pm\mathbf{r}_1)$, hence we have $\mathbf{A}_2(t) = \mathbf{A} - \mathbf{u}_1(t)\mathbf{u}_1^T(t)\mathbf{A}\mathbf{v}_1(t)\mathbf{v}_1^T(t) \to \mathbf{A} - \mathbf{l}_1\mathbf{l}_1^T\mathbf{A}(\pm\mathbf{r})(\pm\mathbf{r}_1^T) = \mathbf{A} - \sigma_1\mathbf{l}_1\mathbf{r}_1^T = \sum_{i=2}^m \sigma_i\mathbf{l}_i\mathbf{r}_i^T$, as $t \to \infty$. Once $A_2(t)$ approaches $\mathbf{A} - \sigma_1\mathbf{l}_1\mathbf{r}_1^T$, the repeated use of Lemma 7.2 gives $(\mathbf{u}_2(t), \mathbf{v}_2(t)) \to (\mathbf{l}_2, \pm\mathbf{r}_2)$ as $t \to \infty$, which is the stability of (7.25a–7.25b) when $i = 2$. Mimicking this process, we can prove the stability of (7.25a–7.25b) when $i = 3,\ldots, m$. This completes the proof of Theorem 7.1.

Remark 7.3 We may expect that (7.25a–7.25b) perform the standard SVD, which is achieved by the following method: if the stable state vectors $\boldsymbol{u}_i$ and $\boldsymbol{v}_i$ of (7.25a–7.25b) satisfy condition$\mathbf{u}_i^T\mathbf{A}\mathbf{v}_i < 0$, then $\boldsymbol{v}_i$ is replaced by $-\boldsymbol{v}_i$.

7.3.3 Simulation Experiments and Conclusions

Two simulations are presented. In simulation 1, SVD of an ill-conditioned matrix is used to evaluate the efficiency of this parallel neural network. In Simulation 2, the solution of a TLS problem is obtained by neural-network-based SVD for non-squared data matrix.

Simulation 1: Consider the following matrix:

$$\mathbf{A} = [\mathbf{u}_0, \ldots, \mathbf{u}_8]\mathrm{diag}(10, 10, 10, 10^{-3}, 10^{-3}, 10^{-3}, 10^{-7}, 10^{-7}, 10^{-7})[\mathbf{v}_0, \ldots, \mathbf{v}_8]^T, \tag{7.32}$$

where $\boldsymbol{u}_i$ and $\boldsymbol{v}_i$ $(i = 0, \ldots, 8)$ are the ith components of 11- and nine-dimensional orthogonal discrete cosine basis functions, respectively, i.e.,

$$\mathbf{u}_i = k_i\left[\cos\left(\frac{1\times i\pi}{22}\right)\cos\left(\frac{3\times i\pi}{22}\right)\cdots\cos\left(\frac{21\times i\pi}{22}\right)\right]^T$$
$$k_i = \begin{cases}\sqrt{1/11} & i = 0\\ \sqrt{2/11} & i \neq 0\end{cases} \tag{7.33a}$$

$$\mathbf{v}_i = k_i\left[\cos\left(\frac{1\times i\pi}{18}\right)\cos\left(\frac{3\times i\pi}{18}\right)\cdots\cos\left(\frac{17\times i\pi}{18}\right)\right]^T$$
$$k_i = \begin{cases}\sqrt{1/9} & i = 0\\ \sqrt{2/9} & i \neq 0\end{cases} \tag{7.33b}$$

The matrix $\boldsymbol{A}$ given by (7.32) has nine nonzero singular values in which the entire three distinct singular values 10, 10^{-3} and 10^{-7} have multiplicity 3. The data matrix is ill conditioned, since its condition number is 10^8[33].

We used three methods to compute the SVD of in (7.32).

The estimation errors are defined as follows.

(1) The proposed algorithm:

$$e_i(k) = 10\log_{10}\left(\left\|\mathbf{A}_i\mathbf{v}_i(k)/(\mathbf{u}_i^T(t)\mathbf{A}_i\mathbf{v}_i(t)) - \mathbf{u}_i(k)\right\|^2 + \left\|\mathbf{A}_i^T\mathbf{u}_i(k)/(\mathbf{u}_i^T(t)\mathbf{A}_i\mathbf{v}_i(t)) - \mathbf{v}_i(k)\right\|^2\right), i = 1, 2, \ldots, m \tag{7.34}$$

(2) The DFCNN [5]:

$$\tilde{e}_i(k) = 10\log_{10}\left(\left\|\mathbf{A}_i\mathbf{v}_i(k) - \mathbf{u}_i(k)(\mathbf{u}_i^T(t)\mathbf{A}_i\mathbf{v}_i(t))\right\|^2 + \left\|\mathbf{A}_i^T\mathbf{u}_i(k) - \mathbf{v}_i(k)(\mathbf{u}_i^T(t)\mathbf{A}_i\mathbf{v}_i(t))\right\|^2\right), i = 1, 2, \ldots, m \tag{7.35}$$

(3) The Moore's neural network [11]:

$$\hat{e}(k) = 10\log_{10}\left(\left\|\mathbf{X}(k)[\mathbf{N}\mathbf{Y}(k)\mathbf{Y}^T(k) - \mathbf{X}^T(k)\mathbf{X}(k)\mathbf{N}]\right\|^2 + \left\|[\mathbf{X}^T(k)\mathbf{X}(k)\mathbf{N} - \mathbf{N}\mathbf{Y}(k)\mathbf{Y}^T(k)]\mathbf{Y}(k)\right\|^2\right). \tag{7.36}$$

In the DFCNN and Moore's neural networks, the suitable fixed time-step length takes $0.49/\sigma_{\max}$, and the variables and parameters in (7.35) and (7.36) can be found in [5] and [11], respectively.

Randomly generating the initial state vectors with unit-norm, the MATLAB software obtains the simulation results. The evolution curves of all the errors against iteration number shown in Fig. 7.2. The proposed algorithm has emerged as a useful neural-based technique, when the underlying matrix is close to a singular matrix.

Simulation 2: Give the following problem of identification for multiple input and single output static system:

$$O(t) = \mathbf{S}^T(t)\mathbf{h}, \tag{7.37}$$

where $h = [-0.3, -0.9, 0.8, -0.7, 0.6]^T$ is system parameter vector; $S(t)$ is a five-dimensional white input vector; $O(t)$ is the output. The input and output samples are corrupted by additive white noise. The unbiased estimate of parameter can be gotten by the total least squares approach [33]. The estimation error is defined as $\rho(k) = 20\log 10(\|\mathbf{h} - \hat{\mathbf{h}}(k)\|)$, where $\hat{\mathbf{h}}(k)$ represents the estimate of $\boldsymbol{h}$. Define the augmented output vector $\mathbf{x}(t) = [\mathbf{S}^T(t), O(t)]$. The data matrix consists of 30 samples of the augmented output vector $\boldsymbol{x}(t)$, where any row vector of the data matrix represent a sample of the augmented output vector.

The evolution curves of $\rho(k)$ against iteration number under cases (SNR = 0.1, 0.25, 0.5) are shown in Fig. 7.3 from which we see that the proposed neural network can find the TLS solution, while the DFCNN cannot.

In this section, a new recurrent neural network for SVD of a non-squared data matrix is proposed. The global convergence of this nonlinear system is studied. The norm of its state vectors is governed by stable ordinary differential equations independent of the data matrix and globally exponentially converges to 1 in the fixed

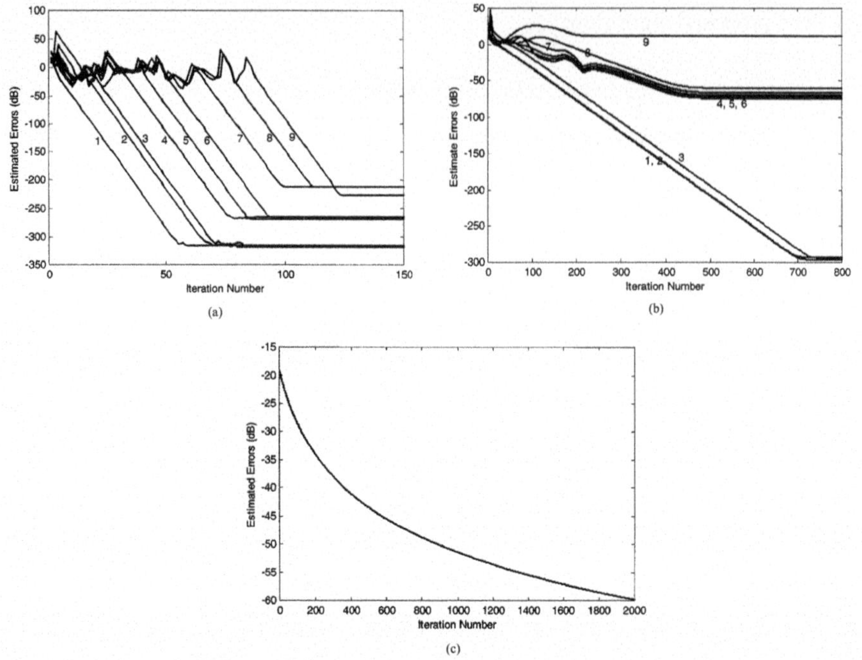

Fig. 7.2 The convergent curves of all the estimation errors against iteration number, where integer 1–9 in figure **a** and **b** shows the estimation errors associated with singular component 1–9. **a** ANN. **b** DFCNN. **c** Moore's approach

decay rate 2. Both theoretical analysis and simulation results show that the time-step length or leaning rate in the iterative algorithm associated with this neural network is independent of the data matrix. The proposed algorithm has also emerged as a useful neural-network-based technique when the data matrix is close to a singular matrix.

7.4 A Fast and Effective Neural Network Algorithm to Perform SVD

In this section, based on a novel information criterion, we first propose a novel dynamical system. Then, we obtain an effective online gradient-based algorithm for PSS tracking from the dynamical system. Moreover, all principal singular values of the PSS can be estimated in a parallel way based on the extracted subspace. Compared with traditional methods, the proposed algorithm can not only track the PSS but also estimate all of the corresponding principal singular values. Compared with Kaiser et al. [1] algorithm, the proposed algorithm can extract multiple principal singular triplets (singular value and singular vectors) in a parallel way. Compared with

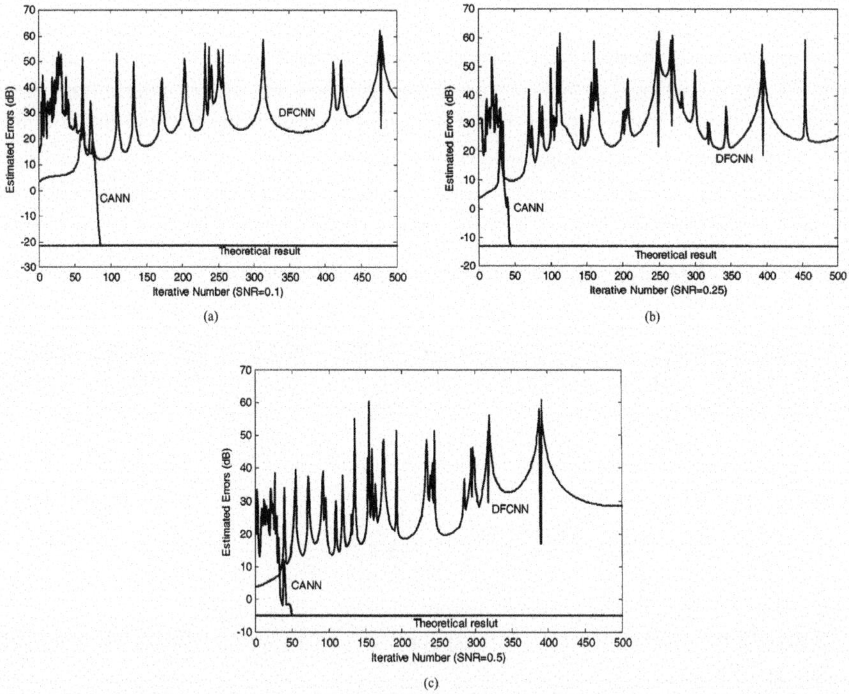

Fig. 7.3 The estimation errors against iteration number

Hasan's [15] algorithm, the proposed algorithm does not involve matrix inversion computation.

7.4.1 Preliminary

The SVD of a $m \times n$ real matrix $\boldsymbol{A}$ is given by $\mathbf{A} = \overline{\mathbf{U}}\overline{\mathbf{S}}\overline{\mathbf{V}}^T + \overline{\mathbf{U}}_2\overline{\mathbf{S}}_2\overline{\mathbf{V}}_2^T$, where $\overline{\mathbf{U}} = [\overline{\mathbf{u}}_1, \overline{\mathbf{u}}_2, \ldots, \overline{\mathbf{u}}_M] \in \Re^{m \times M}$ denotes the matrix composed of the left principal singular vectors, $\overline{\mathbf{S}} = \text{diag}(\sigma_1, \sigma_2, \ldots, \sigma_M) \in \Re^{M \times M}$ denotes the matrix with the principal singular values on its diagonal, and $\overline{\mathbf{V}} = [\overline{\mathbf{v}}_1, \overline{\mathbf{v}}_2, \ldots, \overline{\mathbf{v}}_M] \in \Re^{n \times M}$ denotes the matrix composed of the right principal singular vectors. These matrices are referred to as the principal portion of the SVD. Thus, $\hat{\mathbf{A}} = \overline{\mathbf{U}}\overline{\mathbf{S}}\overline{\mathbf{V}}^T$ is the best rank-M approximation of $\boldsymbol{A}$, where $M \leq p = \min\{m, n\}$. Moreover, $(\overline{\mathbf{u}}_j, \overline{\mathbf{v}}_j, \sigma_j)$ is called the jth singular triplet of the cross-covariance matrix $\boldsymbol{A}$. Furthermore, $\overline{\mathbf{U}}_2 = [\overline{\mathbf{u}}_{M+1}, \overline{\mathbf{u}}_{M+2}, \ldots, \overline{\mathbf{u}}_p] \in \Re^{m \times (p-M)}, \overline{\mathbf{S}}_2 = diag(\sigma_{M+1}, \sigma_{M+2}, \ldots, \sigma_p) \in \Re^{(p-M) \times (p-M)}$, and $\overline{\mathbf{V}}_2 = [\overline{\mathbf{v}}_{M+1}, \overline{\mathbf{v}}_{M+2}, \ldots, \overline{\mathbf{v}}_p] \in \Re^{n \times (p-M)}$ correspond to the minor portion of the SVD. Left and right singular vectors are normal, i.e., $\|\overline{\mathbf{u}}_j\| = \|\overline{\mathbf{v}}_j\| = 1, \forall j$, and mutually orthogonal. Thus, $\overline{\mathbf{U}}^T\overline{\mathbf{U}} = \overline{\mathbf{V}}^T\overline{\mathbf{V}} = \mathbf{I}_M$

and $\overline{\mathbf{U}}_2^T\overline{\mathbf{U}}_2 = \overline{\mathbf{V}}_2^T\overline{\mathbf{V}}_2 = \mathbf{I}_{p-M}$. Moreover, we assume that the singular values are ordered and mutually different with respect to their absolute values such that $|\sigma_1| > \cdots |\sigma_M| > |\sigma_{M+1}| \geq \cdots \geq |\sigma_p|$. In the following, all considerations (e.g., concerning fixed points) depend on the principal portion of the SVD in the domain $\{(\overline{\mathbf{U}}, \overline{\mathbf{V}})|\overline{\mathbf{U}}^T\mathbf{A}\overline{\mathbf{V}} > 0\}$ only, thus $\sigma_j > 0\ \forall j \leq p$.

7.4.2 Novel Information Criterion and Algorithm

For a data matrix $\mathbf{A} \in \Re^{m\times n}$, and given $\mathbf{U} \in \Re^{m\times r}$ and $\mathbf{V} \in \Re^{n\times r}$ in the domain $\{(\mathbf{U}, \mathbf{V})|\mathbf{U}^T\mathbf{A}\mathbf{V} > 0\}$, we present a novel information criterion for PSS tracking of $\boldsymbol{A}$ as follows:

$$\begin{aligned} &\min\{J(\mathbf{U}, \mathbf{V})\} \\ &J(\mathbf{U}, \mathbf{V}) = -\frac{1}{2}\mathrm{tr}(\mathbf{U}^T\mathbf{A}\mathbf{V}\mathbf{D}) - \mathrm{tr}\left(\frac{\mathbf{U}^T\mathbf{A}\mathbf{V}}{\|\mathbf{U}\|\|\mathbf{V}\|}\right) + \frac{1}{2}\mathrm{tr}(\mathbf{U}^T\mathbf{U} + \mathbf{V}^T\mathbf{V}), \end{aligned} \tag{7.38}$$

where $\boldsymbol{D}$ is a diagonal matrix whose eigenvalues are all positive.

Based on the gradient of $J(\boldsymbol{U}, \boldsymbol{V})$ respect to $\boldsymbol{U}$ and $\boldsymbol{V}$, we can obtain a dynamical system as follows:

$$\dot{\mathbf{U}} = \mathbf{A}\mathbf{V}\mathbf{D} + \frac{\mathbf{A}\mathbf{V}}{\|\mathbf{U}\|\|\mathbf{V}\|} - \frac{\mathbf{U}\mathbf{U}^T\mathbf{A}\mathbf{V}}{\|\mathbf{U}\|^3\|\mathbf{V}\|} - \mathbf{U} \tag{7.39}$$

$$\dot{\mathbf{V}} = \mathbf{A}^T\mathbf{U}\mathbf{D} + \frac{\mathbf{A}^T\mathbf{U}}{\|\mathbf{U}\|\|\mathbf{V}\|} - \frac{\mathbf{V}\mathbf{U}^T\mathbf{A}\mathbf{V}}{\|\mathbf{U}\|\|\mathbf{V}\|^3} - \mathbf{V} \tag{7.40}$$

Clearly, we can obtain an adaptive algorithm with normalized steps from the above dynamical system as follows:

$$\begin{aligned} \tilde{\mathbf{U}}(k+1) = \mathbf{U}(k) + \eta\left\{\frac{\mathbf{A}(k+1)\mathbf{V}(k)(\mathbf{D}+\mathbf{I}_r)}{\|\mathbf{U}(k)\|\|\mathbf{V}(k)\|}\right\} \\ - \mathbf{U}(k)\left[\frac{\mathbf{U}^T(k)\mathbf{A}(k+1)\mathbf{V}(k)}{\|\mathbf{U}(k)\|^3\|\mathbf{V}(k)\|} + \mathbf{I}_r\right] \end{aligned} \tag{7.41}$$

$$\mathbf{U}(k+1) = \frac{\tilde{\mathbf{U}}(k+1)}{\|\mathbf{U}(k+1)\|} \tag{7.42}$$

$$\begin{aligned} \tilde{\mathbf{V}}(k+1) = \mathbf{V}(k) + \eta\left\{\frac{\mathbf{A}^T(k+1)\mathbf{U}(k)(\mathbf{D}+\mathbf{I}_r)}{\|\mathbf{U}(k)\|\|\mathbf{V}(k)\|}\right\} \\ - \mathbf{V}(k)\left[\frac{\mathbf{U}^T(k)\mathbf{A}(k+1)\mathbf{V}(k)}{\|\mathbf{U}(k)\|\|\mathbf{V}(k)\|^3} + \mathbf{I}_r\right] \end{aligned} \tag{7.43}$$

$$\mathbf{V}(k+1) = \frac{\tilde{\mathbf{V}}(k+1)}{\|\mathbf{V}(k+1)\|}, \tag{7.44}$$

where k is the time step, and η is the learning rate.

Note that in the above algorithm two normalized steps (7.42) and (7.44) are added to ensure convergence. Since $\boldsymbol{U}(k)$ and $\boldsymbol{V}(k)$ are normalized at each step, we have $\|\mathbf{U}(k)\| = \|\mathbf{V}(k)\| = 1, \quad \forall k > 0$. Substituting $\|\mathbf{U}(k)\| = \|\mathbf{V}(k)\| = 1, \quad \forall k > 0$ into (7.41–7.44) and ignoring the normalize step, it yields

$$\begin{aligned}\mathbf{U}(k+1) = \mathbf{U}(k) + \eta\{&\mathbf{A}(k+1)\mathbf{V}(k)(\mathbf{D}+\mathbf{I}_r)\} \\ &- \mathbf{U}(k)\left[\mathbf{U}^T(k)\mathbf{A}(k+1)\mathbf{V}(k) + \mathbf{I}_r\right]\end{aligned} \tag{7.45}$$

$$\begin{aligned}\mathbf{V}(k+1) = \mathbf{V}(k) + \eta\{&\mathbf{A}^T(k+1)\mathbf{U}(k)(\mathbf{D}+\mathbf{I}_r)\} \\ &- \mathbf{V}(k)\left[\mathbf{U}^T(k)\mathbf{A}(k+1)\mathbf{V}(k) + \mathbf{I}_r\right],\end{aligned} \tag{7.46}$$

where $\boldsymbol{I}_r$ is an identity matrix of dimension r. Since $\boldsymbol{D}$ and $\boldsymbol{I}_r$ are known, the above algorithm can be simplified to

$$\begin{aligned}\mathbf{U}(k+1) = \mathbf{U}(k) + \eta\{&\mathbf{A}(k+1)\mathbf{V}(k)\mathbf{D}\} \\ &- \mathbf{U}(k)\left[\mathbf{U}^T(k)\mathbf{A}(k+1)\mathbf{V}(k) + \mathbf{I}_r\right]\end{aligned} \tag{7.47}$$

$$\begin{aligned}\mathbf{V}(k+1) = \mathbf{V}(k) + \eta\{&\mathbf{A}^T(k+1)\mathbf{U}(k)\mathbf{D}\} \\ &- \mathbf{V}(k)\left[\mathbf{U}^T(k)\mathbf{A}(k+1)\mathbf{V}(k) + \mathbf{I}_r\right],\end{aligned} \tag{7.48}$$

where $\boldsymbol{D}$ is a diagonal matrix whose eigenvalues are all larger than 1. And the corresponding dynamical system yields

$$\dot{\mathbf{U}} = \mathbf{AVD} - \mathbf{U}\left[\mathbf{U}^T\mathbf{AV} + \mathbf{I}_r\right] \tag{7.49}$$

$$\dot{\mathbf{V}} = \mathbf{A}^T\mathbf{UD} - \mathbf{V}\left[\mathbf{U}^T\mathbf{AV} + \mathbf{I}_r\right] \tag{7.50}$$

It is worth noting that in the simplified system (7.49) and (7.50), the stationary point relates to the singular values of $\boldsymbol{A}$. This means that the state matrix pencil ($\boldsymbol{U}(k)$, $\boldsymbol{V}(k)$) of the proposed algorithm (7.47) and (7.48) converges to a stationary point ($\boldsymbol{U}^*$, $\boldsymbol{V}^*$) that relates to the first r principal singular values of $\boldsymbol{A}$. In this case, the first r principal singular values of $\boldsymbol{A}$ can be estimated via the extracted PSS.

Particularly, if $r = 1$, then the matrices $\boldsymbol{U}$ and $\boldsymbol{V}$ reduce to vectors $\mathbf{u} \in \Re^{m\times 1}$ and $\mathbf{v} \in \Re^{n\times 1}$, respectively, and the matrix $\boldsymbol{D}$ reduces to a positive real number $d(>1)$. In this case, we obtain an algorithm for principal singular triplet (PST) extraction as

$$\begin{aligned}\mathbf{u}(k+1) = \mathbf{u}(k) + \eta[d\mathbf{A}(k+1)\mathbf{v}(k) \\ - \mathbf{u}(k)(\mathbf{u}^T(k)\mathbf{A}(k+1)\mathbf{v}(k)+1)]\end{aligned} \tag{7.51}$$

$$\begin{aligned}\mathbf{v}(k+1) = \mathbf{v}(k) + \eta[d\mathbf{A}^T(k+1)\mathbf{u}(k) \\ - \mathbf{v}(k)(\mathbf{u}^T(k)\mathbf{A}(k+1)\mathbf{v}(k)+1)]\end{aligned} \tag{7.52}$$

which corresponds to the dynamical system

$$\dot{\mathbf{u}} = d\mathbf{A}\mathbf{v} - \mathbf{u}(\mathbf{u}^T\mathbf{A}\mathbf{v}+1) \tag{7.53}$$

$$\dot{\mathbf{v}} = d\mathbf{A}^T\mathbf{u} - \mathbf{v}(\mathbf{u}^T\mathbf{A}\mathbf{v}+1) \tag{7.54}$$

This is applicable if only the first PST estimation is demanded in some practical applications.

Next, we discuss the influence of parameters of the proposed algorithm. There are two parameters in our algorithm. The first one is the learning rate η, and the second one is the matrix $\boldsymbol{D}$ (or value d). Actually, like most iteration algorithms, it is hard to choose a very suitable learning rate for the proposed algorithm. In general, for convergence of iteration algorithms, a larger learning rate can lead to fast convergence, but algorithms may suffer from the oscillation or divergence. However, a very small learning rate will result in a slow convergence speed and degrade the performance of algorithms. Since it is hard to analyze the influence of $\boldsymbol{D}$ (or d) in theory, we perform a simple simulation to show the influence of d in the proposed PST extraction algorithm (see Sect. 7.4.4.2, and similar situation also holds for each element of $\boldsymbol{D}$ in the proposed PSS tracking algorithm). From the simulation result, we find that the algorithm is prone to diverge with larger d, and under the precondition that the algorithm is convergent, the parameter d has little influence on the convergence speed.

7.4.3 Convergence Analysis

The convergence property of the system (7.49) and (7.50) can be established as in the following theorem.

Theorem 7.2 *Assume that* $\boldsymbol{U}(k)$ *and* $\boldsymbol{V}(k)$ *are the solutions of dynamical system* (7.49) *and* (7.50) *for all* $k \geq 0$ *in the domain* $\{(\mathbf{U}(k), \mathbf{V}(k)) | \mathbf{U}^T(k)\mathbf{A}\mathbf{V}(k) > 0\}$. *Assume also that* $\mathbf{U}(0) \in \Re^{m\times r}$ and $\mathbf{V}(0) \in \Re^{n\times r}$ *are full-ranked and bounded, where* $0 < r \leq p$. *Denote* $\mathbf{P} = \lim_{k\to\infty} \mathbf{U}^T(k)\mathbf{U}(k)$, $\mathbf{Q} = \lim_{k\to\infty} \mathbf{V}^T(k)\mathbf{V}(k)$ *and* $\mathbf{B} = \lim_{k\to\infty} \mathbf{U}^T(k)\mathbf{A}\mathbf{V}(k)$. *Then,*

The dynamical system (7.49) *and* (7.50) *has equilibrium points as follows*:

$$\hat{\mathbf{U}}^* = \hat{\mathbf{U}}_r(\mathbf{D} - \hat{\mathbf{S}}_r^{-1})^{1/2} \tag{7.55}$$

$$\hat{\mathbf{V}}^* = \hat{\mathbf{V}}_r(\mathbf{D} - \hat{\mathbf{S}}_r^{-1})^{1/2}, \tag{7.56}$$

where $\hat{\mathbf{U}}_r$ and $\hat{\mathbf{V}}_r$ consist of arbitrary r singular vectors of $\overline{\mathbf{U}}$ and $\overline{\mathbf{V}}$ as introduced in $\mathbf{A} = \overline{\mathbf{U}}\overline{\mathbf{S}}\overline{\mathbf{V}}^T + \overline{\mathbf{U}}_2\overline{\mathbf{S}}_2\overline{\mathbf{V}}_2^T$, respectively, and $\hat{\mathbf{S}}$ consists of r corresponding singular values of $\overline{\mathbf{S}}$.

Algorithm (7.47) *and* (7.48) *converges to the global minimum point of* (7.49) *and* (7.50)

$$\mathbf{U}^* = \overline{\mathbf{U}}_r\left(\mathbf{D} - \overline{\mathbf{S}}_r^{-1}\right)^{1/2} \tag{7.57}$$

$$\mathbf{V}^* = \overline{\mathbf{V}}_r\left(\mathbf{D} - \overline{\mathbf{S}}_r^{-1}\right)^{1/2} \tag{7.58}$$

with probability 1 as k $\rightarrow \infty$, *where* $\overline{\mathbf{S}}_r = \mathrm{diag}(\sigma_1, \ldots, \sigma_r)$, *and* $\overline{\mathbf{U}}_r$ *and* $\overline{\mathbf{V}}_r$ *consist of the first r columns of* $\overline{\mathbf{U}}$ *and* $\overline{\mathbf{V}}$ *corresponding to singular values* $\sigma_1,\ldots,\sigma_r$, *respectively. And all other stationary points are saddle (unstable) points of* (7.49) *and* (7.50). *Thus,* $\boldsymbol{U}^*$ *and* $\boldsymbol{V}^*$ *produce an orthogonal basis of the left and right PSS, respectively.*

$\boldsymbol{P}$, $\boldsymbol{Q}$, *and* $\boldsymbol{B}$ *are diagonal and* $\boldsymbol{P} = \boldsymbol{Q}$. *The first r principal singular values of* $\boldsymbol{A}$ *are the diagonal elements of the matrix* $\boldsymbol{P}^{-1}\boldsymbol{B}$ *or* $(\boldsymbol{D}^{-}\boldsymbol{P})^{-1}$.

Proof The proof of (a) can be done by solving $\dot{\mathbf{U}} = 0$ and $\dot{\mathbf{V}} = 0$ in (7.49) and (7.50).

Pre-multiplying the left side of (7.49) and (7.50) by $\boldsymbol{U}^T$ and $\boldsymbol{V}^T$, respectively, we have

$$\mathbf{U}^T\dot{\mathbf{U}} = \mathbf{U}^T\mathbf{AVD} - \mathbf{U}^T\mathbf{U}\left[\mathbf{U}^T\mathbf{AV} + \mathbf{I}_r\right] \tag{7.59}$$

$$\mathbf{V}^T\dot{\mathbf{V}} = \mathbf{V}^T\mathbf{A}^T\mathbf{UD} - \mathbf{V}^T\mathbf{V}\left[\mathbf{U}^T\mathbf{AV} + \mathbf{I}_r\right] \tag{7.60}$$

Denote $\tilde{\mathbf{U}} = \mathbf{U}^T\mathbf{U}$, $\tilde{\mathbf{V}} = \mathbf{V}^T\mathbf{V}$, and $\tilde{\mathbf{A}} = \mathbf{U}^T\mathbf{AV}$ in the equilibrium point (where $\dot{\mathbf{U}} = 0$ and $\dot{\mathbf{V}} = 0$). Then we have that

$$\tilde{\mathbf{A}}\mathbf{D} = \tilde{\mathbf{U}}(\tilde{\mathbf{A}} + \mathbf{I}_r) \tag{7.61}$$

$$\tilde{\mathbf{A}}^T\mathbf{D} = \tilde{\mathbf{V}}(\tilde{\mathbf{A}} + \mathbf{I}_r) \tag{7.62}$$

which means that

$$\tilde{\mathbf{A}}\tilde{\mathbf{A}}^{-T} = \tilde{\mathbf{U}}\tilde{\mathbf{V}}^{-1} \tag{7.63}$$

Let $\boldsymbol{R}$ be a matrix defined by $\mathbf{R} = \tilde{\mathbf{U}}\tilde{\mathbf{V}}^{-1}$, and then

$$\tilde{\mathbf{A}} = \mathbf{R}\tilde{\mathbf{A}}^T \tag{7.64}$$

Equation (7.64) yields

$$\tilde{\mathbf{A}}(\mathbf{I} - \mathbf{R}^T) + (\mathbf{I} - \mathbf{R})\tilde{\mathbf{A}}^T = 0 \tag{7.65}$$

Since $\tilde{\mathbf{U}}$ and $\tilde{\mathbf{V}}$ are positive definite, then the eigenvalues of $\boldsymbol{R}$ are all real numbers with corresponding real eigenvectors. Thus, assume that λ is an eigenvalue of $\boldsymbol{R}^T$ with associative eigenvector z, then $\mathbf{R}^T\mathbf{z} = \lambda\mathbf{z}$. Pre- and post-multiplying the left and right side of (7.65) by z^T and z, respectively, give

$$\mathbf{z}^T\tilde{\mathbf{A}}\mathbf{z}(1 - \lambda) + \mathbf{z}^T\tilde{\mathbf{A}}^T\mathbf{z}(1 - \lambda) = 0 \tag{7.66}$$

and hence

$$(1 - \lambda)\mathbf{z}^T(\tilde{\mathbf{A}} + \tilde{\mathbf{A}}^T)\mathbf{z} = 0 \tag{7.67}$$

Since $\tilde{\mathbf{A}} + \tilde{\mathbf{A}}^T$ is positive definite in the domain $\{(\mathbf{U}(k), \mathbf{V}(k))|\mathbf{U}^T(k)\mathbf{A}\mathbf{V}(k) > 0\}$, it follows that $\lambda = 1$, i.e., each eigenvalue of $\boldsymbol{R}$ is equal to 1. The eigenvalues of $\tilde{\mathbf{U}}\tilde{\mathbf{V}}^{-1} = \mathbf{R}$ are same as those of $\tilde{\mathbf{U}}^{1/2}\tilde{\mathbf{V}}^{-1}\tilde{\mathbf{U}}^{1/2}$. Since $\tilde{\mathbf{U}}^{1/2}\tilde{\mathbf{V}}^{-1}\tilde{\mathbf{U}}^{1/2} = \mathbf{I}$ and hence $\tilde{\mathbf{U}} = \tilde{\mathbf{V}}, \mathbf{R} = \mathbf{I}$.

It is known that each equilibrium point of (7.49) and (7.50) must span an r-dimensional left and right subspace, respectively, thus each equilibrium point must be [15]

$$\hat{\mathbf{U}}^* = \hat{\mathbf{U}}_r\mathbf{X} \tag{7.68}$$

$$\hat{\mathbf{V}}^* = \hat{\mathbf{V}}_r\mathbf{Y}, \tag{7.69}$$

where $\boldsymbol{X}$ and $\boldsymbol{Y}$ are unknown diagonal matrices. In the vicinity of stationary point $(\hat{\mathbf{U}}^*, \hat{\mathbf{V}}^*)$, it holds that

$$\tilde{\mathbf{U}} = \mathbf{U}^T\mathbf{U} = \hat{\mathbf{U}}^{*T}\hat{\mathbf{U}}^* = \mathbf{X}^T\hat{\mathbf{U}}_r^T\hat{\mathbf{U}}_r\mathbf{X} = \mathbf{X}^T\mathbf{X} \tag{7.70}$$

While $\boldsymbol{X}$ is a diagonal matrix, thus $\mathbf{X} = \tilde{\mathbf{U}}^{1/2}$ and similarly $\mathbf{Y} = \tilde{\mathbf{V}}^{1/2} = \tilde{\mathbf{U}}^{1/2}$. We have

$$\tilde{\mathbf{A}} = \mathbf{U}^T\mathbf{A}\mathbf{V} = \mathbf{X}^T\hat{\mathbf{U}}_r^T\mathbf{A}\hat{\mathbf{V}}_r\mathbf{Y} = \mathbf{X}^T\hat{\mathbf{S}}_r\mathbf{Y} \tag{7.71}$$

While $\boldsymbol{X}$, $\boldsymbol{Y}$, and $\hat{\mathbf{S}}_r$ are all diagonal, it yields that

$$\mathbf{X}^T\hat{\mathbf{S}}_r\mathbf{Y} = \mathbf{X}^T\mathbf{Y}\hat{\mathbf{S}}_r = \tilde{\mathbf{U}}\hat{\mathbf{S}}_r. \tag{7.72}$$

thus $\tilde{\mathbf{A}} = \tilde{\mathbf{U}}\hat{\mathbf{S}}_r$. Substituting $\tilde{\mathbf{A}} = \tilde{\mathbf{U}}\hat{\mathbf{S}}_r$ into (7.61), it yields

$$\tilde{\mathbf{U}}\hat{\mathbf{S}}_r\mathbf{D} = \tilde{\mathbf{U}}(\tilde{\mathbf{U}}\hat{\mathbf{S}}_r + \mathbf{I}_r) \tag{7.73}$$

which means that

$$\tilde{\mathbf{U}} = \mathbf{D} - \hat{\mathbf{S}}_r^{-1}, \tag{7.74}$$

Similarly, we have that

$$\tilde{\mathbf{V}} = \mathbf{D} - \hat{\mathbf{S}}_r^{-1}. \tag{7.75}$$

Thus, the dynamical system (7.49) and (7.50) has equilibrium points as

$$\hat{\mathbf{U}}^* = \hat{\mathbf{U}}_r\left(\mathbf{D} - \hat{\mathbf{S}}_r^{-1}\right)^{1/2}, \tag{7.76}$$

$$\hat{\mathbf{V}}^* = \hat{\mathbf{V}}_r\left(\mathbf{D} - \hat{\mathbf{S}}_r^{-1}\right)^{1/2} \tag{7.77}$$

The proof of (b) can be done by calculating the eigenvalues of the Jacobian matrix of the system (7.49) and (7.50) and applications of some propositions in Appendix A [24]. The system is stable if all of the eigenvalues of the Jacobian are negative. However, the Hessian matrix is a high-dimensional complicated matrix that is hard to analyze. Thus, in this section, the proof of (b) is omitted due to the complexity. Based on the results of (a) and (b), we conclude that

$$\mathbf{P} = \mathbf{Q} = \hat{\mathbf{U}}^{*T}\hat{\mathbf{U}}^* = \mathbf{D} - \overline{\mathbf{S}}_r^{-1} \tag{7.78}$$

$$\mathbf{B} = (\mathbf{D} - \overline{\mathbf{S}}_r^{-1})^{T/2}\overline{\mathbf{S}}_r(\mathbf{D} - \overline{\mathbf{S}}_r^{-1})^{1/2} = (\mathbf{D} - \overline{\mathbf{S}}_r^{-1})\overline{\mathbf{S}}_r. \tag{7.79}$$

Thus, $\boldsymbol{P}$, $\boldsymbol{Q}$, and $\boldsymbol{B}$ are all diagonal. Moreover, we can deduce that

$$\mathbf{P}^{-1}\mathbf{B} = (\mathbf{D} - \overline{\mathbf{S}}_r^{-1})^{-1}(\mathbf{D} - \overline{\mathbf{S}}_r^{-1})\overline{\mathbf{S}}_r = \overline{\mathbf{S}}_r \tag{7.80}$$

$$(\mathbf{D} - \mathbf{P})^{-1} = (\overline{\mathbf{S}}_r^{-1})^{-1} = \overline{\mathbf{S}}_r, \tag{7.81}$$

where $\overline{\mathbf{S}}_r$ is a diagonal matrix who contains the first r principal singular values of $\boldsymbol{A}$ in its diagonal elements. Because $\boldsymbol{D}$ and $\boldsymbol{P}$ are diagonal matrices, it is easy to get the inversion of $(\boldsymbol{D} - \boldsymbol{P})$.

This completes the proof of Theorem 7.2.

7.4.4 Experiments and Conclusions

In this section, we provide two experiments to illustrate the performance of the proposed algorithm. In the first experiment, SVD of an ill-conditioned matrix is used to evaluate the efficiency of the proposed parallel neural network algorithm (7.47) and (7.48). The second experiment mainly shows the efficiency of the proposed algorithm for PST extraction. In all experiments, the results are an average of $L =$ 100 independent runs. Note that in Experiments 1 and 2, the performance of right state matrix(vector) is similar to that of left state matrix (vector), thus they are not shown in figures. In all experiments, we compare the performance of the proposed algorithm with that of some well-known algorithms, i.e., algorithms proposed by Feng et al. [14], Hasan [15] and Kong et al. [16].

7.4.4.1 Experiment 1: Principal Singular Subspace Tracking

Generate randomly the nine-dimensional Gaussian white sequence $\boldsymbol{y}(k)$, whereas $\boldsymbol{x}(k) = \boldsymbol{A}\boldsymbol{y}(k)$, where $\boldsymbol{A}$ is given by the following:

$$\mathbf{A} = [\mathbf{u}_0, \ldots, \mathbf{u}_8] \cdot \mathbf{A}_0 \cdot [\mathbf{v}_0, \ldots, \mathbf{v}_8]^T, \tag{7.82}$$

where

$$\mathbf{A}_0 = \text{diag}(10, 10, 10, 10^{-3}, 10^{-3}, 10^{-3}, 10^{-7}, 10^{-7}, 10^{-7}), \tag{7.83}$$

and $\boldsymbol{u}_i$ and $\boldsymbol{v}_i (i = 0,\ldots, 8)$ are the ith components of 11- and 9-dimensional orthogonal discrete cosine basis function, respectively.

In theory, the matrix $\boldsymbol{A}$ given by (7.82) has nine nonzero singular values in which the entire three distinct singular values 10, 10^{-3} and 10^{-7} have multiplicity 3. The data matrix is ill conditioned, since its condition number is 10^8. In this experiment, we choose $r = 3$ while $10 >> 10^{-3}$, that is to say, the PSS spanned by the first 3 principal singular components will be tracked.

In order to measure the convergence speed and precision of learning algorithm, we compute the direction cosine between the state matrix, i.e., $\boldsymbol{U}(k)$ and $\boldsymbol{V}(k)$, and the true principal singular subspace, i.e., $\overline{\mathbf{U}}_r$ and $\overline{\mathbf{V}}_r$, at the k-th update as follows:

$$\text{DC}(\mathbf{U}(k)) = \left\|\mathbf{U}^T(k) \cdot \overline{\mathbf{U}}_r\right\| / \|\mathbf{U}(k)\| \left\|\overline{\mathbf{U}}_r\right\| \tag{7.84}$$

$$\text{DC}(\mathbf{V}(k)) = \left\|\mathbf{V}^T(k) \cdot \overline{\mathbf{V}}_r\right\| / \|\mathbf{V}(k)\| \left\|\overline{\mathbf{V}}_r\right\| \tag{7.85}$$

Also, we give the following index parameter

$$\text{Dist}(\mathbf{U}(k)) = \left\|\mathbf{U}^T(k)\mathbf{U}(k) - \text{diag}(\mathbf{U}^T(k)\mathbf{U}(k))\right\| \tag{7.86}$$

$$\mathrm{Dist}(\mathbf{V}(k)) = \left\| \mathbf{V}^T(k)\mathbf{V}(k) - \mathrm{diag}(\mathbf{V}^T(k)\mathbf{V}(k)) \right\|, \tag{7.87}$$

which denotes the deviation of a state matrix from the orthogonality. Here, diag($\boldsymbol{X}$) $= \boldsymbol{Z}$ is a diagonal matrix such that $z_{i,i} = x_{i,i}$. Clearly, if direction cosines DC($\boldsymbol{U}(k)$) and DC($\boldsymbol{V}(k)$) converge to 1, then the state matrices $\boldsymbol{U}(k)$ and $\boldsymbol{V}(k)$ must approach the direction of the left and right PSS, respectively. If Dist($\boldsymbol{U}(k)$) and Dist($\boldsymbol{V}(k)$) converge to zero, then it means that $\boldsymbol{U}(k)$ and $\boldsymbol{V}(k)$ produce an orthogonal basis of the left PSS and right PSS, respectively.

Moreover, the sample standard deviation of direction cosine and that of norm estimation of $\boldsymbol{U}(k)$ and $\boldsymbol{V}(k)$ are also calculated to measure the numerical stability of algorithms. The sample standard deviation of direction cosine are defined by the following:

$$\mathrm{DD}(\mathbf{U}(k)) = \sqrt{\frac{1}{L-1}\sum_{j=1}^{L}[\mathrm{DC}_j(\mathbf{U}(k)) - \overline{\mathrm{DC}}(\mathbf{U}(k))]^2} \tag{7.88}$$

$$\mathrm{DD}(\mathbf{V}(k)) = \sqrt{\frac{1}{L-1}\sum_{j=1}^{L}[\mathrm{DC}_j(\mathbf{V}(k)) - \overline{\mathrm{DC}}(\mathbf{V}(k))]^2}, \tag{7.89}$$

where $\mathrm{DC}_j(\boldsymbol{U}(k))$ and $\mathrm{DC}_j(\boldsymbol{V}(k))$ are the direction cosine of the jth independent run ($j = 1, 2, \ldots, L$) and $\overline{\mathrm{DC}}(\boldsymbol{U}(k))$ and $\overline{\mathrm{DC}}(\boldsymbol{V}(k))$ are the average of $L = 100$ independent runs. And the sample standard deviation of norm estimation are defined by the following:

$$\mathrm{DN}(\mathbf{U}(k)) = \sqrt{\frac{1}{L-1}\sum_{j=1}^{L}[\left\|(\mathbf{U}_j(k))\right\| - \left\|\overline{\mathbf{U}}(k)\right\|]^2} \tag{7.90}$$

$$\mathrm{DN}(\mathbf{V}(k)) = \sqrt{\frac{1}{L-1}\sum_{j=1}^{L}[\left\|(\mathbf{V}_j(k))\right\| - \left\|\overline{\mathbf{V}}(k)\right\|]^2}, \tag{7.91}$$

where $\left\|\mathbf{U}_j(k)\right\|$ and $\left\|\mathbf{V}_j(k)\right\|$ are the norm of $\boldsymbol{U}(k)$ and $\boldsymbol{V}(k)$ in the jth independent run ($j = 1, 2, \ldots, L$), respectively. And $\left\|\overline{\mathbf{U}}(k)\right\|$ and $\left\|\overline{\mathbf{V}}(k)\right\|$ are the average of $L =$ 100 independent runs.

In this experiment, we set the learning rate as $\gamma = 0.002$ for all algorithms and set $\boldsymbol{D} =$ diag(2.5456, 3.7235, 1.9371) for our algorithm and Hasan's algorithm, where $\boldsymbol{D}$ is randomly generated. Moreover, the initial value $\boldsymbol{U}(0)$ and $\boldsymbol{V}(0)$ are randomly generated.

Figure 7.4 shows the direction cosine between $\boldsymbol{U}(k)$ and $\overline{\mathbf{U}}_r$, where $\boldsymbol{U}(k)$ is generated by each aforementioned algorithm. It is found that all algorithms can extract the

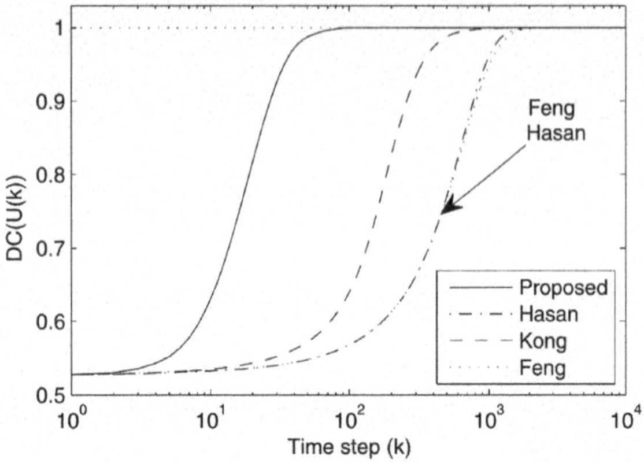

Fig. 7.4 Direction cosine between $\boldsymbol{U}(k)$ and $\overline{\mathbf{U}}_r$

PSS of the cross-covariance matrix effectively, and the proposed one has the highest convergence speed.

Figure 7.5 shows the norm of each column of state matrix $\boldsymbol{U}(k)$, which is generated by the proposed algorithm. Each column of $\boldsymbol{U}(k)$ converges to a length that relates to the corresponding principal singular value, i.e., $\left\|\mathbf{U}_p(k)\right\| \rightarrow \sqrt{\mathbf{D}_{p,p} - \overline{\mathbf{S}}_{p,p}^{-1}}$ as $k \rightarrow \infty (p = 1, 2, \ldots, \mathrm{r})$, and $\left\|\mathbf{v}_p(k)\right\| = \left\|\mathbf{u}_p(k)\right\|$, where $\boldsymbol{X}_{p,p}$ denotes the pth diagonal element of matrix $\boldsymbol{X}$. This indicates that each principal singular value can be estimated from the norm of each column of state matrix $\boldsymbol{U}(k)$ and(or) $\boldsymbol{V}(k)$, i.e.,

$$\sigma_p = \lim_{k \to \infty} \frac{1}{\mathbf{D}_{p,p} - \left\|\mathbf{u}_p(k)\right\|^2}$$

Note that in order to show the singular value estimation clearly, we calculate them from the norm of each column of the state matrix. Actually, in practical applications, all singular values are estimated in a parallel way.

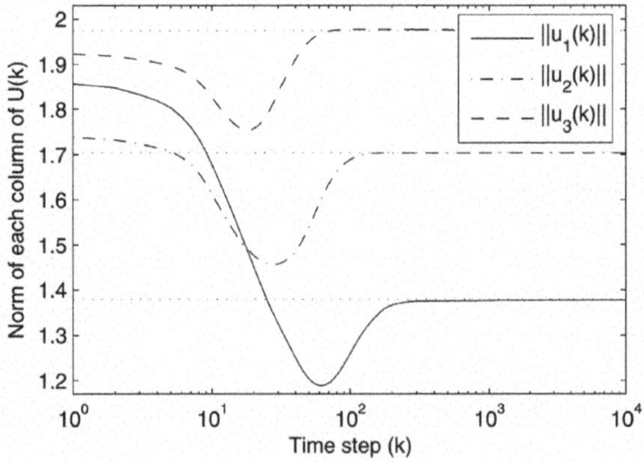

Fig. 7.5 Norm of vector $\boldsymbol{u}1(k)$, $\boldsymbol{u}2(k)$, and $\boldsymbol{u}3(k)$, where $\boldsymbol{U}(k) = [\boldsymbol{u}1(k), \boldsymbol{u}2(k), \boldsymbol{u}3(k)]$ is generated by proposed algorithm

Fig. 7.6 Norm of vector $\boldsymbol{u}1(k)$, $\boldsymbol{u}2(k)$ and $\boldsymbol{u}3(k)$, where $\boldsymbol{U}(k) = [\boldsymbol{u}1(k), \boldsymbol{u}2(k), \boldsymbol{u}3(k)]$ is generated by Hasan's algorithm

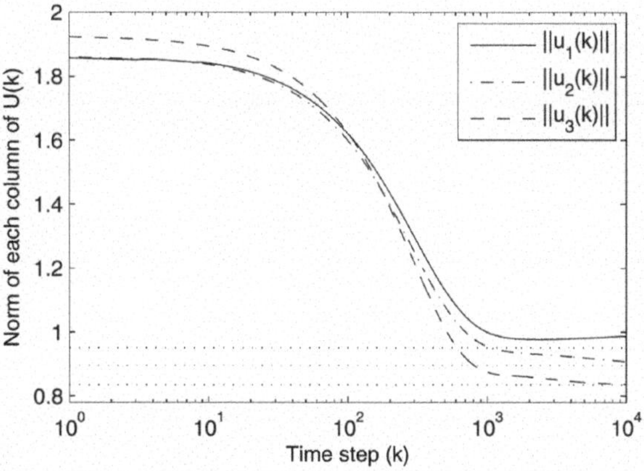

Similar to Figs. 7.5, 7.6 shows the norm of each column of state matrix $\boldsymbol{U}(k)$ which is generated by Hasan's algorithm. Each column of $\boldsymbol{U}(k)$ converges to a length that relates to the corresponding principal singular value, i.e.,

$$\left\|\mathbf{u}_p(k)\right\| \to \sqrt{1 - \mathbf{D}_{p,p}\overline{\mathbf{S}}_{p,p}^{-1}} ask \to \infty(0 < p \leq r),$$

and $\left\|\mathbf{v}_p(k)\right\| = \left\|\mathbf{u}_p(k)\right\|$.This indicates that each principal singular value can be estimated from the norm of each column of state matrix $\boldsymbol{U}(k)$ or(and) $\boldsymbol{V}(k)$, i.e.,

$$\sigma_p = \lim_{k\to\infty} \frac{\mathbf{D}_{p,p}}{1 - \left\|\mathbf{u}_p(k)\right\|^2}$$

Both of the proposed algorithm and Hasan's algorithm can extract the principal singular values, however, the proposed algorithm converges faster and extract the principal singular value more accurately than Hasan's algorithm, as shown in Figs. 7.5 and 7.6.

Figure 7.7 shows the norm of $\boldsymbol{U}(k)$ which are generated by Kong's algorithm and Feng's algorithm. It can be found that in Kong's algorithm the norm of $\boldsymbol{U}(k)$ converges to a non-unit length ($\sqrt{r}$ as discussed in [16]) while that of Feng's algorithm converges to the unit length.

by Kong's algorithm and Feng's algorithm, respectively.

Figure 7.8 shows the deviation of $\boldsymbol{U}(k)$ from the orthogonality, respectively. All curves converge to zero means that $\boldsymbol{U}(k)$ produce the orthonormal basis of the left PSS. We can also find that the proposed algorithm has the highest convergence speed.

Moreover, Figs. 7.9 and 7.10 show the sample standard deviation of direction cosine and norm estimation, respectively. From Figs. 7.9 and 7.10 we find that the proposed algorithm has better numerical stability than other algorithms. Based on

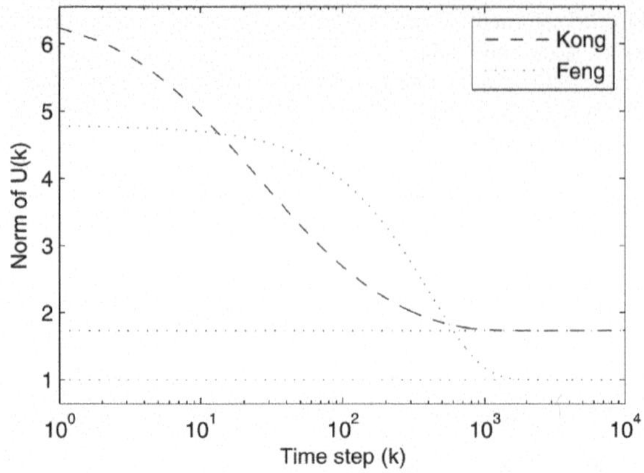

Fig. 7.7 7N or m of $\boldsymbol{U}(k)$, where $\boldsymbol{U}(k)$ are generated

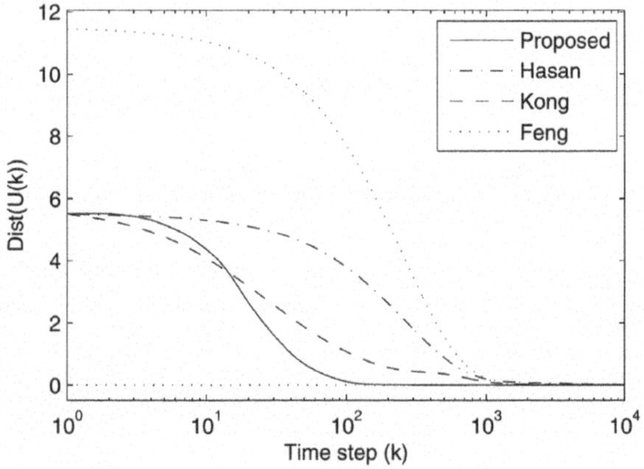

Fig. 7.8 The deviation of $U(k)$ from the orthogonality

the experiment results, we conclude that the proposed algorithm has the best performance, because it has the highest convergence speed, the best accuracy of orthogonality, and the best numerical stability. What's more, all principal singular values can be estimated based on the norm of the estimated state matrix.

7.4.4.2 Experiment 2: Principal Singular Triplet Extraction

The PST extraction algorithm (7.51–7.52) is a special case of the PSS tracking algorithm (7.47–7.48) for extracting the principal singular triplet according to the largest singular value. In this case, this experiment compares the performance of the proposed PST extraction algorithm with other algorithms aforementioned in Experiment 1 for the first PST extraction.

Generate randomly the nine-dimensional Gaussian white sequence $\boldsymbol{y}(k)$, whereas $\boldsymbol{x}(k) = \boldsymbol{A}\boldsymbol{y}(k)$, where $\boldsymbol{A}$ is given by the following:

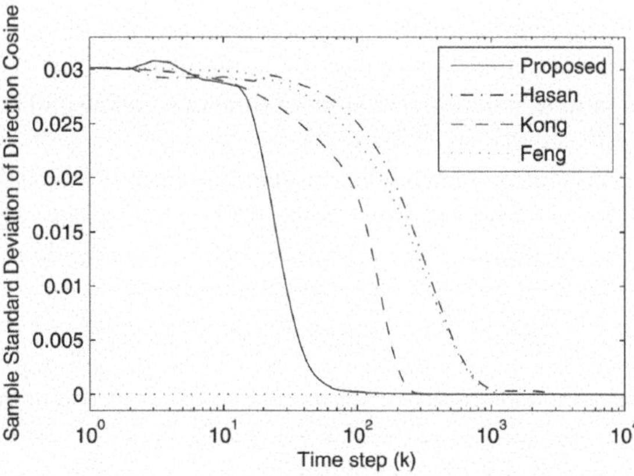

Fig. 7.9 The sample standard deviation of direction cosine

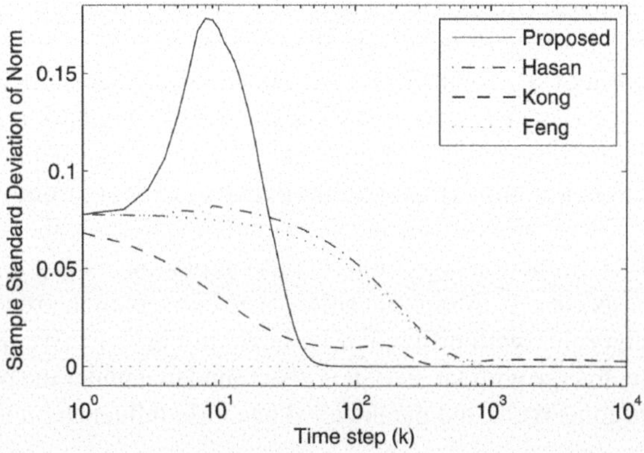

Fig. 7.10 The sample standard deviation of norm estimation

$$\mathbf{A} = [\mathbf{u}_0, \ldots, \mathbf{u}_8] \cdot \mathrm{randn}(9, 9) \cdot [\mathbf{v}_0, \ldots, \mathbf{v}_8]^T, \tag{7.92}$$

where $\boldsymbol{u}_i$ and $\boldsymbol{v}_i$ ($i = 0,\ldots, 8$) are the ith components of 11- and 9-dimensional orthogonal discrete cosine basis function, respectively. In this case, the cross-covariance matrix $\boldsymbol{A}$ has distinct eigenvalues.

In PST extraction, the direction cosines at the k-th update are defined as follows:

$$\mathrm{DC}(\mathbf{u}(k)) = \left\| \mathbf{u}^T(k) \cdot \bar{\mathbf{u}}_1 \right\| / \|\mathbf{u}(k)\| \|\bar{\mathbf{u}}_1\| \tag{7.93}$$

$$\mathrm{DC}(\mathbf{v}(k)) = \left\| \mathbf{v}^T(k) \cdot \bar{\mathbf{v}}_1 \right\| / \|\mathbf{v}(k)\| \|\bar{\mathbf{v}}_1\| \tag{7.94}$$

Clearly, if the direction cosines (7.93) and (7.94) converge to 1, then the state vector $\boldsymbol{u}(k)$ and $\boldsymbol{v}(k)$ must approach the direction of the true left and right principal singular vectors, respectively.

The sample standard deviations of direction cosine and norm of $\boldsymbol{u}(k)$ and $\boldsymbol{v}(k)$ are also calculated to measure the numerical stability of all algorithms. The definition of sample standard deviation of direction cosine and norm estimation in this case are similar to that in (7.88–7.91).

In this experiment, the learning rate is set as $\gamma = 0.01$ for all learning rules. We set $d = 1.8973$ for the proposed algorithm and Hasan's algorithm, where d is randomly generated. The initial value $\boldsymbol{u}(0)$ and $\boldsymbol{v}(0)$ are randomly generated.

Figure 7.11 shows the direction cosine between $\boldsymbol{u}(k)$ and $\bar{\mathbf{u}}_1$, where $\boldsymbol{u}(k)$ is generated by each aforementioned algorithm. Figure 7.12 shows the norm of state vectors $\boldsymbol{u}(k)$ for all algorithms. As Fig. 7.11 shows, all algorithms can effectively extract the principal singular vectors of a matrix. It is found that all algorithms have similar convergence speed for PST extraction. As shown in Fig. 7.12, in vector length estimation, the proposed algorithm converges faster than the others. What's more, the proposed algorithm and Hasan's algorithm converge to a length that relate to the first principal singular value ($\sqrt{d - 1/\sigma_1}$ for the proposed algorithm and $\sqrt{1 - d/\sigma_1}$ for Hasan's algorithm). Thus, the principal singular value can be estimated based on the norm of $\boldsymbol{u}(k)$ and (or) $\boldsymbol{v}(k)$ in the proposed algorithm and Hasan's algorithm. Moreover, Figs. 7.13 and 7.14 show the sample standard deviation of direction cosine and norm, respectively. From Figs. 7.13 and 7.14 we find that the proposed algorithm has better numerical stability than the other algorithms.

Here, we perform an extra simulation to show the influence of parameter d in our PST extraction algorithm. In this simulation, we use $d = 2$, $d = 5$, $d = 8$ and $d = 10$, respectively, in our algorithm. The convergence process of direction cosine of our algorithm with different d is shown in Fig. 7.15. We find that the algorithm is prone to diverge with larger d. It is also found that under the precondition that the algorithm is convergent, the parameter d has little influence on the convergence speed.

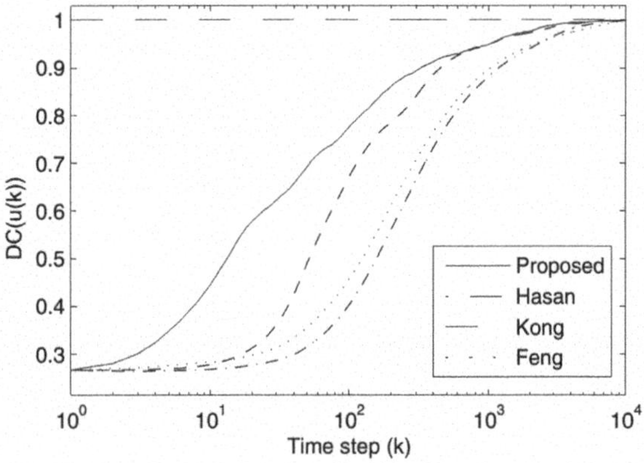

Fig. 7.11 Direction cosine between $\boldsymbol{u}(k)$ and $\bar{\mathbf{u}}_1$

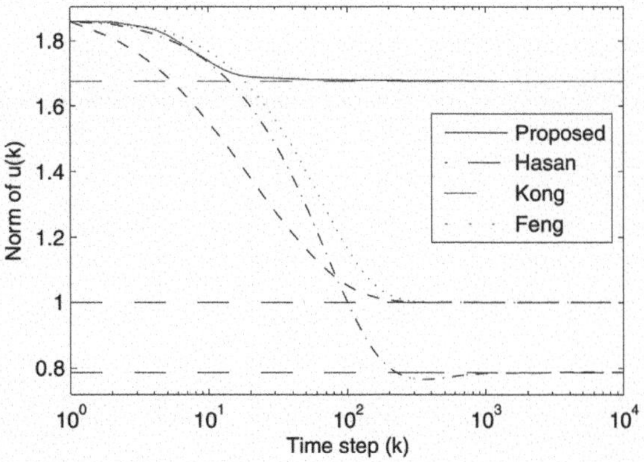

Fig. 7.12 Norm of $u(k)$

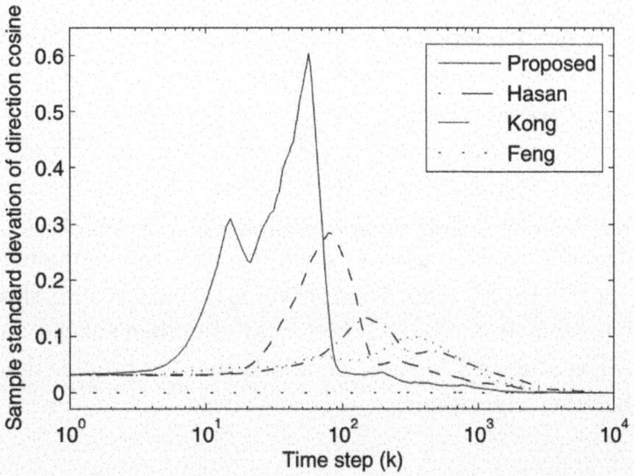

Fig. 7.13 Sample standard deviation of direction cosine

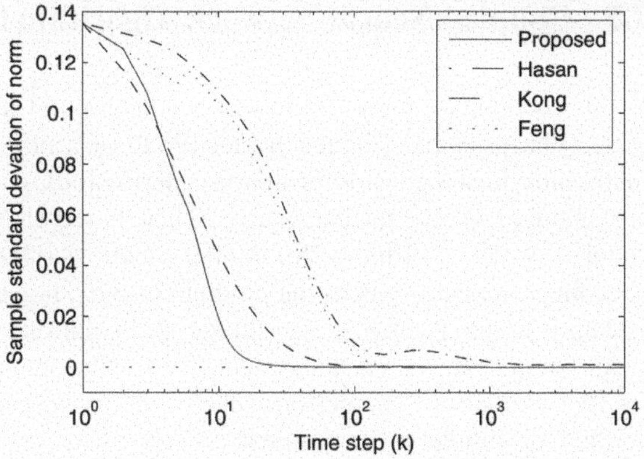

Fig. 7.14 Sample standard deviation of norm estimation

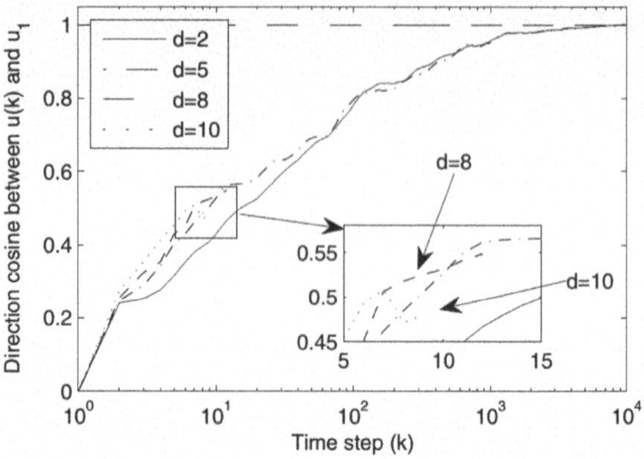

Fig. 7.15 Convergence process of direction cosine of our algorithm with different *d*

7.4.4.3 Conclusion

In this section, we propose an effective algorithm to track the PSS of a data matrix. In the proposed algorithm, the left and right PSS of a data matrix are tracked in a parallel way. Moreover, all principal singular values according to the PSS can also be estimated based on the extracted left or(and) right subspace. Thus, the proposed algorithm can be used to track the PSS and estimate all of the corresponding principal singular values. Moreover, experiment results show that the proposed algorithm converges faster and has better stability than similar algorithms. Based on the results of Experiments 1–2, we conclude that the proposed algorithm performs better than the other algorithms.

7.5 Coupled Neural Network Algorithm for PST Extraction

In this section, we propose a novel coupled neural network learning algorithm to extract the principal singular triplet (PST) of a non-square data matrix [25]. We firstly introduce a novel information criterion (NIC), in which the stationary points are singular triplet of the data matrix. Then, based on Newton's method, we obtain a coupled system of ordinary differential equations (ODEs) from the NIC. The ODEs have the same equilibria as the gradient of NIC, however, only the first PST of the system is stable (which is also the desired solution), and all others are (unstable) saddle points. Based on the system, we finally obtain a fast and stable algorithm for PST extraction. The performance of the proposed algorithm will be tested and verified via simulation experiments.

7.5.1 Novel Information Criterion and Its Coupled System

We know that gradient-based algorithms can be derived by maximizing the variance of the projected data or minimizing the reconstruction error based on an information criterion. Thus, it is required that the stationary points of the information criterion must be attractors. However, the gradient-based method is not suitable for the NIC since the first PST of the NIC is a saddle point. Different from the gradient method, Newton's method has the beneficial property that it turns even saddle points into attractors which guarantees the stability of the resulting learning rules [20]. In this case, the learning rule can be derived to an information criterion which is neither subject to minimization nor to maximization [1]. Moreover, Newton's method has higher convergence speed than gradient method. In this section, a coupled system is derived from an NIC based on Newton's method. The NIC is presented as follows:

$$p = \mathbf{u}^T \mathbf{A}\mathbf{v} - \frac{1}{2}\sigma \mathbf{u}^T \mathbf{u} - \frac{1}{2}\sigma \mathbf{v}^T \mathbf{v} + \sigma \tag{7.95}$$

The gradient of (7.95) is determined through

$$\nabla p = \left[\left(\frac{\partial p}{\partial \mathbf{u}} \right)^T, \left(\frac{\partial p}{\partial \mathbf{v}} \right)^T, \frac{\partial p}{\partial \sigma} \right]^T \tag{7.96}$$

which has the components

$$\frac{\partial p}{\partial \mathbf{u}} = A\mathbf{v} - \sigma \mathbf{u} \tag{7.97}$$

$$\frac{\partial p}{\partial \mathbf{v}} = A^T \mathbf{u} - \sigma \mathbf{v} \tag{7.98}$$

$$\frac{\partial p}{\partial \sigma} = -\frac{1}{2}\mathbf{u}^T \mathbf{u} - \frac{1}{2}\mathbf{v}^T \mathbf{v} + 1 \tag{7.99}$$

It is clear that the stationary points of (7.95) are also the singular triplets of $\boldsymbol{A}$, and we can also conclude that $\mathbf{u}^T \mathbf{A}\mathbf{v} = \sigma$ and $\mathbf{u}^T \mathbf{u} = \mathbf{v}^T \mathbf{v} = 1.\mathbf{u}^T \mathbf{u} = \mathbf{v}^T \mathbf{v} = 1$.

In Newton descent, the gradient is pre-multiplied by the inverse Hessian. The Hessian $\mathbf{H} = \nabla \nabla^T p$ of (7.95) is

$$\mathbf{H} = \begin{pmatrix} -\sigma \mathbf{I}_m & \mathbf{A} & -\mathbf{u} \\ \mathbf{A}^T & -\sigma \mathbf{I}_n & -\mathbf{v} \\ -\mathbf{u}^T & -\mathbf{v}^T & 0 \end{pmatrix}, \tag{7.100}$$

where $\boldsymbol{I}_{\mathrm{m}}$ and $\boldsymbol{I}_{\mathrm{n}}$ are identity matrices of dimension m and n, respectively. In the following, we determine an approximation for the inverse Hessian in the vicinity of the PST.

Newton's method requires an inversion on the Hessian (7.100) in the vicinity of the stationary point, here the PST. The inversion of the Hessian is simplified by a similarity transformation, using the orthogonal matrix

$$\mathbf{T} = \begin{pmatrix} \overline{\mathbf{U}} & \mathbf{0}_{mn} & \mathbf{0}_m \\ \mathbf{0}_{nm} & \overline{\mathbf{V}} & \mathbf{0}_n \\ \mathbf{0}_m^T & \mathbf{0}_n^T & 1 \end{pmatrix}, \tag{7.101}$$

Here, $\overline{\mathbf{U}} = [\overline{u}_1, \ldots, \overline{u}_m]$ and $\overline{\mathbf{V}} = [\overline{v}_1, \ldots, \overline{v}_n]$ are matrices which contain all left and right singular vectors in their columns, respectively. For the computation of the transformed matrix $\mathbf{H}^* = \mathbf{T}^T\mathbf{H}\mathbf{T}$, we exploit $\mathbf{A}^T\overline{\mathbf{U}} = \overline{\mathbf{V}}\overline{\mathbf{S}}^T$ and $\mathbf{A}\overline{\mathbf{V}} = \overline{\mathbf{U}}\overline{\mathbf{S}}$, where $\overline{\mathbf{S}}$ is a $n \times n$ matrix whose first N diagonal elements are the singular values arranged in decreasing order, and all remaining elements of $\overline{\mathbf{S}}$ are zero. Moreover, in the vicinity of the stationary point $(\overline{\mathbf{u}}_1, \overline{\mathbf{v}}_1, \overline{\sigma}_1)$, we can approximate $\overline{\mathbf{u}}^T\mathbf{u} \approx e_m$ and $\overline{\mathbf{v}}^T\mathbf{v} \approx e_n$, where e_m and e_n are unit vectors of the specified dimensions with a 1 as the first element. We obtain

$$\mathbf{H}_0^* = \begin{pmatrix} -\sigma\mathbf{I}_m & \overline{\mathbf{S}} & -\mathbf{e}_m \\ \overline{\mathbf{S}}^T & -\sigma\mathbf{I}_n & -\mathbf{e}_n \\ -\mathbf{e}_m^T & -\mathbf{e}_n^T & 0 \end{pmatrix}, \tag{7.102}$$

Next, by approximating $\sigma \approx \overline{\sigma}_1$ and by assuming $|\overline{\sigma}_1| \geq |\overline{\sigma}_j|, \forall j = 2, \ldots, M$, the expression $\overline{\mathbf{S}}$ can be approximated as $\overline{\mathbf{S}} = \sigma\sigma^{-1}\overline{\mathbf{S}} \approx \sigma\mathbf{e}_m\mathbf{e}_n^T$. In this case, (6.103) yields

$$\mathbf{H}^* = \begin{pmatrix} -\sigma\mathbf{I}_m & \sigma\mathbf{e}_m\mathbf{e}_n^T & -\mathbf{e}_m \\ \sigma\mathbf{e}_n\mathbf{e}_m^T & -\sigma\mathbf{I}_n & -\mathbf{e}_n \\ -\mathbf{e}_m^T & -\mathbf{e}_n^T & 0 \end{pmatrix}. \tag{7.103}$$

As introduced in [21], suppose that an invertible matrix of size $(j+1) \times (j+1)$ is of the form

$$\mathbf{R}_{j+1} = \begin{pmatrix} \mathbf{R}_j & \mathbf{r}_j \\ \mathbf{r}_j^T & \rho_j \end{pmatrix}, \tag{7.104}$$

where $\mathbf{r}_j$ is a vector of size j and ρ_j is a real number. If $\mathbf{R}_j$ is invertible and $\mathbf{R}_j^{-1}$ is known, then $\mathbf{R}_{j+1}^{-1}$ can be determined from

$$\mathbf{R}_{j+1}^{-1} = \begin{pmatrix} \mathbf{R}_j^{-1} & \mathbf{0}_j \\ \mathbf{0}_j^T & 0 \end{pmatrix} + \frac{1}{\beta_j}\begin{pmatrix} \mathbf{b}_j\mathbf{b}_j^T & \mathbf{b}_j \\ \mathbf{b}_j^T & 1 \end{pmatrix}, \tag{7.105}$$

where $\mathbf{b}_j = -\mathbf{R}_j^{-1}\mathbf{r}_j$ and $\beta_j = \rho_j + \mathbf{r}_j^T\mathbf{b}_j$. Here, it is obvious that $r_j = \left[\mathbf{e}_m^T, \mathbf{e}_n^T\right]^T$, $\rho = 0$ and

$$\mathbf{R}_j = \begin{pmatrix} -\sigma\mathbf{I}_m & \sigma\mathbf{e}_m\mathbf{e}_n^T \\ \sigma\mathbf{e}_n\mathbf{e}_m^T & -\sigma\mathbf{I}_n \end{pmatrix}. \quad (7.106)$$

Based on the method introduced in [22] and [23], we obtain

$$\mathbf{R}_j^{-1} = \begin{pmatrix} -\sigma\mathbf{I}_m & \sigma\mathbf{e}_m\mathbf{e}_n^T \\ \sigma\mathbf{e}_n\mathbf{e}_m^T & -\sigma\mathbf{I}_n \end{pmatrix}^{-1} = -\sigma^{-1}\begin{pmatrix} (\mathbf{I}_m - \mathbf{e}_m\mathbf{e}_m^T)^{-1} & \mathbf{e}_m\mathbf{e}_n^T(\mathbf{I}_n - \mathbf{e}_n\mathbf{e}_n^T)^{-1} \\ \mathbf{e}_n\mathbf{e}_m^T(\mathbf{I}_m - \mathbf{e}_m\mathbf{e}_m^T)^{-1} & (\mathbf{I}_n - \mathbf{e}_n\mathbf{e}_n^T)^{-1} \end{pmatrix}. \quad (7.107)$$

It is found that $\mathbf{I}_m - \mathbf{e}_m\mathbf{e}_m^T$ and $\mathbf{I}_n - \mathbf{e}_n\mathbf{e}_n^T$ are singular and thus we add a penalty term $\varepsilon \approx 1$ to (7.107), then it yields

$$\begin{aligned} \mathbf{R}_j^{-1} &\approx -\sigma^{-1}\begin{pmatrix} (\mathbf{I}_m - \varepsilon\mathbf{e}_m\mathbf{e}_m^T)^{-1} & \mathbf{e}_m\mathbf{e}_n^T(\mathbf{I}_n - \varepsilon\mathbf{e}_n\mathbf{e}_n^T)^{-1} \\ \mathbf{e}_n\mathbf{e}_m^T(\mathbf{I}_m - \varepsilon\mathbf{e}_m\mathbf{e}_m^T)^{-1} & (\mathbf{I}_n - \varepsilon\mathbf{e}_n\mathbf{e}_n^T)^{-1} \end{pmatrix} \\ &\approx -\sigma^{-1}\begin{pmatrix} \mathbf{I}_m + \eta\mathbf{e}_m\mathbf{e}_m^T & (1+\eta)\mathbf{e}_m\mathbf{e}_n^T \\ (1+\eta)\mathbf{e}_n\mathbf{e}_m^T & \mathbf{I}_n + \eta\mathbf{e}_n\mathbf{e}_n^T \end{pmatrix}, \end{aligned} \quad (7.108)$$

where $\eta = \varepsilon/(1-\varepsilon)$. Thus,

$$\mathbf{b}_j = -\mathbf{R}_j^{-1}\mathbf{r}_j = -2(1+\eta)\sigma^{-1}[\mathbf{e}_m^T, \mathbf{e}_n^T]^T, \quad (7.109)$$

$$\beta_j = \rho_j + \mathbf{r}_j^T\mathbf{b}_j = 4(1+\eta)\sigma^{-1}. \quad (7.110)$$

Substituting (7.108–7.110) into (7.105), it yields

$$(\mathbf{H})^{*-1} = \mathbf{R}_{j+1}^{-1} \approx \begin{pmatrix} -\sigma^{-1}(\mathbf{I}_m - \mathbf{e}_m\mathbf{e}_m^T) & \mathbf{0}_{mn} & -\frac{1}{2}\mathbf{e}_m \\ \mathbf{0}_{mn} & -\sigma^{-1}(\mathbf{I}_n - \mathbf{e}_n\mathbf{e}_n^T) & -\frac{1}{2}\mathbf{e}_n \\ -\frac{1}{2}\mathbf{e}_m^T & -\frac{1}{2}\mathbf{e}_n^T & \frac{\sigma}{4(1+\eta)} \end{pmatrix} \quad (7.111)$$

From the inverse transformation $\mathbf{H}^{-1} = \mathbf{T}\mathbf{H}^{*-1}\mathbf{T}^T$ and by approximating $\frac{\sigma}{4(1+\sigma)} = \frac{1}{4}\sigma(1-\varepsilon) \approx 0$, we get

$$\mathbf{H}^{-1} \approx \begin{pmatrix} -\sigma^{-1}(\mathbf{I}_m - \mathbf{u}\mathbf{u}_m^T) & \mathbf{0}_{mn} & -\frac{1}{2}\mathbf{u} \\ \mathbf{0}_{mn} & -\sigma^{-1}(\mathbf{I}_n - \mathbf{v}\mathbf{v}_n^T) & -\frac{1}{2}\mathbf{v} \\ -\frac{1}{2}\mathbf{u}^T & -\frac{1}{2}\mathbf{v}^T & 0 \end{pmatrix}, \quad (7.112)$$

where we approximated the unknown singular vectors by $\overline{\mathbf{u}}_1 \approx \mathbf{u}$ and $\overline{\mathbf{v}}_1 \approx \mathbf{v}$, and $\mathbf{0}_{mn}$ is a zero matrix of size $m \times n$, and $\mathbf{0}_{nm}$ denotes its transpose.

The Newton's method for SVD is defined as follows:

$$\begin{bmatrix} \dot{\mathbf{u}} \\ \dot{\mathbf{v}} \\ \dot{\sigma} \end{bmatrix} = -\mathbf{H}^{-1}\nabla p = - - \mathbf{H}^{-1} \begin{bmatrix} \frac{\partial p}{\partial \mathbf{u}} \\ \frac{\partial p}{\partial \mathbf{v}} \\ \frac{\partial p}{\partial \sigma} \end{bmatrix} \tag{7.113}$$

By applying the gradient of p in (7.97–7.98) and the inverse Hessian (7.112) to (7.113), we obtain a system of ODEs

$$\dot{\mathbf{u}} = \sigma^{-1}\mathbf{A}\mathbf{v} - \sigma^{-1}\mathbf{u}\mathbf{u}^T\mathbf{A}\mathbf{v} - \frac{1}{2}\mathbf{u} + \frac{3}{4}\mathbf{u}\mathbf{u}^T\mathbf{u} - \frac{1}{4}\mathbf{u}\mathbf{v}^T\mathbf{v}, \tag{7.114}$$

$$\dot{\mathbf{v}} = \sigma^{-1}\mathbf{A}^T\mathbf{u} - \sigma^{-1}\mathbf{v}\mathbf{u}^T\mathbf{A}\mathbf{v} - \frac{1}{2}\mathbf{v} + \frac{3}{4}\mathbf{v}\mathbf{v}^T\mathbf{v} - \frac{1}{4}\mathbf{v}\mathbf{u}^T\mathbf{u}, \tag{7.115}$$

$$\dot{\sigma} = \mathbf{u}^T\mathbf{A}\mathbf{v} - \frac{1}{2}\sigma(\mathbf{u}^T\mathbf{u} + \mathbf{v}^T\mathbf{v}). \tag{7.116}$$

It is straight forward to show that this system of ODEs has the same equilibria as the gradient of p. However, only the first PST is stable.

7.5.2 *Experiments and Conclusions*

7.5.2.1 Experiment

In this experiment, we conduct a simulation to test the ability of PST extraction of proposed algorithm, and also compare the performance with that of some other algorithms, i.e., the coupled algorithm [1], two non-coupled algorithms[14, 16]. Here, the vector data sequence is generated by $\mathbf{x}(k) = \mathbf{A} \cdot \mathbf{y}(k)$, where $\boldsymbol{A}$ is given by $\mathbf{A} = [\mathbf{u}_0, \ldots, \mathbf{u}_8] \cdot \mathrm{randn}(9, 9) \cdot [\mathbf{v}_0, \ldots, \mathbf{v}_8]^T$, where $\mathbf{u}_i$ and $\mathbf{v}_i (i = 0,\ldots,8)$ are the ith components of 11-dimensional and 9-dimensional orthogonal discrete cosine basis functions, respectively. In this simulation, we let $\mathbf{y}_t \in \Re^{9\times 1}$ be Gaussian, spatially and temporally white, and randomly generated. In order to measure the convergence speed and precision of learning algorithms, we compute the direction cosine between the state vector, i.e., $\boldsymbol{u}(k)$ and $\boldsymbol{v}(k)$, and the true principal singular vector, i.e., $\overline{\mathbf{u}}_1$ and $\overline{\mathbf{v}}_1$, at the k-th update:

$$DC(\mathbf{u}(k)) = \frac{\left|\mathbf{u}^T(k) \cdot \overline{\mathbf{u}}_1\right|}{\|\mathbf{u}(k)\| \cdot \|\overline{\mathbf{u}}_1\|}, \tag{7.117}$$

$$DC(\mathbf{v}(k)) = \frac{\left|\mathbf{v}^T(k) \cdot \overline{\mathbf{v}}_1\right|}{\|\mathbf{v}(k)\| \cdot \|\overline{\mathbf{v}}_1\|}. \tag{7.118}$$

Clearly, if direction cosine (7.117) and (7.118) converge to 1, then state vector $\boldsymbol{u}(k)$ and $\boldsymbol{v}(k)$ must approach the direction of the true left and right singular vectors, respectively. For coupled algorithms, we define the left and right singular error at the k-th update:

$$\varepsilon_L(k) = \left\| \sigma(k)^{-1} \mathbf{A}^T \mathbf{u}(k) - \mathbf{v}(k) \right\| \tag{7.119}$$

$$\varepsilon_R(k) = \left\| \sigma(k)^{-1} \mathbf{A} \mathbf{v}(k) - \mathbf{u}(k) \right\| \tag{7.120}$$

If these two singular errors converge to 0, then the singular value estimate $\sigma(k)$ must approach the true singular value as $k \to \infty$. In this experiment, the learning rate is chosen as $\gamma = 0.02$ for all rules. The initial value $\boldsymbol{u}(0)$ and $\boldsymbol{v}(0)$ are set to be orthogonal to $\overline{\mathbf{u}}_1$ and $\overline{\mathbf{v}}_1$. Experiment results are shown in Figs. 7.16, 7.17 and 7.18.

From Fig. 7.16a and b, it is observed that all algorithms can effectively extract both of the left and right principal singular vectors of a cross-correlation matrix

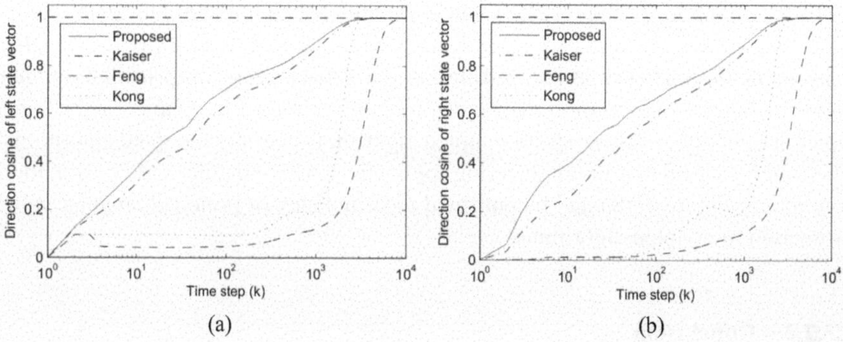

Fig. 7.16 Direction cosine of left and right singular vectors estimation **a** Direction cosine between $\boldsymbol{u}(k)$ and $\overline{\mathbf{u}}_1$ **b** Direction cosine between $\boldsymbol{v}(k)$ and $\overline{\mathbf{v}}_1$

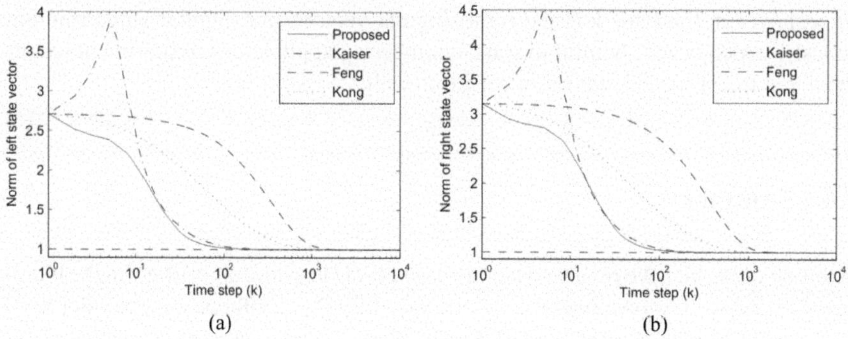

Fig. 7.17 Norm of left and right state vector **a** Norm of left state vector **b** Norm of right state vector

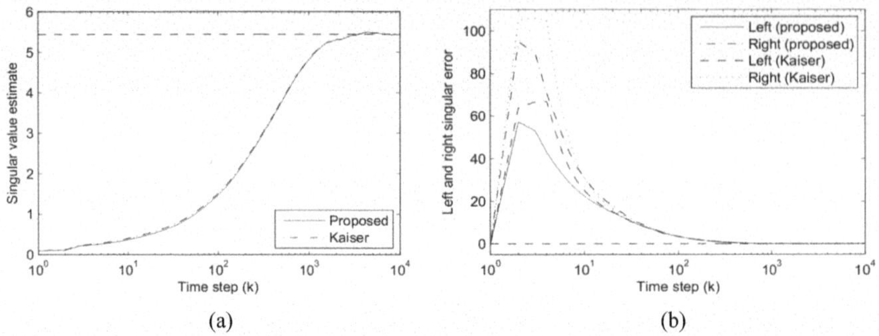

Fig. 7.18 Principal singular value estimation and their errors **a** Principal singular value estimation **b** Estimation errors

between two data streams. The coupled algorithms have higher convergence speed than Feng's algorithm. Compared with Kong's algorithm, the coupled algorithms have similar convergence speed in the whole process but higher convergence speed in the beginning steps. In Fig. 7.17a and b, we find that all left and right state vectors of all algorithms converge to a unit length, and coupled algorithms have higher convergence speed than non-coupled algorithms. What is more, the principal singular value of the cross-correlation matrix can also be estimated in coupled algorithms, which is actually an advantage of coupled algorithms over non-coupled algorithms. This is very helpful in some engineering applications when singular values estimation is required. Figure 7.18a and b confirmed the efficiency of principal singular value estimation of coupled algorithms.

7.5.2.2 Conclusion

In this section, a novel CSVD algorithm is presented by finding the stable stationary point of an NIC via Newton's method. The proposed algorithm can solve the speed-stability problem and thus perform much better than non-coupled algorithms. CSVD algorithms can track the left and right singular vectors and singular value simultaneously, which is very helpful in some engineering applications. Experiment results show that the proposed algorithm performs well.

7.6 Summary

In this chapter, we have reviewed neural-based SVD learning algorithms in the literature. Several well-known neural-network-based SVD algorithms, e.g., parallel SVD learning algorithms on double Stiefel manifol, cross-associative neural network for SVD, coupled SVD of a cross-covariance matrix, are reviewed and introduced. Then, three neural-based SVD algorithms proposed by us, including a neural network for SVD of non-squared data matrix, a fast and effective neural-based algorithm to perform SVD, and a coupled neural-based algorithm for PST extraction, are presented in detail. The stability and convergence of the proposed algorithms are theoretically analyzed. Simulation experiments results show that the effectiveness and superiority of the proposed algorithms. The proposed SVD algorithms can provide a fast and flexible method for many potential real TLS applications.

References

1. A. Kaiser, W. Schenck, R. Moller, Coupled singular value decomposition of a cross-covariance matrix. Int. J. Neural Syst. **20**(4), 293–318 (2010)
2. A. Cochocki, R. Unbehauen, *Neural Networks for Optimization and Signal Processing* (John Wiley, New York, 1993)
3. A.L. Yuille, D.M. Kammen, D.S. Cohen, Quadrature and the development of orientation selective cortical cells by Hebb rules. Biol. Cybern.Cybern. **61**(3), 183–194 (1989)
4. D.Z. Feng, Z. Bao, X.D. Zhang, A cross-associative neural network for SVD of non-squared data matrix in signal processing. IEEE Trans. Neural Netw.Netw. **12**(5), 1215–1221 (2001)
5. K.I. Diamantaras, S.Y. Kung, Cross-correlation neural network models. IEEE Trans. Signal Process. **42**(11), 3218–3223 (1994)
6. J.R. Bunch, C.P. Nielsen, Updating the singular value decomposition. Numerical Mathem. **31**(2), 111–129 (1978)
7. P. Comon, G.H. Golub, Tracking a few extreme singular values and vectors in signal processing. Proc. IEEE **78**(8), 1327–1343 (1990)
8. W. Ferzali, J. G. Proakis, Adaptive SVD algorithm for covariance matrix eigenstructure computation. in *Proceedings of the 1990 International Conference on Acoustics, Speech, and Signal Processing* (Albuquerque, NM, IEEE, 1990), pp. 2615–2618
9. M. Moonen, P. Van Dooren, J. Vandewalle, A singular value decomposition updating algorithm for subspace tracking. SIAM J. Matrix Anal. Appl. **13**(4), 015–1038 (1992)
10. U. Helmke, J.B. Moore, Singular-value decomposition via gradient and self-equivalent flows. Linear Algebra Appl. **169**, 223–248 (1992)
11. J.B. Moore, R.E. Mahony, U. Helmke, Numerical gradient algorithms for eigenvalue and singular value calculations. SIAM J. Matrix Anal. Appl. **15**(3), 881–902 (1994)
12. G. Hori, A general framework for SVD flows and joint SVD flows. in *Proceedings of the 2003 IEEE International Conference on Acoustics, Speech, and Signal Processing* (Hong Kong, IEEE, II-693–6, 2003)
13. D.Z. Feng, Z. Bao, W.X. Shi, Cross-correlation neural network models for the smallest singular component of general matrix. Signal Process. **64**(3), 333–346 (1998)
14. D.Z. Feng, X.D. Zhang, Z. Bao, A neural network learning for adaptively extracting cross-correlation features between two high-dimensional data streams. IEEE Trans. Neural Networks **15**(6), 1541–1554 (2004)
15. M. A. Hasan, A logarithmic cost function for principal singular component analysis. in *Proceedings of the 2008 IEEE International Conference on Acoustics, Speech, and Signal Processing* (Las Vegas, NV, IEEE, 2008), pp. 1933–1936
16. X.Y. Kong, H.G. Ma, Q.S. An, Q. Zhang, An effective neural learning algorithm for extracting cross-correlation feature between two highdimensional data streams. Neural. Process. Lett. **42**(2), 459–477 (2015)
17. L. Cheng, Z.G. Hou, M. Tan, A simplified neural network for linear matrix inequality problems. Neural. Process. Lett. **29**(3), 213–230 (2009)

18. L. Cheng, Z.G. Hou, Y. Lin, M. Tan, W.C. Zhang, F.X. Wu, Recurrent neural network for non-smooth convex optimization problems with application to the identification of genetic regulatory networks. IEEE Trans. Neural Netw.Netw. **22**(5), 714–726 (2011)
19. Q.S. Liu, T.W. Huang, J. Wang, One-layer continuous-and discretetime projection neural networks for solving variational inequalities and related optimization problems. IEEE Trans. Neural Netw. Learn. Syst. **25**(7), 1308–1318 (2014)
20. R. Moller, A. Konies, Coupled principal component analysis. IEEE Trans. Neural Netw.Netw. **15**(1), 214–222 (2004)
21. B. Noble, J.W. Daniel, *Applied Linear Algebra (Third edition)* (Prentice Hall, Englewood Cliffs, NJ, 1988)
22. H. Hotelling, Some new methods in matrix calculation. Ann. Math. Stat. **14**(1), 1–34 (1943)
23. H. Hotelling, Further points on matrix calculation and simultaneous equations. Ann. Math. Stat. **14**(4), 440–441 (1943)
24. X.W. Feng, X.Y. Kong, D.H. Xu, J.Q. Qin, A fast and effective principal singular subspace tracking algorithm. Neurocomputing **267**, 201–209 (2017)
25. X.W. Feng, X.Y. Kong, H.G. Ma, Coupled cross-correlation neural network algorithm for principal singular triplet extraction of a cross-covariance matrix. IEEE/CAA J. Automatica Sinica **3**(2), 149–156 (2016)
26. T.D. Sanger, Two iterative algorithms for computing the singular value decomposition from input/output samples. ed. by J. D. Cowan, G. Tesauro, J. Alspector, *Advances in Neural Information Processing Systems*, vol 6 (Morgan Kaufmann, 1994), pp. 144–151
27. S. Fiori, Singular value decomposition learning on double Stiefel manifolds. Int. J. Neural Syst. **13**(2), 1–16 (2003)
28. A. Weingessel, K. Hornik, SVD algorithms: APEX-like versus subspace methods. Neural Proc. Letter **5**, 177–184 (1997)
29. A. Weingessel, *An Analysis of Learning Algorithms in PCA and SVD Neural Networks*. PhD thesis, Technische University at Wien (1999)
30. U. Helmke, J.B. Moore, Singular-value decomposition via gradient and self- equivalent flows. Linear Algebra and its Appl. **169**, 223–248 (1992)
31. R.W. Brockett, Dynamical systems that sort lists, diagonalize matrices and solve linear programming problems. Linear Algebra Appl. **146**, 79–91 (1991)
32. B. Yang, Projection approximation subspace tracking. IEEE Trans. Signal Process. **43**(1), 95–107 (1995)
33. G.H. Golub, C.F. Van Loan, *Matrix Computations* (The Johns Hopkins University Press, Baltimore, MD, 1989)

Chapter 8
Neural-Based TLS Algorithms

8.1 Introduction

The TLS problem can be solved by using direct and iterative methods [1, 2]. The neural approaches can be considered iterative methods, and they have lower computational complexity compared with other iterative methods, which make them more suitable in real-time application [2]. There are three neural ways of solving TLS problem: (1) One is a neural network for the SVD [3–5], which finds the right singular vector associated with the smallest singular value of the argumented input data matrix of by minimizing a cost function with a gradient learning, and a subsequent normalization is needed. (2) Another is a linear neuron for the MCA [6–14], which finds the eigenvector associated with the smallest eigenvalue of the correlation matrix of augmented input data by minimizing the Rayleigh quotient with a gradient learning, and a subsequent normalization ($\boldsymbol{x}_{n+1} = -1$) is needed. (3) The third is linear neural network acting directly on the $\boldsymbol{x}_{n+1} = -1$ hyperplane (TLS NN) [17–24].

The existing TLS NNs are the hopfield-like neural network of Luo [15, 16], the principal limit of which is the fact that it is linked to the dimensions of the data matrix and cannot be used without structural changes for other TLS problems. Others are the linear neurons [17–24] which is correct enough for small gains. The linear neuron [17, 18] is associated with a constrained anti-Hebbian learning law, it is correct enough for small gains and, above all, for weight norms much smaller than 1. The linear neurons [19, 20] are associated with different learning laws to deal with OLS, TLS, and DLS problems. For each algorithm, an appropriate cost energy is devised, which contains the classical minimization plus regularization terms for ill-conditioned problems and nonlinear functions for the robustness of the method. The TLS EXIN neuron [21] is a new neural approach for linear parameter estimation, which permits the use of such powerful algorithms as SCG and BFGS. The linear neuron [22] is a modified gradient algorithm based on TLS EXIN, capable of resolving the total least squares (TLS)

X. Kong and D. Feng, *Efficient Online Learning Algorithms for Total Least Square Problems*, Engineering Applications of Computational Methods 21,
https://doi.org/10.1007/978-981-97-1765-1_8

problem present in this kind of estimation. In [23], both the TLS and mixed LS–TLS framework with gradient algorithms for unbiased equation-error adaptive IIR filtering is presented, and a mixed LS-TLS as a QR-decomposition-based algorithm for unbiased, equation-error adaptive IIR filtering is introduced [24].

Despite the large number of neural algorithms of solving the TLS problem proposed to date, there are few algorithms that possess the following desirable properties: (1) The weight vector norm of algorithm does not diverge nor exist the sudden divergence. (2) The algorithm is self-stabilizing such that the sign of the weight vector length change is independent of presented input vector. (3) The algorithm is good for larger learning factor or under higher noise environments. Considering the practical application as online system identification or fault diagnosis for higher noise environment, it is necessary to find an efficient neural approach with the above properties.

The objective of this chapter is to find more satisfactory self-stabilizing TLS neural approach, which is applied to the parameter estimation of an adaptive FIR filter for system identification in the presence of additive noise in both input and output data. This chapter is organized as follows: In Sect. 8.2, the review of neural-based TLS algorithms is given. In Sect. 8.3, a new self-stabilizing of TLS algorithm [25] is presented, and the self-stabilizing and stability analysis of the proposed algorithm are given. Finally, Sect. 8.4 is the conclusion of this chapter.

8.2 Review of Neural-Based TLS Algorithms

8.2.1 The Hopfield-Like Neural Network of Luo, Li, and He

This network [15, 16] is made up of $3(m + n) + 2$ neurons (m and n are the dimensions of the data matrix) grouped in a main network and in four subnetworks; the output of the main network gives the TLS solution, and the available data matrix and observation vector are taken directly as the interconnections and the bias current of the network. Thus, the principal limit of this network is the fact that it is linked to the dimensions of the data matrix and cannot be used without structural changes for other TLS problems. The authors demonstrate the stability of the network for *batch* operation: that is, working on all the equations together (as opposed to *sequential* or *online* operation, which updates at every equation presentation). The initial conditions cannot be null. The network is based on an analog circuit architecture which has continuous-time dynamics. The authors apply the network to the TLS linear prediction frequency estimation problem [2].

8.2.2 The Linear Neuron of Gao, Ahmad, and Swamy

This is a single linear neuron associated with a constrained anti-Hebbian learning law [17, 18], which follows from the linearization of $E_{\text{TLS}}(\boldsymbol{x})$, so it is correct enough for small gains and, above all, for weight norms much smaller than 1. The weight vector gives the TLS solution after the learning phase. It has been applied to adaptive FIR and IIR parameter estimation problems. In the first problem, after learning, the output of the neuron gives the error signal of the adaptive filter; this is a useful property because in a great many adaptive filter applications, the error signal has the same importance as the filter parameters and other signal quantities. From now on, this neuron will be termed TLS GAO [2].

The simplest neural model is a linear unit as shown in Fig. 8.1, where the neuron receives a set $n+1$ scalar-valued input $\xi_1(t), ..., \xi_n(t)$, and $\xi_{n+1}(t)$ through $n+1$ synaptic connection with coupling weight $\psi_1(t), ..., \psi_n(t)$, and $\psi_{n+1}(t)$, respectively. Write the input vector $\boldsymbol{\xi}(t) = [\xi_1(t), ..., \xi_n(t), \xi_{n+1}(t)]^T$, the weight vector $\boldsymbol{\psi}(t) = [\psi_1(t), ..., \psi_n(t), \psi_{n+1}(t)]^T$, the output signal $\eta(t)$ of the neuron is given by $\eta(t) = \boldsymbol{\psi}(t)^T\boldsymbol{\xi}(t)$. The constrained anti-Hebbian algorithm, TLS GAO, is given by

$$\boldsymbol{\psi}(t+1) = \boldsymbol{\psi}(t) - \mu\eta(t)[\boldsymbol{\xi}(t) + \boldsymbol{\psi}(t)\xi_{n+1}(t)]. \tag{8.1}$$

By substituting $\eta(t) = \boldsymbol{\psi}(t)^T\boldsymbol{\xi}(t)$ into (8.1), the TLS GAO algorithm becomes

$$\boldsymbol{\psi}(t+1) = \boldsymbol{\psi}(t) - \mu[\boldsymbol{\xi}(t)\boldsymbol{\xi}(t)^T\boldsymbol{\psi}(t) + \boldsymbol{\psi}(t)^T\boldsymbol{\xi}(t)\xi_{n+1}(t)\boldsymbol{\psi}(t)] \tag{8.2}$$

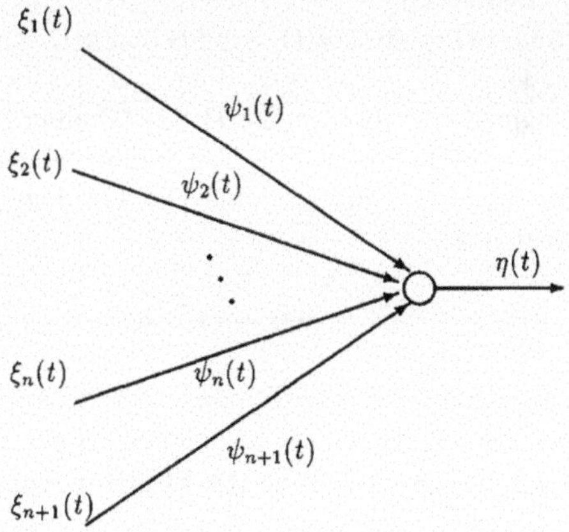

Fig. 8.1 Linear neuron unit

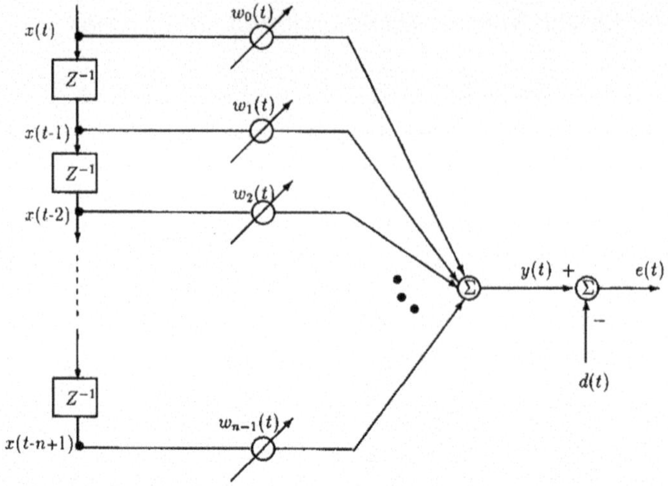

Fig. 8.2 Adaptive FIR filter

8.2.2.1 The Adaptive FIR Filtering Application of TLS GAO

An adaptive FIR filter is shown in Fig. 8.2, where $\mathbf{x}(t) = [x(t), x(t-1), ..., x(t-n+1)]^T$ is the input vector, $\mathbf{w}(t) = [w_0(t), w_1(t), ..., w_{n-1}(t)]^T$ is the weight vector, and $d(t)$ is the desired output of the filter. The actual output signal of the adaptive filter is given by $y(t) = \mathbf{w}^T(t)\mathbf{x}(t)$, and the error signal of the adaptive filter is given by $e(t) = y(t) - d(t)$.

Letting $\boldsymbol{\xi}(t) = [\mathbf{x}^T(t), d(t)]^T$,$[\psi_1(t), ..., \psi_n(t)]^T = \mathbf{w}(t)$ and $\psi_{n+1}(t) = -1$, then from $\eta(t) = \boldsymbol{\psi}(t)^T\boldsymbol{\xi}(t)$, it holds that $\eta(t) = \mathbf{w}^T(t)\mathbf{x}(t) - d(t) = e(t)$. Thus, the error signal $e(t)$ of the adaptive filter, which is the output signal $\eta(t)$ of the neuron, can be obtained directly from the TLS GAO algorithm. Substituting the established relation $\eta(t) = e(t)$ into the TLS GAO algorithm (8.1), and using the first n components of the vector $\boldsymbol{\xi}(t)$ and $\boldsymbol{\psi}(t)$, the TLS GAO algorithm for the adaptive FIR filter can be written as

$$\mathbf{w}(t+1) = \mathbf{w}(t) - \mu e(t)[\mathbf{x}(t) + d(t)\mathbf{w}(t)]. \tag{8.3}$$

8.2.2.2 The Adaptive IIR Filtering Application of TLS GAO

The output of an IIR adaptive filter with the equation-error formulation is given by

$$y(t) = \sum_{i=1}^{N-1} a_i(t)d(t-i) + \sum_{i=0}^{M-1} b_i(t)x(t-i), \tag{8.4}$$

where $\{x(t)\}$ and $\{d(t)\}$ are, respectively, the measured input sequence and output sequence from the unknown system. The error signal of the adaptive IIR filter is defined as $e(t) = y(t) - d(t)$.

Let $\varphi(t) = [d(t-1), ..., d(t-N+1), x(t), ..., x(t-M+1)]^T$, $\boldsymbol{\theta}(t) = [a_1(t), ..., a_{N-1}(t), b_0(t), ..., b_{M-1}(t)]^T$.

Thus, (8.4) can be written as $y(t) = \varphi(t)^T \boldsymbol{\theta}(t)$. Taking $[\varphi(t)^T, d(t)]^T$ as the input vector and $[\boldsymbol{\theta}(t)^T, -1]^T$ as the weight vector of the neuron, the output of the neuron can be obtained from $\eta(t) = \boldsymbol{\psi}(t)^T \boldsymbol{\xi}(t)$ as $\eta(t) = \varphi^T(t)\boldsymbol{\theta}(t) - d(t) = e(t)$. Thus, the TLS GAO algorithm for the adaptive IIR filter can be written as

$$\boldsymbol{\theta}(t+1) = \boldsymbol{\theta}(t) - \mu e(t)[\varphi(t) + d(t)\boldsymbol{\theta}(t)] \quad (8.5)$$

or the following algorithm as

$$\theta_i(t+1) = \theta_i(t) - \mu e(t)[\varphi_i(t) + d(t)\theta_i(t)],\ i = 1, ..., N + M - 1. \quad (8.6)$$

The algorithm provides a very simple adaptive scheme of finding the TLS solution for the IIR filter parameters.

8.2.3 The Linear Neurons of Cichocki and Unbehauen

This is a linear neurons with different learning laws to deal with OLS, TLS, and DLS problems [19, 20]. For each algorithm, an appropriate cost energy to be minimized is devised, which contains the classical minimization [e.g., the minimization of E_{TLS} $(\boldsymbol{x})$ for the TLS] plus regularization terms for ill-conditioned problems and nonlinear functions for the robustness of the method. Then the system of differential equations describing the gradient flow of this energy is implemented in an analog network for the continuous-time learning law and in a digital network for the discrete-time learning law. The analog network consists of analog integrators, summers, and multipliers; the network is driven by independent source signals (zero-mean random, high frequency, uncorrelated *iid*) multiplied by the incoming data a_{ij}, $b_i (i = 1,2,...,m; j = 1,2,...,n)$ from $[\boldsymbol{A}; \boldsymbol{b}]$. The artificial neuron, with an on-chip adaptive learning algorithm, allows both complete processing of the input information simultaneously and the sequential strategy. In the digital neuron, the difference equations of the gradient flow are implemented by CMOS switched-capacitor (SC) technology. The neurons for the TLS do not work on the exact gradient flow, but on its linearization. It gives the same learning law of TLS GAO for a particular choice of the independent signals.

8.2.4 The TLS EXIN Neural Network and Its Modified Version

The TLS EXIN neuron is presented in [21] as a neural approach for linear parameter estimation. In particular, it is able to implement the TLS technique, thus coping with noisy observation vector and data matrix. It permits the use of such powerful algorithms as SCG and BFGS. All these properties are proved mathematically and not only numerically as is often the case for other neuronal approaches. Furthermore, the use of a sigmoidal activation function would improve the TLS EXIN robustness to outliers.

The MCA EXIN learning rule [2, 21] can be used to solves the TLS problem:

$$\psi(t+1) = \psi(t) - \frac{\alpha(t)y(t)}{\psi^T(t)\psi(t)}[\xi(t) - \frac{y(t)\psi(t)}{\psi^T(t)\psi(t)}]. \tag{8.7}$$

Replacing $\psi(t)$ with $[\mathbf{w}^T(t); -1]^T$, $\xi(t)$ with $[\mathbf{a}_i^T; b_i]^T$, i being the index of the row of $[A; \mathrm{b}]$ input to the neuron at the time t, and taking only the first n components yields

$$\mathbf{w}(t+1) = \mathbf{w}(t) - \alpha(t)\frac{(\mathbf{w}^T(t)\mathbf{a}_i - b_i)}{1+\mathbf{w}^T(t)\mathbf{w}(t)}\mathbf{a}_i + \alpha(t)\left[\frac{(\mathbf{w}^T(t)\mathbf{a}_i - b_i)}{(1+\mathbf{w}^T(t)\mathbf{w}(t))}\right]^2 \mathbf{w}(t). \tag{8.8}$$

This is the TLS EXIN learning law. The TLS EXIN neuron is a linear unit with n inputs (vector $\boldsymbol{a}_i$), n weights (vector $\boldsymbol{w}$), one output (scalar $y_i = \boldsymbol{w}^T\boldsymbol{a}_i$) and one training error [scalar $\boldsymbol{w}^T\boldsymbol{a}_i - b_i$]. With this typology, the training is considered as supervised, b_i being the target.

About the input of the TLS EXIN neuron, two cases have to be considered:

1. The equations (rows of $[\boldsymbol{a}; B]$) are not known in advance and therefore are fed directly at their presentation.
2. The equations (rows of $[\boldsymbol{A}; b]$) are known in advance and are fed either in a random sequence or in a cyclical order. In both cases, the stochastic approximation is valid, and the learning process is called online or sequential.

8.2.5 The LS-TLS Neural Algorithm of Bruce and Williamson

Steepest descent gradient algorithms for unbiased equation-error adaptive IIR filtering are analyzed collectively for both the TLS and mixed LS-TLS framework in [23]. These algorithms have a monic normalization that allows for a direct filtering implementation. The algorithms can converge to the desired filter coefficient vector. A region of initialization is given under which the algorithm converges to the desired solution, which is derived using interlacing relationships between the eigenvalues of the data correlation matrices and their respective Schur complements. In addition, a mixed LS-TLS is introduced in [24] as a QR-decomposition-based algorithm for unbiased, equation-error adaptive IIR filtering. The algorithm is based on casting adaptive IIR filtering into a mixed LS-TLS framework. This formulation is shown to be equivalent to the minimization of the mean square equation error subject to a unit norm constraint on the denominator parameter vector. An efficient implementation of the mixed LS-TLS solution is achieved through the use of back substitution and inverse iteration. Unbiasedness of the system parameter estimates is established for the mixed LS-TLS solution in the case of uncorrelated output noise, and the algorithm is shown to converge to this solution.

Consider the system identification problem, as shown in Fig. 8.3. Here is to estimate the unknown causal stable system using an adaptive IIR filter in an equation-error setting. The input drives the unknown system with impulse response $\{h(k)\}$ to generate the output $v(k)$. We note that neither is observable. Instead, only their noisy versions $x(k)$ and $y(k)$ are available. Let the z-transform of $h(k)$ be denoted as $H(z)$. Further, let $H(z)$ be given by

$$H(z) = \frac{b_0 + b_1 z^{-1} + \cdots + b_M z^{-M}}{1 - a_1 z^{-1} - \cdots a_N z^{-N}} = \frac{B(z)}{1 - A(z)}. \tag{8.9}$$

Here is to seek to determine parameter weights at time $k\{c_i(k), d_j(k)\}, 0 \le i \le M, 1 \le j \le N$ such that $C(z,k) = c_0(k) + c_1(k)z^{-1} + \cdots + c_M(k)z^{-M}$, $D(z,k) = d_1(k)z^{-1} + \cdots + d_N(k)z^{-N}$ model $B(z)$ and $A(z)$, respectively.

Scenario 1: Input and output noise of the same variance

Adaptive algorithm:

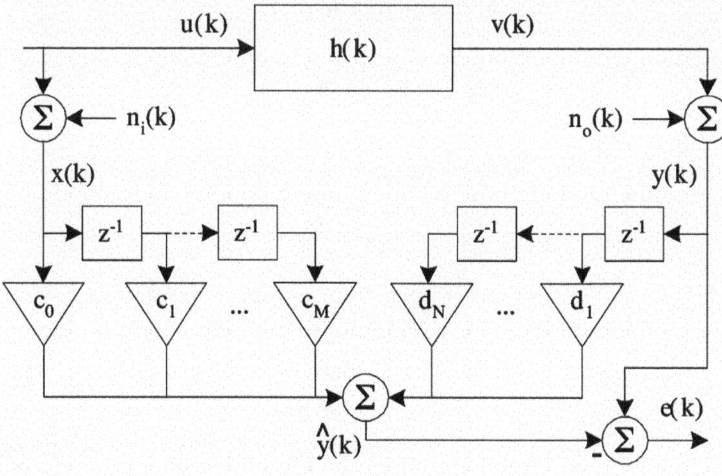

Fig. 8.3 Equation-error filtering diagram

$$\boldsymbol{\theta}(k+1) = \boldsymbol{\theta}(k) + \mu \frac{(1+\boldsymbol{\theta}^T(k)\boldsymbol{\theta}(k))e(k)\boldsymbol{\Phi}(k) + e^2(k)\boldsymbol{\theta}(k)}{(1+\boldsymbol{\theta}^T(k)\boldsymbol{\theta}(k))^2} \tag{8.10}$$

Scenario 2: Output noise only
Adaptive algorithm:

$$\boldsymbol{\theta}(k+1) = \boldsymbol{\theta}(k) + \mu \frac{(1+\boldsymbol{\theta}_1^T(k)\boldsymbol{\theta}_1(k))e(k)\boldsymbol{\Phi}(k) + e^2(k)\begin{bmatrix}\boldsymbol{\theta}_1(k)\\ 0\end{bmatrix}}{(1+\boldsymbol{\theta}_1^T(k)\boldsymbol{\theta}_1(k))^2} \tag{8.11}$$

Scenario 3: Disparate input and output noises

In certain filtering situations, significant input and output noise is present, but these noises are of non-equal variance. Consider the case where both $n_i(k)$ and $n_o(k)$ are white with variances of $\sigma_{n_i}^2$ and $\sigma_{n_o}^2$, respectively. Let $\alpha = \sigma_{n_o}^2/\sigma_{n_i}^2$. Set up the scaled composite regressor vectors such that the input and output noises have equal variance. With the aforementioned modifications, Scenario 3 becomes Scenario 1.

Scenario 4: Input noise only

This scenario is the dual of Scenario 2 in that the input portion of the data is noisy. About the details, See [23].

In Eqs. (8.10) and (8.11),

$$\boldsymbol{\theta}_1(k) = \left[d_1(k) \cdots d_N(k)\right],$$

$$\boldsymbol{\theta}_2(k) = \left[c_0(k) \cdots c_M(k)\right],$$

$$\boldsymbol{\theta}(k) = [\boldsymbol{\theta}_1^T(k), \boldsymbol{\theta}_1^T(k)]^T,$$

$$\boldsymbol{\Phi}(k) = [\mathbf{Y}^T(k), \mathbf{X}^T(k)]^T,$$

$$\mathbf{X}(k) = \left[x(k) \cdots x(k-M)\right]^T,$$

$$\mathbf{Y}(k) = \left[y(k-1) \cdots y(k-N)\right]^T$$

$$e(k) = y(k) - \boldsymbol{\theta}^T(k)\boldsymbol{\Phi}(k).$$

8.3 A Self-stabilizing Neural Algorithm for TLS Filtering

8.3.1 *The TLS Linear Neuron with a Self-stabilizing Algorithm*

In this section, a neural approach for solving the TLS problem is presented. It is based on a linear neuron with a self-stabilizing neural algorithm, capable of resolving the TLS problem present in the parameter estimation of an adaptive FIR filter for system identification, where noisy errors affect not only the observation vector but also the data matrix. The learning rule is analyzed mathematically, and the condition to guarantee the stability of algorithm is obtained. The computer simulations are given to illustrate that the neural approach is self-stabilizing and considerably outperforms the existing TLS methods when a larger learning factor is used or the signal–noise-rate (SNR) is lower.

Consider the adaptive filter when the input and output observation data are both corrupted by noise, and denote the input and output observation series at k as $\{[\tilde{x}(k), \tilde{d}(k)]|k = 1, 2, \cdots, N\}, \tilde{x}(k) = x(k) + n_i(k), \tilde{d}(k) = d(k) + n_o(k)$, where $n_i(k)$ and $n_o(k)$ are additive noises. Denote the filter weight vector as $\mathbf{w}(k) = [w_1(k), w_2(k), ..., w_n(k)]^T$ and the input vector at k as $\tilde{\mathbf{x}}(k) = [\tilde{x}(k), \tilde{x}(k-1), ..., \tilde{x}(k-n+1)]^T$. Then the filter output at k can be written as $y(k) = \tilde{\mathbf{x}}^T(k)\mathbf{w}(k)$, and the output error is $\varepsilon(k) = y(k) - \tilde{d}(k)$. Denote the augmented input vector as $\mathbf{z}(k) = [\tilde{\mathbf{x}}^T(k), \tilde{d}(k)]^T$ and the augmented weighted vector as $\overline{\mathbf{w}}(k) = [\mathbf{w}^T(k), -1]^T$. Then the output error can be written as $\varepsilon(k) = \mathbf{z}^T(k)\overline{\mathbf{w}}(k)$.

Define the TLS cost function as follows

$$E_{\mathrm{TLS}} = \frac{\varepsilon^2}{\overline{\mathbf{w}}^T(k)\overline{\mathbf{w}}(k)} = \frac{(\mathbf{z}^T(k)\overline{\mathbf{w}}(k))^2}{\overline{\mathbf{w}}^T(k)\overline{\mathbf{w}}(k)} = \frac{(\mathbf{z}^T(k)\overline{\mathbf{w}}(k))^2}{1 + \mathbf{w}^T(k)\mathbf{w}(k)} \tag{8.12}$$

which is the Rayleigh quotient of the augmented weighted vector $\overline{\mathbf{w}}(k)$ constrained to the TLS hyperplane (i.e.,$w_{n+1}(k) = -1$).

An adaptive algorithm can then be obtained from (8.12) as:

$$\overline{\mathbf{w}}(k+1) = \overline{\mathbf{w}}(k) - \frac{\alpha(k)\varepsilon(k)}{\overline{\mathbf{w}}^T(k)\overline{\mathbf{w}}(k)}(\mathbf{z}(k) - \frac{\varepsilon(k)\overline{\mathbf{w}}(k)}{\overline{\mathbf{w}}^T(k)\overline{\mathbf{w}}(k)}), \tag{8.13}$$

where $\alpha(k)$ is the learning rate, which controls the stability and rate of convergence of the algorithm. Equation (8.13) is the MCA EXIN algorithm in [2, 9]. Its TLS EXIN algorithm is given by

$$\begin{aligned} w(k+1) = w(k) - \alpha(k)\frac{[w^T(k)\tilde{\mathbf{x}}(k) - \tilde{d}(k)]}{1 + w^T(k)w(k)}\tilde{\mathbf{x}}(k) \\ + \alpha(k)\left[\frac{w^T(k)\tilde{\mathbf{x}}(k) - \tilde{d}(k)}{1 + w^T(k)w(k)}\right]^2 w(k), \end{aligned} \tag{8.14}$$

In [2, 9, 22], Eq. (8.13) is analyzed in detail and it is concluded that the algorithm is the best TLS neuron in terms of stability (no finite time divergence), speed, and accuracy. In [22], a modified gradient algorithm was proposed, and the algorithm has the best accuracy and the best convergence performance when a larger learning factor is used or SNR is lower in linear system.

However, from the analysis [22], the weight norm of the above two neurons can be obtained as follows:

$$\|\overline{\mathbf{w}}(k+1)\|_2^2 = \begin{cases} \|\overline{\mathbf{w}}(k)\|_2^2 + \frac{\alpha^2(k)}{4\|\overline{\mathbf{w}}(k)\|_2^2}\|\mathbf{z}(k)\|_2^4 \sin^2 2\theta_{\mathbf{z}\overline{\mathbf{w}}} \ \ for \text{ EXIN algorithm} \\ \|\overline{\mathbf{w}}(k)\|_2^2 + \frac{2\alpha(k)\varepsilon^2(k)}{\|\overline{\mathbf{w}}(k)\|_2^4}(\|\overline{\mathbf{w}}(k)\|_2^2 - 1) + O(\alpha^2(k)) \ \ for \ [22] \ a\lg orithm. \end{cases} \tag{8.15}$$

From (8.15), we can see that the squared modulus of the augmented weight vector in EXIN algorithm always increases with an oscillatory behavior, and the squared modulus of the augmented weight vector in [22] is divergent when the initial value $\|\mathbf{W}(k)\|_2$ is larger than one. In the TLS problem, the initial value of the squared modulus is larger than or equal to one. Then $\|\overline{\mathbf{w}}(k)\|_2 > 1$ is true.

In order to avoid the possible divergence and preserve the good performance as much as possible, based on the EXIN algorithm and by adding a restriction on norm of weight, a modified algorithm, called as a self-stabilizing EXIN algorithm, is proposed in [25] as follows:

$$\overline{\mathbf{w}}(k+1) = \overline{\mathbf{w}}(k) - \frac{\alpha(k)}{\|\overline{\mathbf{w}}(k)\|^2}[\varepsilon(k)\mathbf{z}(k) - \frac{(\varepsilon^2(k) + 1 - \|\overline{\mathbf{w}}(k)\|^2)}{\|\overline{\mathbf{w}}(k)\|^2}\overline{\mathbf{w}}(k)]. \tag{8.16}$$

From (8.16), we can obtain a self-stabilizing TLS algorithm as follows

$$\begin{aligned} w(k+1) = w(k) - \alpha(k)\frac{[w^T(k)\tilde{x}(k) - \tilde{d}(k)]\tilde{x}(k)}{1 + w^T(k)w(k)} \\ + \alpha(k)\frac{[(w^T(k)\tilde{x}(k) - \tilde{d}(k))^2 + w^T(k)w(k)]}{(1 + w^T(k)w(k))^2}w(k). \end{aligned} \tag{8.17}$$

The presented TLS neuron is a linear unit with n inputs, n weights, one output and one training error, and the presented algorithm is a modified gradient algorithm, which can be used for the NN weights estimation where the input and output observation data are both corrupted by noise.

For convenience of derivation and analysis, the self-stabilizing and stability analysis of the proposed algorithm (8.17) is completed by analyzing the algorithm (8.16).

8.3.2 Self-stabilizing and Stability

8.3.2.1 The Self-stabilizing Analysis

From (8.16), the temporal derivative $D(k)$ of the squared weight vector norm is:

$$\begin{aligned} D(k) &= \frac{1}{2}\frac{d\bar{w}^T(t)\bar{w}(t)}{\mathrm{d}t} = \bar{w}^T(k)\Delta\bar{w}(k) = \bar{w}^T(k)(\bar{w}(k+1) - \bar{w}(k)) \\ &= -[\varepsilon^2(k) - (\varepsilon^2(k) + 1 - \|\bar{w}(t)\|^2)]/(\|\bar{w}(t)\|^2) \\ &= \frac{1}{\|\bar{w}(t)\|^2}[1 - \|\bar{w}(t)\|^2] = \begin{cases} > 0 \ \|\bar{w}(t)\| < 1 \\ < 0 \ \|\bar{w}(t)\| > 1 \\ = 0 \ \|\bar{w}(t)\| = 1. \end{cases} \end{aligned} \tag{8.18}$$

From (8.18), it is clear that the sign of the temporal derivative $D(k)$ is independent of $\varepsilon(k)$, but depends only on the sign of $(1 - \|\overline{\mathbf{w}}(k)\|^2)$. This means that Algorithm (8.16) has self-stabilizing property. Thus, the norm of weight vector may stably approach the equilibrium point 1, as $k \to \infty$. This also renders our algorithm to have satisfactory one-tending property.

8.3.2.2 The Convergence Analysis

According to the stochastic approximation theory [26, 27], it can be shown that if some conditions are satisfied [9], Eq. (8.16) can be effectively represented by Eq. (8.19), i.e., their asymptotic paths are close with a large probability, and eventually, the solution of Eq. (8.16) tends with probability one to the uniformly asymptotically stable solution of the ODE.

$$\frac{d\overline{\mathbf{w}}(t)}{dt} = -\frac{1}{\|\overline{\mathbf{w}}(t)\|^2}[\mathbf{R}\overline{\mathbf{w}}(t) - \frac{(\overline{\mathbf{w}}^T(t)\mathbf{R}\overline{\mathbf{w}}(t) + 1 - \|\overline{\mathbf{w}}(t)\|^2)}{\|\overline{\mathbf{w}}(t)\|^2}\overline{\mathbf{w}}(t)]. \tag{8.19}$$

From a computational point of view, the most important conditions are the following:

(1) $\mathbf{z}(t)$ is zero mean, stationary, and bounded with probability one.
(2) $\alpha(t)$ is a decreasing sequence of positive scalars.
(3) $\sum_t \alpha(t) = \infty$;
(4) $\sum_t \alpha^p(t) < \infty$ *for some* p; .
(5) $\sum_{t\to\infty} \sup[(1/\alpha(t)) - (1/\alpha(t-1))] < \infty$.

The asymptotic property of (8.19) can be ensured by the following theorem.

Theorem 6.1 *Let* $\mathbf{R}$ *be a positive semi-definite matrix,* λ_{n+1} *and* $\mathbf{v}_{n+1}$ *be respectively its smallest eigenvalue and associated normalized eigenvector with nonzero last* *component. If the initial weight vector* $\overline{\mathbf{w}}(0)$ *satisfies* $\overline{\mathbf{w}}^T(0)\mathbf{v}_{n+1} \neq 0$ *and* λ_{n+1} *is single, then* $\lim_{t\to\infty} \overline{\mathbf{w}}(t) = \pm\mathbf{v}_{n+1}$ *holds.*

The proof of Theorem 6.1 is similar to that of Theorem 1 in [22]. For the details, refer to [22]. Only the difference is given below.

From a series of consequence, we have

$$\lim_{t\to\infty} \overline{\mathbf{w}}(t) = \lim_{t\to\infty}\left[\sum_{i=1}^{n+1} f_i(t)\mathbf{v}_i\right] = f_{n+1}(t)\mathbf{v}_{n+1}. \tag{8.20}$$

Equation (8.20) shows that $\overline{\mathbf{w}}(t)$ tends to the direction of the eigenvector associated with the smallest eigenvalue of the autocorrelation matrix of the augmented input data. From (8.20), it follows that

$$\lim_{t\to\infty} \|\overline{\mathbf{w}}(t)\| = \lim_{t\to\infty} \|f_{n+1}(t)\mathbf{v}_{n+1}\| = \lim_{t\to\infty} \|f_{n+1}(t)\|. \tag{8.21}$$

Furthermore, from (8.21), it holds that

$$\begin{aligned}
\frac{\mathrm{d}\bar{\boldsymbol{w}}^T(t)\bar{\boldsymbol{w}}(t)}{\mathrm{d}t} &= -\frac{2}{\|\bar{\boldsymbol{w}}(t)\|^4}[\|\bar{\boldsymbol{w}}(t)\|^2\bar{\boldsymbol{w}}^T(t)\boldsymbol{R}\bar{\boldsymbol{w}}(t) - (\bar{\boldsymbol{w}}^T(t)\boldsymbol{R}\bar{\boldsymbol{w}}(t) \\
&\quad + 1 - \|\bar{\boldsymbol{w}}(t)\|^2)\|\bar{\boldsymbol{w}}(t)\|^2] \\
&= -\frac{2}{\|\bar{\boldsymbol{w}}(t)\|^2}[\bar{\boldsymbol{w}}^T(t)\boldsymbol{R}\bar{\boldsymbol{w}}(t) - (\bar{\boldsymbol{w}}^T(t)\boldsymbol{R}\bar{\boldsymbol{w}}(t) + 1 - \|\bar{\boldsymbol{w}}(t)\|^2)] \\
&= 2[\frac{1}{\|\bar{\boldsymbol{w}}(t)\|^2} - 1] \\
&= \begin{cases} > 0 \ \|\bar{\boldsymbol{w}}(t)\| < 1 \\ < 0 \ \|\bar{\boldsymbol{w}}(t)\| > 1 \\ = 0 \ \|\bar{\boldsymbol{w}}(t)\| = 1. \end{cases}
\end{aligned} \tag{8.22}$$

Because the initial value of the squared modulus is larger than or equal to one in the TLS problem, and from (8.22), we have $\lim_{t\to\infty} \|\overline{\mathbf{w}}(t)\| = 1$. Thus we have $\lim_{t\to\infty} f_{n+1}(t) = \pm 1$, which gives $\lim_{t\to\infty} \overline{\mathbf{w}}(t) = \pm\mathbf{v}_{n+1}$.

This completes the proof.

8.3.2.3 The Divergence Analysis

In [2, 9], Cirrincione et al. found a sudden divergence for Luo algorithm. Sudden divergence is very adverse for practical applications. Does our proposed algorithm have a sudden divergence? In this section, we will study algorithm (8.16) in detail.

In [9], Corollary 19 gave simple convergence behavior for the solutions of a gradient flow as follows. The solutions of a gradient flow have a particularly simple convergence behavior, i.e., there are no periodic solutions, strange attractors or any chaotic behaviors. Based on the above corollary, dynamic behaviors of the weight vector for Luo, OJAn and MCA EXIN are described, and MCA divergences for Luo, OJAn, MCA EXIN, OJA, and OJA+ learning laws are proved. Following the above analysis method, the divergence analysis of the proposed algorithm is performed as follows.

Averaging the behavior of the weight vector after the first critical point has been reached ($t \geq t_0$), it follows that [9]:

$$\overline{\mathbf{w}}(t) = \|\overline{\mathbf{w}}(t)\|\mathbf{v}_{n+1}, \quad \forall t \geq t_0 \tag{8.23}$$

where $\mathbf{v}_{n+1}$ is the unit eigenvector associated with the smallest eigenvalue of $\boldsymbol{R}$.

From (8.22), we can obtain that $d\|\overline{\mathbf{w}}(t)\|^2/dt = 2(1/\|\overline{\mathbf{w}}(t)\|^2 - 1)$. Assuming that the MC direction has been approached, the equation can be approximated as

$$\frac{\mathrm{d}p}{\mathrm{d}t} = 2(\frac{1}{p} - 1), \tag{8.24}$$

where $p = \|\overline{\mathbf{w}}(t)\|^2$. Define the time instant at which the MC direction is approached as t_0 and the corresponding value of the squared weight modulus as p_0. The solution of (8.24) is given by

$$\begin{cases} p + \ln|p - 1| = p_0 + \ln|p_0 - 1| - 2(t - t_0) \text{ if } p_0 \neq 1 \\ p = p_0 \text{ if } p_0 = 1. \end{cases} \tag{8.25}$$

Figure 8.4a shows these results for different values of p_0. Figure 8.4b shows the divergence of the squared weight modulus for a few algorithms [9].

From the experimental results, it can be seen that the norm of the weight vector of proposed algorithm tends to one and the sudden divergence does not happen in a finite time. However, the norms of the weight vector in OJA, OJAn, OJA+, LUO, EXIN are divergent.

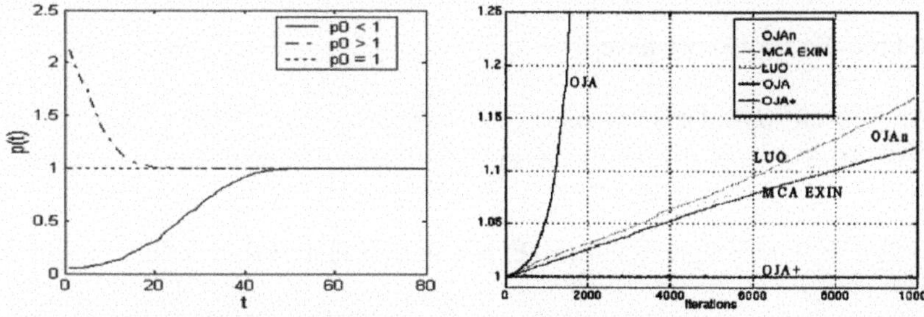

(a) The algorithm (8.16) (b) the EXIN, OJAn, LUO,OJA, OJA+ algorithms

Fig. 8.4 Asymptotic behavior of the ODE for different values of the initial conditions

8.3.2.4 The Temporal Behavior Analysis via the Stochastic Discrete-Time Method

The above analysis is based on the stochastic approximation theory [26, 27]. The obtained results are then approximations on some conditions. The purpose of this section is to analyze the temporal behavior of our TLS neurons and the relation between the dynamic stability and learning rate, by using mainly the stochastic discrete-time (SDT) system following the approach in [2, 9].

Define

$$r'(k) = \frac{\left|\overline{\mathbf{w}}^T(k+1)\mathbf{z}(k)\right|^2}{\|\overline{\mathbf{w}}(k+1)\|^2},\ r(k) = \frac{\left|\overline{\mathbf{w}}^T(k)\mathbf{z}(k)\right|^2}{\|\overline{\mathbf{w}}(k)\|^2}, \tag{8.26}$$

$$\rho(\alpha) = \frac{r'}{r} \geq 0,\ p = \|\overline{\mathbf{w}}(k)\|^2,\ u = \varepsilon^2(k). \tag{8.27}$$

The two scalars r' and r represent, respectively, the squared perpendicular distance from the input $z(k)$ to the data fitting hyperplane whose normal is given by the weight and passes through the origin, after and before the weight increment. Recalling the definition of MC, we should have $r' \leq r$. If this inequality is not valid, it means that the learning law increases the estimation error due to disturbances caused by noisy data. When this disturbance is too large, it will make $\overline{\mathbf{w}}(t)$ deviate drastically from the normal learning, which may result in divergence or fluctuations (implying an increased learning time).

Next, we will prove a theorem which provides the stability condition for algorithm (8.16).

Theorem 6.2 *Under definitions* (8.26, 8.27)*, if.*

$$0 < \alpha(k) < \frac{2p^2}{p\|\mathbf{z}(k)\|^2 - 2(u + 1 - p)} \ \Lambda\ p\|\mathbf{z}(k)\|^2 - 2(u + 1 - p) > 0\ ,$$

then $r' \leq r$, which implies algorithm (8.16) *is stable.*

Proof From Eq. (8.16), we have.

$$\begin{aligned}
\bar{w}^T(k+1)z(k) &= \varepsilon(k) \\
&\quad - \frac{\alpha(k)}{\|\bar{w}(k)\|^4}[\|\bar{w}(k)\|^2\varepsilon(k)\|z(k)\|^2 \\
&\quad - (\varepsilon^2(k) + 1 - \|\bar{w}(k)\|^2)\varepsilon(k)] \\
&= \varepsilon(k)(1 - \frac{\alpha(k)}{\|\bar{w}(k)\|^4}[\|\bar{w}(k)\|^2\|z(k)\|^2 \\
&\quad - (\varepsilon^2(k) + 1 - \|\bar{w}(k)\|^2)]).
\end{aligned} \tag{8.28}$$

From (8.16), it holds that

$$\begin{aligned}\|\bar{w}(k+1)\|^2 &= \bar{w}^T(k+1)\bar{w}(k+1)\\ &= \|\bar{w}(k)\|^2 - \frac{2\alpha(k)}{\|\bar{w}(k)\|^4}(\|\bar{w}(k)\|^2\varepsilon^2(k)\\ &\quad - (\varepsilon^2(k)+1-\|\bar{w}(k)\|^2)\|\bar{w}(k)\|^2)\\ &\quad + \frac{\alpha^2(k)}{\|\bar{w}(k)\|^8}(\|\bar{w}(k)\|^4\varepsilon^2(k)\|z(k)\|^2\\ &\quad - 2\|\bar{w}(k)\|^2\varepsilon^2(k)(\varepsilon^2(k)\\ &\quad + 1 - \|\bar{w}(k)\|^2) + (\varepsilon^2(k)+1-\|\bar{w}(k)\|^2)^2\|\bar{w}(k)\|^2). \end{aligned} \tag{8.29}$$

Therefore,

$$\begin{aligned}\rho(\alpha) = \frac{r'(k)}{r(k)} &= \frac{(\overline{\mathbf{w}}^T(k+1)\mathbf{z}(k))^2}{\|\overline{\mathbf{w}}(k+1)\|^2}\frac{\|\overline{\mathbf{w}}(k)\|^2}{(\varepsilon(k))^2}\\ &= \frac{(1-\frac{\alpha(k)}{\|\overline{\mathbf{w}}(k)\|^4}[\|\overline{\mathbf{w}}(k)\|^2\|\mathbf{z}(k)\|^2 - (\varepsilon^2(k)+1-\|\overline{\mathbf{w}}(k)\|^2)])^2}{1-\frac{2\alpha(k)}{\|\overline{\mathbf{w}}(k)\|^2}(1-\frac{1}{\|\overline{\mathbf{w}}(k)\|^2}) + \frac{\alpha^2(k)}{\|\overline{\mathbf{w}}(k)\|^2}E}\\ &= \frac{p^2(1-\alpha(k)q)^2}{p^2 - 2\alpha(k)(p-1) + \alpha^2(k)pE}, \end{aligned} \tag{8.30}$$

where $q = (1/p^2)(\|\mathbf{z}(k)\|^2 p - (u+1-p))$ and $E = (1/p^3)(up\|\mathbf{z}(k)\|^2 - 2u(u+1-p) + (u+1-p)^2)$.

Then, $\rho(\alpha) < 1$(dynamic stability)if and only if

$$\begin{aligned}p^2(1-\alpha(k)q)^2 &< p^2 - 2\alpha(k)(p-1) + \alpha^2(k)\frac{1}{p^2}(up\|z(k)\|^2\\ &\quad - 2u(u+1-p) + (u+1-p)^2). \end{aligned} \tag{8.31}$$

Notice that $u/p = \|\mathbf{z}(k)\|^2\cos^2\theta_{\mathbf{z}\overline{\mathbf{w}}}$, where $\theta_{\mathbf{z}\overline{\mathbf{w}}}$ is the angle between the augmented input vector and the augmented weight vector.

From (8.31), it holds that

$$\alpha(k)\|\mathbf{z}(k)\|^2 p\sin^2\theta_{\mathbf{z}\overline{\mathbf{w}}}(\alpha(k)(2u+2-2p-p\|\mathbf{z}(k)\|^2) + 2p^2) > 0. \tag{8.32}$$

The dynamic stability condition is then

$$0 < \alpha(k) < \frac{2p^2}{p\|\mathbf{z}(k)\|^2 - 2(u+1-p)} \wedge p\|\mathbf{z}(k)\|^2 - 2(u+1-p) > 0. \tag{8.33}$$

The second condition implies the absence of the negative instability. It can be rewritten as

$$\cos^2 \theta_{\mathbf{z}\overline{\mathbf{w}}} \leq \frac{1}{2} + \frac{1}{\|\mathbf{z}(k)\|^2} \frac{p-1}{p}. \tag{8.34}$$

In reality, the second condition is included in the first one. For the case $0 < \alpha_b \leq \gamma < 1$, it holds

$$\begin{aligned} \cos^2 \theta_{\mathbf{z}\overline{\mathbf{w}}} &\leq \frac{1}{2} + \frac{1}{\|\mathbf{z}(k)\|^2} \frac{p-1}{p} - \frac{p}{\gamma \|\mathbf{z}(k)\|^2} \\ &\leq \frac{1}{2} + \frac{1}{\|\mathbf{z}(k)\|^2} - \frac{2}{\|\mathbf{z}(k)\|^2 \sqrt{\gamma}} = \Upsilon, \end{aligned} \tag{8.35}$$

which is more restrictive than (8.34). Figure 8.5 shows this condition, where

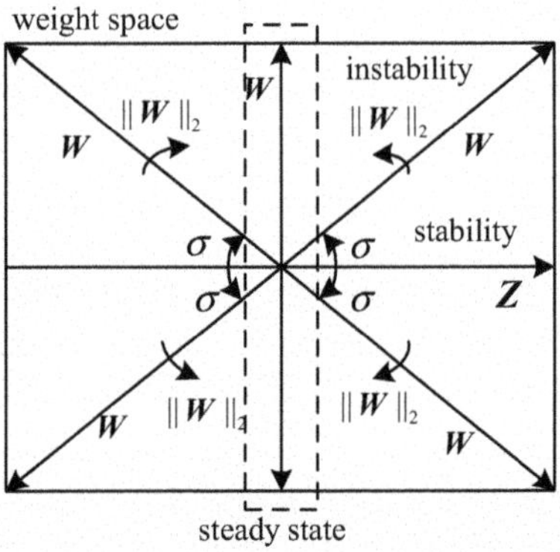

Fig. 8.5 Stability subspace of the weight space with regard to the input vector (two-dimensional space) of proposed algorithm

$\sigma = \arccos \sqrt{\Upsilon}$. From (8.35), we can see that the decrease of γ increases the stability, and the angle $\theta_{\mathbf{z}\overline{\mathbf{w}}}$ is not relevant to the change of weight vector modulus. From Fig. 8.5, it is apparent that in the transient (in general low $\theta_{\mathbf{z}\overline{\mathbf{w}}}$), there are less fluctuations and this is beneficial to the stability. From (8.35), it can be concluded that the angle $\theta_{\mathbf{z}\overline{\mathbf{w}}}$ between the weight vector $\overline{\mathbf{w}}(k)$ and the input vector $\mathbf{z}(k)$ is equal to or smaller than 45° when the proposed algorithm is convergent.

This completes the proof.

8.3.3 *Simulation of Algorithm*

In this section, several simulation experiments are conducted to illustrate the convergence and stability of the proposed algorithm in a TLS filtering. Since the OJAm algorithm [12], and TLS EXIN algorithm [21] have better performance than other TLS algorithms, we compare performance of the proposed TLS algorithm (4.58) with these algorithms.

All the learning curves below are obtained by averaging over 30 independent experiments. Assume that the SNR of the input signal is equal to the rate of the output signal. The additive noise is independent zero-mean white noise. The input signal is zero-mean white Gaussian noise with variance 1. In the simulation, the above algorithms are used for a linear system identification problem. The linear system is given by $\boldsymbol{h} = [-0.3, -0.9, 0.8, -0.7, 0.6, 0.2, -0.5, 1.0, -0.7, 0.9, -0.4]^T$, and its normalized one is $\bar{\mathbf{h}} = \mathbf{h}/\|\mathbf{h}\|$. Their convergence is compared under different noise environment and by using different learning factors. Define the learning error of the weight vector as: $Error(k) = 10\log(\|\bar{\mathbf{h}} - \hat{\mathbf{h}}(k)\|^2)$, where $\hat{\mathbf{h}}$ is the estimated weight vector. Figures 6.6–6.8 show respectively the learning curves for SNR = 20, SNR = 10 dB and SNR = 5 dB with different learning factors in the first simulation.

These learning curves indicate that, for FIR filters with larger number of coefficients, algorithm (8.17) has excellent convergence and stable accuracy, and the norm of weight of the algorithm converges to one. Compared with TLS EXIN and OJAm

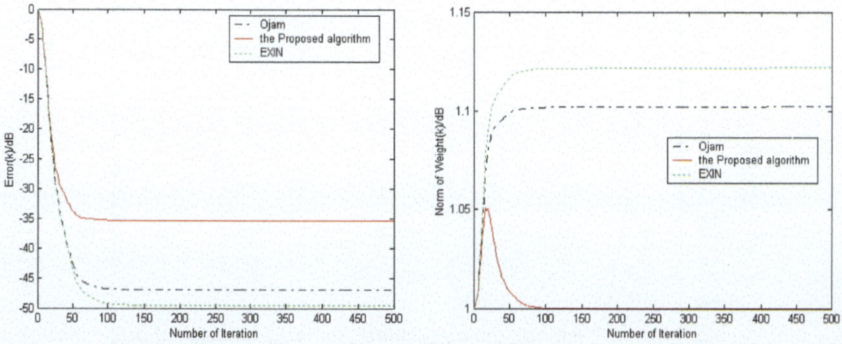

(a) (b)

Fig. 8.6 Learning curve at $\mu = 0.05$, 20 dB **a** error of the weight vector **b** norm of the weight vector

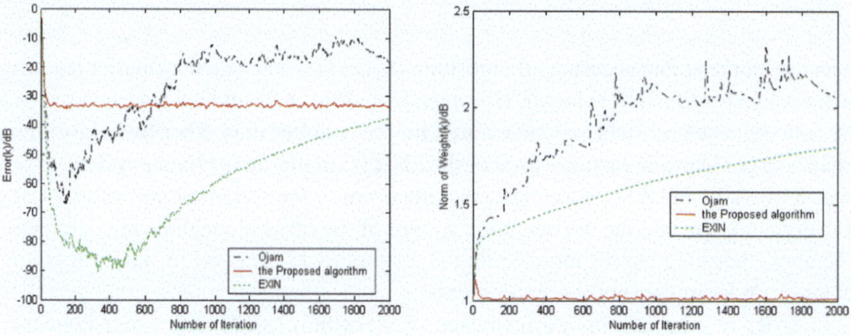

(a) (b)

Fig. 8.7 Learning curve at $\mu = 0.1$, 10 dB **a** error of the weight vector **b** norm of the weight vector

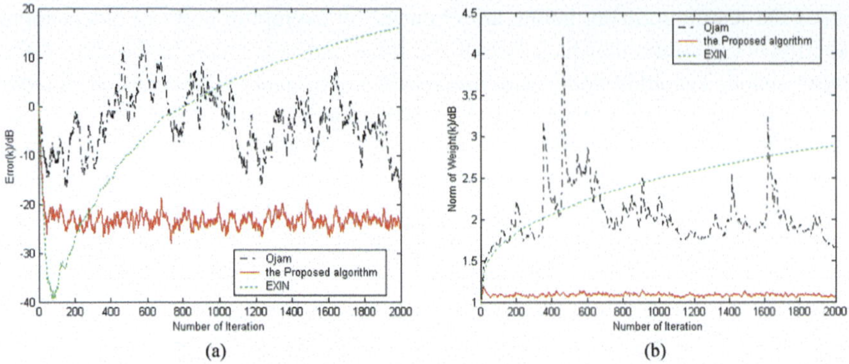

Fig. 8.8 Learning curve at $\mu = 0.1$, 5 dB **a** error of the weight vector **b** norm of the weight vector

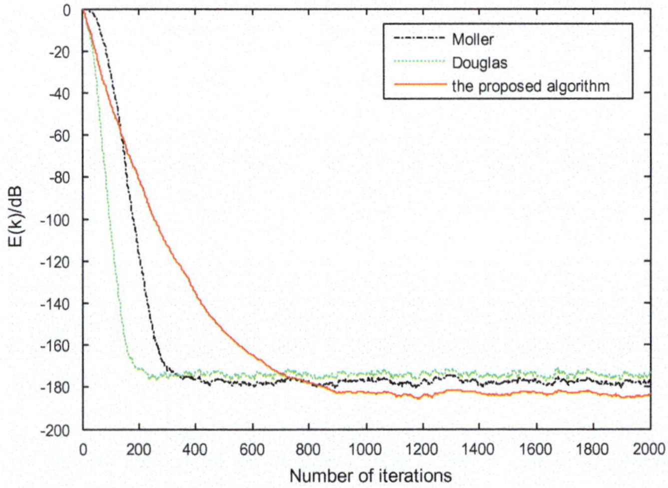

Fig. 8.9 Error of the weight vector ($\mu = 0.01$, 20 dB)

algorithms, the stable accuracy of algorithm (8.17) is worse when a smaller learning factor is used or the SNR is larger. However, from Figs. 8.7a and 8.8a, we can see that the convergence and stable accuracy are obviously better than other two algorithms when a larger learning factor is used or the SNR is smaller in the linear system. From Figs. 8.6b, 8.7b and 8.8b, we can see that the norm of the weight of algorithm (4.58) converges to one, and the norms of the weight of the other algorithms are divergent. The above learning curves indicate that algorithm (8.17) is good for larger learning factor or under higher noise environments.

In order to compare the performance of algorithm (8.17) with other congener self-stabilizing algorithms [11, 13], another simulation experiment is also performed. The performance comparison between algorithm (8.17) and the algorithms [11, 13] is shown in Figs. 8.9–8.12, and only the learning error of the weight vector is shown in these figures. It should be noted that normalization operation of the weight vector is performed in Figs. 8.9–8.12, and other conditions are the same as in the above simulations.

Figure 8.9 shows that algorithm (8.17) has slower convergence speed than Moller and Douglas algorithms when a smaller learning factor is used or the SNR is larger. From Figs. 8.10–8.12, we can see that algorithm (8.17) has the best convergence and stable accuracy compared with Moller and Douglas algorithms when a larger learning factor is used or the SNR is lower in the linear system. The learning curves in Figs. 8.10–8.12 also indicate that algorithm (8.17) is good for larger learning factor or under higher noise environments compared with other congener self-stabilizing algorithms.

A linear neuron with a self-stabilizing TLS algorithm is presented for the parameter estimation of adaptive FIR filters for system identification where noisy errors affect not only the observation vector but also the data matrix. Compared with other TLS algorithms, the proposed neural approach is self-stabilizing and considerably outperforms the existing TLS methods when a larger learning factor is used or the signal–noise-rate (SNR) is lower.

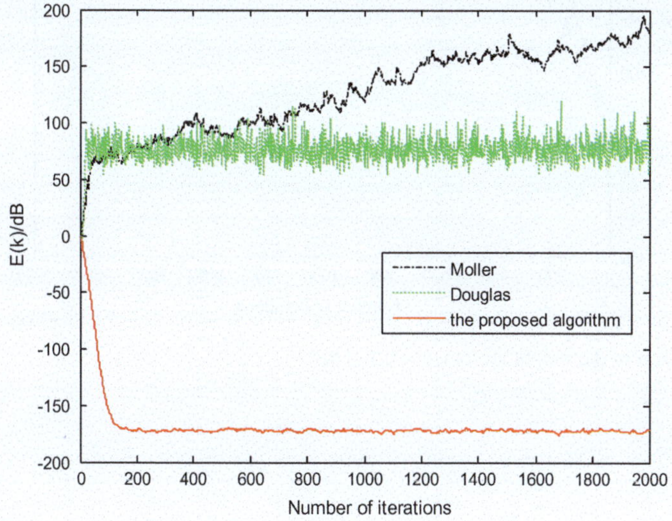

Fig. 8.10 Error of the weight vector ($\mu = 0.1$, 20 dB)

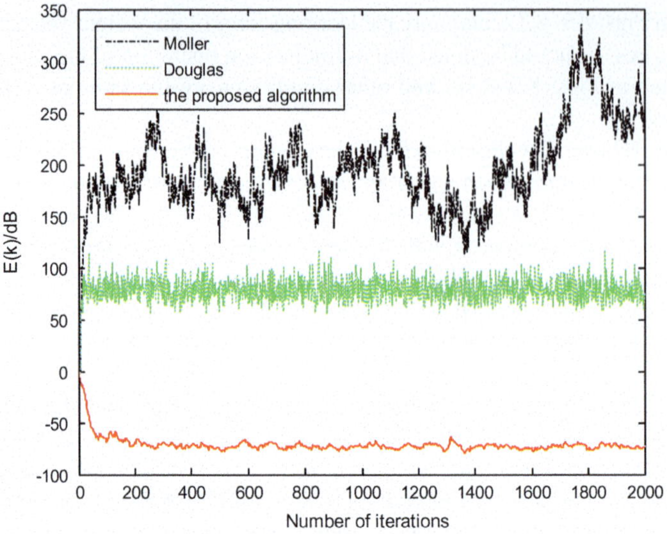

Fig. 8.11 Error of the weight vector ($\mu = 0.2$, 10 dB)

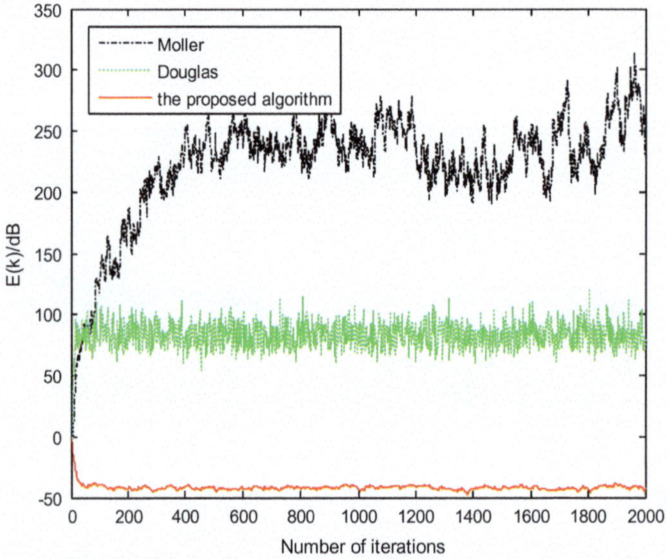

Fig. 8.12 Error of the weight vector ($\mu = 0.1$, 5 dB)

8.4 Conclusion

In this chapter, several neural network-based TLS algorithms, e.g., the Hopfield-like neural network of Luo, the linear neuron of Gao, the linear neurons of Cichocki, the TLS EXIN neural network, the LS-TLS neural algorithm of Bruce, etc., have been reviewed. A linear neuron with self-stabilizing TLS algorithm for FIR filtering is presented, and their performances, e.g., convergence, self-stabilizing, etc. are mathematically proved and numerically verified.

References

1. S. Van Huffel, J. Vandewalle, *The Total Least Squares Problems: Computational Aspects and Analysis* (Frontiers in Applied Mathematics. SIAM, Philadelphia, 1991)
2. G. Cirrincione, M. Cirrincione, Linear system identification using the TLS EXIN neuron. Neurocomputing **28**, 53–74 (1999)
3. A.L. Yuille, D.M. Kammen, D.S. Cohen, Quadrature and the development of orientation selective cortical cells by Hebb rules. Biol. Cybern. **61**(3), 183–194 (1989)
4. D.Z. Feng, Z. Bao, X.D. Zhang, A cross-associative neural network for SVD of non-squared data matrix in signal processing. IEEE Trans. Neural Netw. **12**(5), 1215–1221 (2001)
5. K.I. Diamantaras, S.Y. Kung, Cross-correlation neural network models. IEEE Trans. Signal Process. **42**(11), 3218–3223 (1994)
6. L. Xu, E. Oja, C. Suen, Modified Hebbian learning for curve and surface fitting. Neural Netw. **5**(3), 441–457 (1992)
7. E. Oja, Principal components, minor components and linear neural networks. Neural Netw. **5**(6), 927–935 (1992)
8. F. Luo, R. Unbehauen, A. Cichocki, A minor component analysis algorithm. Neural Netw. **10**(2), 291–297 (1997)
9. G. Cirrincione, M. Cirrincione, The MCA EXIN neuron for the minor component analysis. IEEE Trans. Neural Netw. **13**(1), 160–187 (2002)
10. D. Peng, Y. Zhang, Y. Xiang, H. Zhang, A globally convergent MC algorithm with an adaptive learning rate. IEEE Trans. Neural Netw. Learn. Syst. **23**(359–365), 2012 (2012)
11. S.C. Douglas, S.Y. Kung, S. Amari, A self-stabilized minor subspace rule. IEEE Signal Process. Lett. **5**(12), 328–330 (1998)
12. D.Z. Feng, W.X. Zheng, Y. Jia, Neural network Learning algorithms for tracking minor subspace in high-dimensional data stream. IEEE Trans. Neural Netw. **16**(3), 513–521 (2005)
13. R. Möller, A self-stabilizing learning rule for minor component analysis. Int. J. Neural Syst. **14**(1), 1–8 (2004)
14. T. Chen, S. Amari, Unified stabilization approach to principal and minor components extraction algorithms. Neural Netw. **14**(10), 1377–1387 (2001)
15. F. Luo, Y. Li, C. He, Neural network approach to the TLS linear prediction frequency estimation problem. Neurocomputing **11**, 31–42 (1996)
16. F. Luo, R. Unbehauen, *Applied neural networks for signal processing* (Cambridge University Press, Cambridge, 1997)
17. K.Q. Gao, M.O. Ahmad, M.N. Swamy, Learning algorithm for total least-squares adaptive signal processing. Electron. Lett. **28**(4), 430–432 (1992)
18. K.Q. Gao, M.O. Ahmad, M.N. Swamy, A constrained anti-Hebbian learning algorithm for total leastsquares estimation with applications to adaptive FIR and IIR filtering. IEEE Trans. Circ. Syst. II: Analog Digit. Sign. Process. **41**(11), 718–729 (1994)

19. A. Cichocki, R. Unbehauen, *Neural Networks for Optimization and Signal Processing* (Wiley, NewYork, 1993)
20. A. Cichocki, R. Unbehauen, Simplified neural networks for solving linear least squares and total least squares problems in real time. IEEE Trans. Neural Netw. **5**(6), 910–923 (1994)
21. G. Cirrincione, M. Cirrincione, Linear system identification using the TLS EXIN neuron. Neurocomputing **28**(1), 53–74 (1999)
22. X.Y. Kong, C.Z. Han, R.X. Wei, Modified gradient algorithm for total least squares filtering. Neurocomputing **70**, 568–576 (2006)
23. B.E. Dunne, G.A. Williamson, Analysis of gradient algorithms for TLS-based adaptive IIR filter. IEEE Trans. Signal Process. **52**(12), 3345–3356 (2004)
24. B.E. Dunne, G.A. Williamson, QR-based TLS and mixed LS-TLS algorithms with applications to adaptive IIR filtering. IEEE Trans. Signal Process. **51**(2), 386–394 (2003)
25. X.Y. Kong, C.H. Hu, C.Z. Han, A self-stabilizing neural algorithm for total least square filtering. Neural Process. Lett. **30**, 257–271 (2009). https://doi.org/10.1007/s11063-009-9122-4
26. H. Kushner, D. Clark, *Stochastic approximation methods for constrained and unconstrained systems* (Springer, New York, 1978)
27. L. Ljung, Analysis of recursive stochastic algorithms. IEEE Trans. Autom. Control **22**, 551–575 (1977)

Chapter 9
TLS Algorithm Under Non-Gaussian Noises

9.1 Introduction

The TLS method is one of the prominent solutions which have been proposed to solve the parameter estimation problem of EIV system [1–3]. Unlike the LS method that merely minimizes the sum of squared output error, the TLS method views the estimation problem as a fitting problem with noise existing in both input and output data and its goal is to minimize the sum of squared "total" perturbation/ error required to fit the input to the output [4].

Both the LS method and the TLS method are based on minimizing the squared (total) errors, which rely on the second-order statistics of the (total) errors and achieve optimality when the probability distribution of the (total) errors is Gaussian [4]. However, in real-world applications, non-Gaussian noise is ubiquitous, e.g., artificial noise in electronic devices, atmospheric noise and lighting spikes [5, 6]. Consequently, the probability distribution of errors becomes non-Gaussian. Thus in order to improve the performance of the parameter estimation in practical applications with non-Gaussian noise, it is necessary to take higher-order statistics of error data into account.

The error entropy is an efficient information-theoretic measure which comprises all the higher-order statistics of the error [7]. A numerical of experiments have demonstrated the superiority of the error entropy criterion (EEC) for signal processing under non-Gaussian noise [8–11]. Accordingly, the EEC-based minimum error entropy (MEE) method has shown its ability to achieve more accurate estimates than mean square error (MSE)-based LS method. Nevertheless, like the LS method, the MEE is designed for data models that only consider noise-free input data [8]. So it is foreseeable that the MEE would not achieve optimal estimates for parameter estimation of EIV system [4].

In order to further improve the estimation performance, by additionally considering the existence of noise in input data, a total error entropy criterion (TEEC) and a corresponding method named minimum total error entropy (MTEE) are proposed

X. Kong and D. Feng, *Efficient Online Learning Algorithms for Total Least Square Problems*, Engineering Applications of Computational Methods 21,
https://doi.org/10.1007/978-981-97-1765-1_9

in [4]. In [4], the system parameters by minimizing the quadratic Renyi's entropy of the "total" error was estimated. A gradient-based iterative solutions to optimize the cost function of MTEE is derived, and the MTEE method for both batch mode and online mode is developed. For the online-mode MTEE, to make the algorithm computationally efficient, the gradient is stochastically approximated. Beyond these, theoretical analysis of the local stability is performed for the batch-mode MTEE. The algorithm is proven to be locally stable around the true value of the parameters under some specific conditions.

The objective of this chapter is to introduce and propose satisfactory NG-TLS method, which is applied to the parameter estimation of an adaptive filters for system identification in presence of non-Gaussian noise in both input and output data. In this chapter, on the basis of the method in [4], several individual scenarios such as input and output contaminated with non-Gaussian noises (NG-TLS), partial input and output contaminated with non-Gaussian noises (NG-LS-TLS), input non-Gaussian noise only, output non-Gaussian noise only (NG-LS), and disparate input and output non-Gaussian noises (D-NG-TLS) are studied, the corresponding algorithms are provided, and simulation experiments and analyses are conducted.

This chapter is organized as follows: In Sect. 9.2, the review of the LS/TLS method under non-Gaussian noises is given. The Sect. 9.3 is devoted to the corresponding algorithms of several individual scenarios such as input and output contaminated with non-Gaussian noises (NG-TLS), partial input and output contaminated with non-Gaussian noises (NG-LS-TLS), input non-Gaussian noise only, output non-Gaussian noise only (NG-LS), and disparate input and output non-Gaussian noises (D-NG-TLS). In Sect. 9.4, some simulations experiments and analyses are conducted. Finally, Sect. 9.5 is the summary of this chapter.

9.2 Review of the LS/TLS Method Under Non-Gaussian Noises

9.2.1 The LS Method Under Non-Gaussian Noise

In this chapter, decorated letter or the notation $\{.\}$ is used to denote a set. Superscript $(\hat{\cdot})$ denote an estimator, $(\cdot)^T$ denote transposition, and $(\cdot)^o$ stands for the real value of parameter. The linear system is described by the equation:

$$y_i = \boldsymbol{x}_i^T \boldsymbol{w}^0, \tag{9.1}$$

where $\boldsymbol{w}^0 \in \Re^{D\times 1}$ is the unknown parameter vector of the system, and the input $\boldsymbol{x}_i \in \Re^{D\times 1}$ and output $y_i \in \Re$ are the realization of a kind of zero-mean random process $\{\boldsymbol{X}_{\text{i}}, Y_{\text{i}}\}$ at time instant i. Here, the input is assumed to be stationary and ergodic.

In the context of linear filtering (or parameter estimation), plenty of work has been conducted under the assumption that the measurement/observation for the output signal $\tilde{y}_i$ is noisy while the measurement for the input signal $\tilde{\mathbf{x}}_i$ is noiseless [4].

With the above assumption, the measurement model can be written as below (it is called the errors-in-output model).

$$\begin{cases} \breve{y}_i = y_i + v_i \\ \breve{\boldsymbol{x}}_i = \boldsymbol{x}_i, \end{cases} \tag{9.2}$$

where $v_i \in \Re$ is the zero-mean output measurement noise of variance σ_{out}^2. The noise is supposed to be stationary and ergodic in time and independent of the input signal $\tilde{\mathbf{x}}_i$.

Assuming that the above model has N measurements $\{(\breve{\mathbf{x}}_i^T, \breve{y}_i), i = 1, 2, ..., N\}$, various algorithms are designed to estimate the parameter vector of the system $\mathbf{w}^o$ based on these data. The MSE criterion-based LS method is a well-known solution to this estimation problem. It focuses on minimizing the sum of squared output error, as follows:

$$\min_{w} \sum_{i=1}^{i=N} \left(e_i^{\text{out}}\right)^2, \tag{9.3}$$

where $e_i^{\text{out}} = \breve{y}_i - \breve{\mathbf{x}}_i^T \mathbf{w}$. This method can yield the best linear unbiased estimate for the above errors-in-output model. It is very efficient under the assumption that the distribution of measurement noise is Gaussian.

However, when the noise v_i is non-Gaussian, which is a common situation in real world, the LS method may not be very efficient since it only considers the second-order statistics of error in estimation [7, 8, 12]. In such a case, several studies have shown that using the error entropy criterion (EEC) in designing cost functions is more beneficial to the estimation than using the MSE criterion [8, 9]. The EEC criterion takes the whole distribution of error into account while the MSE criterion is only based on the second-order statistics of error, which is insufficient to capture the information of a non-Gaussian distribution. Clearly, the MEE method is an estimation method which utilize the EEC. For the above estimation problem, the objective of the MEE method is to minimize the Renyi's entropy of the output error, as below,

$$\min_{w} H_\alpha\left(e^{\text{out}}\right) = H_\alpha\left(\breve{y} - \breve{\boldsymbol{x}}^T \boldsymbol{w}\right), \tag{9.4}$$

where $e_i^{\text{out}} = \breve{y}_i - \breve{\boldsymbol{x}}_i^T \boldsymbol{w}$, α is the order of the Renyi's entropy, and it is usually set to be 2 [8, 12]. Although the Renyi's entropy of other proper values of α can also be used to measure entropy, the quadratic Renyi's entropy $H_{\alpha=2}$ has advantages in computational complexity and feasibility [4]. It can be easily estimated based on the samples of output errors $\{e_i^{\text{out}}, i = 1, 2, \ldots, N\}$, by using the Parzen windows method [13].

The non-Gaussian noise signal is usually a non-homogeneous Poisson process with the time-varying density function, and it is possibility to regard the entropy as the manifestation of signal randomness at given moment. The probability density function of the estimation error using the Parzen window function method is

$$\hat{p}\left(e^{\text{out}}\right) = \left(\frac{1}{N}\right)\sum_{i=1}^{N} G_\sigma\left(e^{\text{out}} - e_i^{\text{out}}\right), \tag{9.5}$$

where the window basis function $G_\sigma(\cdot)$ is the kernel function with variance σ^2.

Compared with the LS method, it is more effective to research the uncertainty of data using the information entropy idea. Since the variance and the probability have not a fixed correspondence when the system is affected by the non-Gaussian noises and the error obey normal distribution. In this case, it is difficult to resolve the parameter estimation problem using the LS method, while the entropy, as the measurement of uncertainty of system state, can provide more generalized information description of system and is suitable for solving the higher-order statistics issues of non-Gaussian noises.

The general definitions of entropy is

$$H(\boldsymbol{w}) = -\sum_{i=1}^{N}\left(p\left(e^{\text{out}}\right)\lg p\left(e^{\text{out}}\right)\right), \tag{9.6}$$

where $p\left(e^{\text{out}}\right)$ is the probability density function of the output errors, $0 \le p\left(e^{\text{out}}\right) \le 1$ and $\sum_{i=1}^{N} p\left(e^{\text{out}}\right) = 1$. Using the Quadratic Renyi's entropy express the error entropy of system, i.e.,

$$\begin{aligned} H_{\alpha=2}\left(e^{\text{out}}\right) &= -\lg\int\left(\hat{p}\left(e^{\text{out}}\right)\right)^2 \mathrm{d}x \\ &= -\lg\int\left(\left(\frac{1}{N}\right)\sum_{i=1}^{N} G_\sigma\left(e^{\text{out}} - e_i^{\text{out}}\right)\right)^2 \mathrm{d}x. \end{aligned} \tag{9.7}$$

The Quadratic Renyi's entropy can be applied into the cases when ones can obtain the data from experiments and cannot know the probability density function of data in advance.

In terms of the structure of entropy, the quadratic Renyi's entropy has more computing efficiency compared with other entropy. The error entropy criterion under non-Gaussian noises is to fit a set of data such that the error entropy between the predictive value of system model and the actual measured value reach the minimum, when the system reach the determinant state.

Set

$$U(\boldsymbol{w}) = \int \left(\left(\frac{1}{N} \right) \sum_{i=1}^{N} G_{\sigma}\left(e^{\text{out}} - e_i^{\text{out}}\right) \right)^2 \mathrm{d}x. \tag{9.8}$$

Under the error entropy criterion, the cost function is denoted by

$$J(\boldsymbol{w}) = \min H_{\alpha=2}\left(e^{\text{out}}\right). \tag{9.9}$$

Since the log function is monotonically increasing, minimizing the above estimator of quadratic Renyi's entropy is equivalent to maximizing the term $U(\boldsymbol{w})$, thus there is $J(\boldsymbol{w}) = \max U(\boldsymbol{w})$. From the characteristic of Gauss function, it is known that the convolution of two Gauss functions is still a Gauss function, and its variance is the sum of the variances of two original functions. Accordingly, the $U(\mathbf{w})$ can be simplified to an equivalent discretized Gauss kernel function, and the cost function can be obtained, i.e.,

$$J(\boldsymbol{w}) = \max U(\boldsymbol{w}) = \max \left[\frac{1}{N^2} \sum_{j=1}^{N} \sum_{i=1}^{N} G_{\sigma\sqrt{2}}\left(e_j^{\text{out}} - e_i^{\text{out}}\right) \right]. \tag{9.10}$$

Based on the steepest-ascent theory, the iterative solution of the optimized $J(\boldsymbol{w})$ is as follows

$$\boldsymbol{w}_{\tau+1} = \boldsymbol{w}_{\tau} + \mu \frac{\partial \tilde{V}_2(\boldsymbol{w})}{\partial \boldsymbol{w}} \Big|_{\boldsymbol{w}_{\tau}}, \tag{9.11}$$

where μ is the step length, $\frac{\partial \hat{V}_2(\boldsymbol{w})}{\partial \boldsymbol{w}}$ is the gradient of the vectors, which is

$$\frac{\partial \hat{V}_2(\boldsymbol{w})}{\partial \boldsymbol{w}} = \frac{1}{\sigma^2 N^2} \sum_{i=1}^{N} \sum_{j=1}^{N} G_{\sigma}\left(\Delta e_{ij}\right)\left(\Delta e_{ij} \Delta \tilde{\mathbf{x}}_{ij}\right), \tag{9.12}$$

where the superscript "out" is no longer used without confusion, and the shorthand notation $\Delta e_{ij}^{\text{out}}$ is used to express $\left(e_j^{\text{out}} - e_i^{\text{out}}\right)$, and $\Delta \tilde{\mathbf{x}}_{ij}$ stands for $\breve{\mathbf{x}}_i - \breve{\mathbf{x}}_j$.

9.2.2 TLS Algorithm with Input and Output of the Same Noise Intensity

The methods described above are based on the assumption that the input data does not contain error and the output data contains error; however, this assumption is often unrealistic since the measurement for the input signal is sometimes as noisy as that for the output signal in real world [14]. Taking the measurement error of both input signal and output signal into consideration, the new measurement model (known as the errors-in-variables (EIV) model) can be written as follows:

$$\begin{cases} \tilde{y}_i = y_i + v_i \\ \tilde{\boldsymbol{x}}_i = \boldsymbol{x}_i + \boldsymbol{u}_i \end{cases}, \tag{9.13}$$

where $\boldsymbol{u}_i \in \Re^{D\times 1}$ is zero-mean input measurement noise with auto-covariance matrix $\sigma_{in}^2\mathbf{I}$ ($\boldsymbol{I}$ denotes the identity matrix of appropriate dimensions). The noise $\boldsymbol{u}_i$ is supposed to be stationary and ergodic in time and independent of v_{i} and $\boldsymbol{x}_{\text{i}}$.

In this case, existing literatures have shown that when the noise intensity at input is comparable to that at output, the methods which also pay attention to the input measurement noise (e.g., the TLS method) outperform the methods which omit the corresponding noise (e.g., the LS method) [14, 15]. Different from the LS method, the classical (scaled) TLS method considers both input and output error in the linear modeling, leading to the following optimization problem [1, 4],

$$\begin{gathered} \min_{w,\{\mathbf{e}_i^{\text{in}}\},\{e_i^{\text{out}}\}} \sum_{i=1}^{i=N} \left(e_i^{\text{out}}\varepsilon_0\right)^2 + \left((\mathbf{e}_i^{\text{in}})^T\boldsymbol{\varepsilon}_1\right)^2 \\ s.t.\ \tilde{y}_i - e_i^{\text{out}} = (\tilde{\mathbf{x}}_i - \mathbf{e}_i^{\text{in}})^T\mathbf{w},\ \forall i \in \{1, ..., N\}, \end{gathered} \tag{9.14}$$

where $\mathbf{e}_i^{\text{in}} \in \Re^{D\times 1}$ is the error/perturbation at input and $\boldsymbol{\varepsilon} = [\varepsilon_0, \boldsymbol{\varepsilon}_1^T]^T = [\varepsilon_0, \varepsilon_1, ..., \varepsilon_D]^T$ is a weighting vector with positive entries. In [1], the authors show that the above optimization problem can be efficiently solved by the SVD, and based on the results of SVD, the above optimization problem is proven to be equivalent to the following minimization problem,

$$\min_{w} \sum_{i=1}^{i=N} \frac{\left|\tilde{y}_i - \tilde{\mathbf{x}}_i^T\mathbf{w}\right|^2}{\mathbf{w}^T\boldsymbol{\Lambda}_1^{-2}\mathbf{w} + \varepsilon_0^{-2}}, \tag{9.15}$$

where $\boldsymbol{\Lambda}_1 = \text{diag}(\varepsilon_1, ..., \varepsilon_D)$ is a non-singular weighting matrix.

In (9.15), the designing of weighting coefficients $\{\varepsilon_0, \varepsilon_1, ..., \varepsilon_D\}$ is quite application-specific. Assume that all the dimensions of the input noise intensity σ_{in}^2 are the same, this is, all $\varepsilon_i, i = 1, 2, ..., D$ are set to be equal (thus $\boldsymbol{\Lambda}_1 = \mathbf{I}$), giving the different dimensions of input error the same weight, and also assume that the input noise intensity σ_{in}^2 equals to the output noise intensity σ_{out}^2. Then the (9.15) can be written as

$$\min_{\mathbf{w}} \frac{\sum_{i=1}^{N} \left|\tilde{y}_i - \tilde{\mathbf{x}}_i^T \mathbf{w}\right|^2}{1 + \mathbf{w}^T \mathbf{w}}. \tag{9.16}$$

By defining

$$e_i^{\text{tot}} = \frac{(\tilde{y}_i - \tilde{\mathbf{x}}_i^T \mathbf{w})}{\sqrt{1 + \mathbf{w}^T \mathbf{w}}}, \tag{9.17}$$

the above minimization problem (9.16) can be expressed as

$$\min_{w} \sum_{i=1}^{i=N} (e_i^{\text{tot}})^2. \tag{9.18}$$

By this method, the cost function of LS (9.3) and the cost function of TLS (9.18) have been written in the same form. In (9.17), e^{tot} considers both the output error and the input error, and also it is used in the total least squares, here it is called "total error." Comparing (9.18) with (9.3), it is obvious that the LS minimizes the sum of squared output error while the TLS method minimizes the sum of squared total error in estimation.

The difference in the meaning of cost function (9.18) from that of cost function (9.3) can be interpreted explicitly from the perspective of geometry. Essentially, the estimation problem is tantamount to finding a D-dimensional hyperplane P_w, defined by

$$y = \mathbf{x}^T \mathbf{w},\ \mathbf{x} \in \Re^{D \times 1},\ y \in \Re,$$

to fit the $(D + 1)$-dimensional data points under specific optimization goals. The optimization goals are different for different estimation methods. In the TLS method, the total error e_i^{tot} is actually the (signed) perpendicular distance from data point $(\tilde{\mathbf{x}}_i^T, \tilde{y}_i)$ to the hyperplane P_w. In contrast, in the LS method, the output error can be viewed as the vertical distances from the data points to a fitting straight line P_w, in the case of $D = 1$. So, in other words, the two estimation methods employ two different distance measures in the design of cost functions.

It is noted that, like the LS method, the TLS method is based on the second-order statistics of the total error. Thus it is no longer efficient when the total error is non-Gaussian [4]. The methods combining total error and information-theoretic measure may outperform the traditional TLS method [4]. Thus, information-theoretic estimation algorithms which consider the existence of measurement noise in the input signal have been developed in [4].

Given the above discussions, under the errors-in-variables model, the system parameters by minimizing the quadratic Renyi's entropy of the total error are estimated, and the algorithm called the minimum total error entropy (MTEE) method was proposed in [4]. Specifically, the cost function is given by

$$J(\mathbf{w}) = H_{\alpha=2}(e^{\text{tot}}) = -\log \int p^2(e^{\text{tot}})de^{\text{tot}}, \tag{9.19}$$

where $p(e^{\text{tot}})$ is the probability distribution function (pdf) of e^{tot}.

Using the Parzen window method, the pdf $p(e^{tot})$ can be estimated by samples of e^{tot}, $\{e_i^{\text{tot}} = (\tilde{y}_i - \tilde{\mathbf{x}}_i^T \mathbf{w})/(\sqrt{\|\mathbf{w}\|^2 + 1}), i = 1, 2, ...N\}$:

$$\hat{p}(e^{\text{tot}}) = \frac{1}{N}\sum_{i=1}^{N} G_\sigma(e^{\text{tot}} - e_i^{\text{tot}}), \tag{9.20}$$

where $G_\sigma(\cdot)$ is a Gaussian kernel function with kernel size σ. Substituting (9.20) into (9.19), a convenient evaluation of the integral operator in the formulation of quadratic Renyi's entropy can be obtained [4], as below

$$\hat{H}_2(e^{\text{tot}}) = -\log\left(\frac{1}{N^2}\sum_{i=1}^{N}\sum_{j=1}^{N} G_{\sigma\sqrt{2}}(e_j^{\text{tot}} - e_i^{\text{tot}})\right). \tag{9.21}$$

Since the function is monotonically increasing, minimizing the above estimator of quadratic Renyi's entropy is equivalent to maximizing the following term

$$\hat{V}_2(e^{\text{tot}}) = \frac{1}{N^2}\sum_{i=1}^{N}\sum_{j=1}^{N} G_{\sigma\sqrt{2}}(e_j^{\text{tot}} - e_i^{\text{tot}}), \tag{9.22}$$

which is an estimator of the so-called quadratic information potential (QIP) [8]. Note that it can be viewed as the time average (or expectation) of $p(e^{\text{tot}})$, that is, $\hat{V}_2(e^{\text{tot}}) = (1/N)\sum_{i=1}^{N} p(e_i^{\text{tot}})$.

Given the new cost function $\overline{J}(\mathbf{w}) = \hat{V}_2(e^{\text{tot}})$($e^{tot}$ is a function of w), the steepest-ascent method is employed to reach its maximum. Take the derivative of (9.22), it holds that [4]

$$\begin{aligned}\nabla\overline{J}(\mathbf{w}) &= \frac{\partial \hat{V}_2(e^{\text{tot}})}{\partial \mathbf{w}} = \frac{\partial\left(\frac{1}{N^2}\sum_{i=1}^{N}\sum_{j=1}^{N} G_\sigma(e_j^{\text{tot}} - e_i^{\text{tot}})\right)}{\partial \mathbf{w}} \\ &= \frac{1}{\sigma^2 N^2}\sum_{i=1}^{N}\sum_{j=1}^{N}\left[G_\sigma(e_j^{\text{tot}} - e_i^{\text{tot}})(e_i^{\text{tot}} - e_j^{\text{tot}}) \times \frac{\partial(e_j^{\text{tot}} - e_i^{\text{tot}})}{\partial \mathbf{w}}\right]\end{aligned} \tag{9.23}$$

where

$$
\begin{aligned}
&\frac{\partial(e_j^{\text{tot}}-e_i^{\text{tot}})}{\partial \mathbf{w}} \\
&=\frac{1}{\mathbf{w}^T\mathbf{w}+1}\Big[-(\tilde{\mathbf{x}}_j\sqrt{\mathbf{w}^T\mathbf{w}+1}+\frac{\mathbf{w}}{\sqrt{\mathbf{w}^T\mathbf{w}+1}}(\tilde{y}_j-\tilde{\mathbf{x}}_j^T\mathbf{w})) \\
&+(\tilde{\mathbf{x}}_i\sqrt{\mathbf{w}^T\mathbf{w}+1}+\tfrac{\mathbf{w}}{\sqrt{\mathbf{w}^T\mathbf{w}+1}}(\tilde{y}_i-\tilde{\mathbf{x}}_i^T\mathbf{w})) \\
&=\frac{1}{\sqrt{\mathbf{w}^T\mathbf{w}+1}}(\tilde{\mathbf{x}}_i-\tilde{\mathbf{x}}_j)+\frac{\mathbf{w}}{\mathbf{w}^T\mathbf{w}+1}(e_i^{\text{tot}}-e_j^{\text{tot}}).
\end{aligned}
\tag{9.24}
$$

To simplify the notation, denote $(e_i^{\text{tot}}-e_j^{\text{tot}})$ as Δe_{ij}^{tot}, denote $(\tilde{\mathbf{x}}_i-\tilde{\mathbf{x}}_j)$ as $\Delta\tilde{\mathbf{x}}_{ij}$ and suppress the superscript "tot" in the following derivation, in the case of no confusion. Using the simplified notations and combining (9.23) and (9.24) yields

$$
\frac{\partial \hat{V}_2(e)}{\partial \mathbf{w}}=\frac{1}{\sigma^2N^2}\cdot\sum_{i=1}^{N}\sum_{j=1}^{N}G_\sigma(\Delta e_{ij})\left(\frac{\Delta e_{ij}^2\mathbf{w}}{\mathbf{w}^T\mathbf{w}+1}+\frac{\Delta e_{ij}\Delta\tilde{\mathbf{x}}_{ij}}{\sqrt{\mathbf{w}^T\mathbf{w}+1}}\right). \tag{9.25}
$$

According to the steepest-ascent method, the iterative solution for optimizing $\overline{J}(\mathbf{w})$ is presented as

$$
\mathbf{w}_{\tau+1}=\mathbf{w}_\tau+\mu\frac{\partial \hat{V}_2(e)}{\partial \mathbf{w}}|_{\mathbf{w}_\tau}, \tag{9.26}
$$

where τ denotes the iteration step, and μ denotes the learning step size.

The above results are obtained in the case of batch mode, where the data points N are fixed during the estimation. In an online scenario, at time instant i, the latest L data points $\{(\tilde{\mathbf{x}}_j,\tilde{\mathbf{y}}_j)|j=i-L+1,...,i\}$ can be used to calculate the latest L samples of e, then estimate the QIP based on the L samples. Furthermore, to reduce calculation cost, the estimate of QIP can be approximated stochastically by dropping the time average in (9.22), leading to the expression [8]

$$
\hat{V}_2(e_i)=\frac{1}{L}\sum_{j=i-L+1}^{i}G_{\sigma\sqrt{2}}(e_j-e_i). \tag{9.27}
$$

Similarly, by using the steepest-ascent method, the adaptive updating rule for the estimate of parameter vector can be obtained:

$$
\mathbf{w}_{i+1}=\mathbf{w}_i+\mu\frac{\partial \hat{V}_2(e_i)}{\partial \mathbf{w}}|_{\mathbf{w}_i}, \tag{9.28}
$$

where

$$
\frac{\partial \hat{V}_2(e_i)}{\partial \mathbf{w}}=\frac{1}{\sigma^2L^2}\cdot\sum_{j=i-L+1}^{i}G_\sigma(\Delta e_{ij})\left(\frac{\Delta e_{ij}^2\mathbf{w}}{\mathbf{w}^T\mathbf{w}+1}+\frac{\Delta e_{ij}\Delta\tilde{\mathbf{x}}_{ij}}{\sqrt{\mathbf{w}^T\mathbf{w}+1}}\right). \tag{9.29}
$$

About the local stability analysis of the above method and the determination of a bound for the step size to ensure local stability, see reference [4].

9.3 Mixed LS-TLS Algorithm Under Non-Gaussian Noises

In this section, we consider some other individual scenarios. In each individual scenario, we describe the particular adaptive algorithm and provide some necessary analyses.

9.3.1 *Partial Input and Output Contaminated with Non-Gaussian Noises*

Assume that partial input signal is contaminated with noises, then $y_i = \boldsymbol{x}_i^T \boldsymbol{w}$ can be written as

$$y_i = \boldsymbol{x}_i^T \boldsymbol{w} \rightarrow y_i = \left[\boldsymbol{x}_{1i}^T \; \boldsymbol{x}_{2i}^T\right] \times \begin{bmatrix} \mathbf{w}_1 \\ \mathbf{w}_2 \end{bmatrix}, \tag{9.30}$$

where both $\boldsymbol{x}_i = \left[\boldsymbol{x}_{1i}^T \; \boldsymbol{x}_{2i}^T\right]^T$ and $\boldsymbol{w} = \left[\boldsymbol{w}_1^T \; \boldsymbol{w}_2^T\right]^T$ are block matrix, and $\boldsymbol{x}_{1i} \in \Re^{D_1 \times 1}$ is not contaminated with noise and $\boldsymbol{x}_{2i} \in \Re^{(D-D_1-1)\times 1}$ is contaminated with noise, $\boldsymbol{w}_1 \in \Re^{D_1 \times 1}$ and $\boldsymbol{w}_2 \in \Re^{(D-D_1-1)\times 1}$ are the unknown parameter vectors of the system. Also, assume that the measurement/observation for the output signal $\tilde{y}_i$ is noisy, then (9.30) can be written as

$$\tilde{y}_i = y_i + v_i = [\mathbf{x}_{1i}^T, \mathbf{x}_{2i}^T] \begin{bmatrix} \boldsymbol{w}_1 \\ \boldsymbol{w}_2 \end{bmatrix} + [\mathbf{u}_{1i}^T, \mathbf{u}_{2i}^T] \begin{bmatrix} \boldsymbol{w}_1 \\ \boldsymbol{w}_2 \end{bmatrix}, \tag{9.31}$$

where $\boldsymbol{u}_{1i} \in \Re^{D_1 \times 1}$ is zero vector, $\boldsymbol{u}_{2i} \in \Re^{(D-D_1-1)\times 1}$ is zero-mean input measurement noise with auto-covariance matrix $\sigma_{in}^2 \mathbf{I}$ ($\boldsymbol{I}$ denotes the identity matrix of appropriate dimensions). The noise $\boldsymbol{u}_{2i}$ is supposed to be stationary and ergodic in time and independent of v_i and $\boldsymbol{x}_{2i}$.

Taking the measurement error of both input signal and output signal into consideration, the EIV model can be written as follows:

$$\begin{cases} \tilde{y}_i = y_i + v_i \\ \tilde{\boldsymbol{x}}_i = \boldsymbol{x}_i + \boldsymbol{u}_i = [\boldsymbol{x}_{1i}^T, \boldsymbol{x}_{2i}^T]^T + [\boldsymbol{u}_{1i}^T, \boldsymbol{u}_{2i}^T]^T \end{cases}. \tag{9.32}$$

The TLS method leads to the following optimization problem,

$$\min_{w,\{e_i^{\text{in}}\},\{e_i^{\text{out}}\}} \sum_{i=1}^{i=N} \left(e_i^{\text{out}}\varepsilon_0\right)^2 + \left(\boldsymbol{e}_i^{\text{in}}\boldsymbol{\varepsilon}_1\right)^2$$

$$s.t.\tilde{y}_i - e_i^{\text{out}} = (\tilde{\boldsymbol{x}}_i^T - (\boldsymbol{e}_i^{\text{in}})^T)\boldsymbol{w}$$

$$= (\tilde{\boldsymbol{x}}_i^T - (\boldsymbol{e}_i^{in})^T)\begin{bmatrix} w_1 \\ w_2 \end{bmatrix}, \forall i \in \{1, ..., N\}, \tag{9.33}$$

where $e_i^{out} \in \Re$ is the error/perturbation at output, and $\mathbf{e}_i^{in} \in \Re^{D\times 1}$ is the error/perturbation at input.

Equation (9.33) is equivalent to the following minimization problem,

$$\min_w \sum_{i=1}^{i=N} \frac{\left|\tilde{y}_i - \tilde{\mathbf{x}}_i^T\mathbf{w}\right|^2}{\mathbf{w}^T\boldsymbol{\Omega}^{-2}\mathbf{w} + 1}, \tag{9.34}$$

where the appropriate choice for $\boldsymbol{\Omega}$ is

$$\boldsymbol{\Omega} = \begin{bmatrix} \mathbf{0} & \\ & \mathbf{I} \end{bmatrix}, \tag{9.35}$$

where $\boldsymbol{\Omega}$ is $D \times D$ matrix with the identity matrix $\boldsymbol{I}$ being $(D-D_1-1)\times(D-D_1-1)$.

By defining

$$e_i^{\text{tot}} = \frac{(\tilde{y}_1 - \tilde{\mathbf{x}}_1\mathbf{w})}{\sqrt{\mathbf{w}^T\boldsymbol{\Omega}^{-2}\mathbf{w} + 1}}, \tag{9.36}$$

the above minimization problem (9.34) can be expressed as

$$\min_w \sum_{i=1}^{i=N} (e_i^{\text{tot}})^2. \tag{9.37}$$

When the total error is non-Gaussian, under the errors-in-variables model, the system parameters by minimizing the quadratic Renyi's entropy of the total error are estimated, and the MTEE can be proposed. Specifically, the cost function is given by

$$J(\mathbf{w}) = H_{\alpha=2}(e^{\text{tot}}) = -\log \int p^2(e^{\text{tot}})de^{\text{tot}}, \tag{9.38}$$

where $p(e^{\text{tot}})$ is the probability distribution function (pdf) of e^{tot}.

Using the Parzen window method, the pdf $p(e^{\text{tot}})$ can be estimated by samples of e^{tot}, $\{e_i^{\text{tot}} = (\tilde{y}_i - \tilde{\mathbf{x}}_i\mathbf{w})/(\sqrt{\mathbf{w}^T\boldsymbol{\Omega}^{-2}\mathbf{w} + 1}), i = 1, 2, ...N\}$:

$$\hat{p}(e^{\text{tot}}) = \frac{1}{N}\sum_{i=1}^{N} G_{\sigma}(e^{\text{tot}} - e_i^{\text{tot}}), \tag{9.39}$$

where $G_{\sigma}(\cdot)$ is a Gaussian kernel function with kernel size σ. Substituting (9.39) into (9.38), a convenient evaluation of the integral operator in the formulation of quadratic Renyi's entropy can be obtained, as below

$$\hat{H}_2(e^{\text{tot}}) = -\log\left(\frac{1}{N^2}\sum_{i=1}^{N}\sum_{j=1}^{N} G_{\sigma\sqrt{2}}(e_j^{\text{tot}} - e_i^{\text{tot}})\right). \tag{9.40}$$

Since the function is monotonically increasing, minimizing the above estimator of quadratic Renyi's entropy is equivalent to maximizing the following term

$$\hat{V}_2(e^{\text{tot}}) = \frac{1}{N^2}\sum_{i=1}^{N}\sum_{j=1}^{N} G_{\sigma\sqrt{2}}(e_j^{\text{tot}} - e_i^{\text{tot}}). \tag{9.41}$$

Given the new cost function $\overline{J}(\mathbf{w}) = \hat{V}_2(e^{\text{tot}})$, the steepest-ascent method is employed to reach its maximum. Take the derivative of (9.41), it holds that

$$\begin{aligned}
\nabla\overline{J}(\mathbf{w}) &= \frac{\partial \hat{V}_2(e^{\text{tot}})}{\partial \mathbf{w}} \\
&= \frac{\partial\left(\frac{1}{N^2}\sum_{i=1}^{N}\sum_{j=1}^{N} G_{\sigma}(e_j^{\text{tot}} - e_i^{\text{tot}})\right)}{\partial \mathbf{w}} \\
&= \frac{1}{\sigma^2 N^2}\sum_{i=1}^{N}\sum_{j=1}^{N}\left[G_{\sigma}(e_j^{\text{tot}} - e_i^{\text{tot}})(e_i^{\text{tot}} - e_j^{\text{tot}}) \times \frac{\partial(e_j^{\text{tot}} - e_i^{\text{tot}})}{\partial \mathbf{w}}\right]
\end{aligned} \tag{9.42}$$

where

$$\begin{aligned}
&\frac{\partial(e_j^{\text{tot}} - e_i^{\text{tot}})}{\partial \mathbf{w}} \\
&= \frac{1}{\mathbf{w}^T\boldsymbol{\Omega}^{-2}\mathbf{w}+1}\left[-(\tilde{\mathbf{x}}_j\sqrt{\mathbf{w}^T\Omega\mathbf{w}+1} + \frac{\mathbf{w}}{\sqrt{\mathbf{w}^T\boldsymbol{\Omega}^{-2}\mathbf{w}+1}}(\tilde{y}_j - \tilde{\mathbf{x}}_j^T\mathbf{w}))\right. \\
&\left.+(\tilde{\mathbf{x}}_i\sqrt{\mathbf{w}^T\boldsymbol{\Omega}^{-2}\mathbf{w}+1} + \frac{\mathbf{w}}{\sqrt{\mathbf{w}^T\boldsymbol{\Omega}^{-2}\mathbf{w}+1}}(\tilde{y}_i - \tilde{\mathbf{x}}_i^T\mathbf{w}))\right. \\
&= \frac{1}{\sqrt{\mathbf{w}^T\boldsymbol{\Omega}^{-2}\mathbf{w}+1}}(\tilde{\mathbf{x}}_i - \tilde{\mathbf{x}}_j) + \frac{\mathbf{w}}{\mathbf{w}^T\boldsymbol{\Omega}^{-2}\mathbf{w}+1}(e_i^{\text{tot}} - e_j^{\text{tot}}) \\
&= \frac{\Delta\tilde{\mathbf{x}}_{ij}}{\sqrt{\mathbf{w}_2^T\mathbf{w}_2+1}} + \frac{\mathbf{w}\Delta e_{ij}^{\text{tot}}}{\mathbf{w}_2^T\mathbf{w}_2+1}.
\end{aligned} \tag{9.43}$$

Using the simplified notations and combining (9.42) and (9.43) yields

$$\frac{\partial \hat{V}_2(e)}{\partial \mathbf{w}} = \frac{1}{\sigma^2 N^2} \cdot \sum_{i=1}^{N} \sum_{j=1}^{N} G_\sigma(\Delta e_{ij}) \left(\frac{\Delta e_{ij}^2 \mathbf{w}}{\mathbf{w}_2^T \mathbf{w}_2 + 1} + \frac{\Delta e_{ij} \Delta \tilde{\mathbf{x}}_{ij}}{\sqrt{\mathbf{w}_2^T \mathbf{w}_2 + 1}} \right). \tag{9.44}$$

According to the steepest-ascent method, the iterative solution for optimizing $\overline{J}(\mathbf{w})$ is presented as

$$\mathbf{w}_{\tau+1} = \mathbf{w}_\tau + \mu \frac{\partial \hat{V}_2(e)}{\partial \mathbf{w}} |_{\mathbf{w}_\tau}, \tag{9.45}$$

where τ denotes the iteration step, and μ denotes the learning step-size.

Referring to the derivation of algorithm (9.29), in an online scenario, the adaptive updating rule for the estimate of parameter vector can be obtained:

$$\mathbf{w}_{i+1} = \mathbf{w}_i + \mu \frac{\partial \hat{V}_2(e_i)}{\partial \mathbf{w}} |_{\mathbf{w}_i}, \tag{9.46}$$

where

$$\frac{\partial \hat{V}_2(e_i)}{\partial \mathbf{w}} = \frac{1}{\sigma^2 L^2} \cdot \sum_{j=i-L+1}^{i} G_\sigma(\Delta e_{ij}) \left(\frac{\Delta e_{ij}^2 \mathbf{w}}{\mathbf{w}_2^T \mathbf{w}_2 + 1} + \frac{\Delta e_{ij} \Delta \tilde{\mathbf{x}}_{ij}}{\sqrt{\mathbf{w}_2^T \mathbf{w}_2 + 1}} \right). \tag{9.47}$$

9.3.2 *Disparate Input and Output Non-Gaussian Noises*

In certain filtering situations, significant input and output noise is present, but these noises are of non-equal variance. Consider the case where both v_i and $\mathbf{u}_i$ are with noise intensity σ_{in}^2 and σ_{out}^2, respectively. Let the output noise to input noise intensity ratio be denoted as $\varepsilon_0 = \sigma_{\text{in}}/\sigma_{\text{out}}$.

The new EIV model under disparate input and output noises can be written as follows:

$$\begin{cases} \tilde{y}_i = y_i + v_i \\ \tilde{\boldsymbol{x}}_i = \boldsymbol{x}_i + \boldsymbol{u}_i \end{cases}, \tag{9.48}$$

where $v_i \in \Re$ is the zero-mean output measurement noise, which is supposed to be stationary and ergodic in time and independent of the input signal, and $\boldsymbol{u}_i \in \Re^{D \times 1}$ is zero-mean input measurement noise, which is supposed to be stationary and ergodic in time and independent of v_i and $\boldsymbol{x}_i$.

The scaled TLS method considers both input and output error in the linear modeling, leading to the following optimization problem,

$$\begin{aligned}&\min_{\mathbf{w},\{\mathbf{e}_i^{\text{in}}\},\{e_i^{\text{out}}\}} \sum_{i=1}^{i=N} \left(e_i^{\text{out}}\varepsilon_0\right)^2 + \left((\mathbf{e}_i^{\text{in}})^T \boldsymbol{\varepsilon}_1\right)^2 \\ &\text{s.t. } (\tilde{y}_i - e_i^{\text{out}}) = (\tilde{\mathbf{x}}_i - \mathbf{e}_i^{\text{in}})^T \mathbf{w},\ \forall i \in \{1, \dots, N\},\end{aligned} \tag{9.49}$$

where $e_i^{\text{out}} \in \Re$ is the error/perturbation at output, $\mathbf{e}_i^{in} \in \Re^{D\times 1}$ is the error/ perturbation at input. The above optimization problem is equivalent to the following minimization problem,

$$\begin{aligned}\min_{w} \sum_{i=1}^{i=N} \frac{\left|\tilde{y}_i - \tilde{\mathbf{x}}_i^T \mathbf{w}\right|^2}{\mathbf{w}^T \boldsymbol{\Lambda}_1^{-2} \mathbf{w} + \varepsilon_0^{-2}} &= \min_{w} \sum_{i=1}^{i=N} \frac{\varepsilon_0^2 \left|\tilde{y}_i - \tilde{\mathbf{x}}_i^T \mathbf{w}\right|^2}{(\mathbf{w}^T (\frac{\varepsilon_0}{\varepsilon_1})^2 \mathbf{w} + 1)} \\ \overset{\alpha=\varepsilon_0/\varepsilon_1}{=} \min_{w} \sum_{i=1}^{i=N} \frac{\alpha^2 \varepsilon_1^2 \left|\tilde{y}_i - \tilde{\mathbf{x}}_i^T \mathbf{w}\right|^2}{(\mathbf{w}^T \alpha^2 \mathbf{w} + 1)} &= \min_{w} \varepsilon_1^2 \sum_{i=1}^{i=N} \frac{\left|\alpha \tilde{y}_i - \tilde{\mathbf{x}}_i^T (\alpha \mathbf{w})\right|^2}{((\alpha \mathbf{w})^T (\alpha \mathbf{w}) + 1)} \\ = \min_{\tilde{w}} \varepsilon_1^2 \sum_{i=1}^{i=N} \frac{\left|\alpha \tilde{y}_i - \tilde{\mathbf{x}}_i^T \tilde{\mathbf{w}}\right|^2}{(\tilde{\mathbf{w}}^T \tilde{\mathbf{w}} + 1)},&\end{aligned} \tag{9.50}$$

where $\boldsymbol{\Lambda}_1 = \text{diag}(\varepsilon_1, \dots, \varepsilon_1)$, and $\tilde{\mathbf{w}} = \alpha \mathbf{w}$.

By defining

$$e_i^{\text{tot}} = \frac{(\alpha \tilde{y}_i - \tilde{\mathbf{x}}_i^T \tilde{\mathbf{w}})}{\sqrt{\tilde{\mathbf{w}}^T \tilde{\mathbf{w}} + 1}}, \tag{9.51}$$

the above minimization problem (9.50) can be expressed as

$$\min_{\tilde{w}} \sum_{i=1}^{i=N} (e_i^{\text{tot}})^2. \tag{9.52}$$

Given the above discussions, under the errors-in-variables model, the system parameters by minimizing the quadratic Renyi's entropy of the total error are estimated, and the MTEE can be obtained. Specifically, the cost function is given by

$$J(\tilde{\mathbf{w}}) = H_{\alpha=2}(e^{\text{tot}}) = -\log \int p^2(e^{\text{tot}}) de^{\text{tot}}, \tag{9.53}$$

where $p(e^{\text{tot}})$ is the probability distribution function (pdf) of e^{tot}.

Using the Parzen window method, the pdf $p(e^{\text{tot}})$ can be estimated by samples of e^{tot}, $\{e_i^{\text{tot}} = (\alpha \tilde{y}_i - \tilde{\mathbf{x}}_i^T \tilde{\mathbf{w}})/(\sqrt{\|\tilde{\mathbf{w}}\|^2 + 1}), i = 1, 2, \dots N\}$:

$$\hat{p}(e^{\text{tot}}) = \frac{1}{N}\sum_{i=1}^{N} G_\sigma(e^{\text{tot}} - e_i^{\text{tot}}), \tag{9.54}$$

where $G_\sigma(\cdot)$ is a Gaussian kernel function with kernel size σ. Substituting (9.54) into (9.53), a convenient evaluation of the integral operator in the formulation of quadratic Renyi's entropy can be obtained as below

$$\hat{H}_2(e^{\text{tot}}) = -\log\left(\frac{1}{N^2}\sum_{i=1}^{N}\sum_{j=1}^{N} G_{\sigma\sqrt{2}}(e_j^{\text{tot}} - e_i^{\text{tot}})\right). \tag{9.55}$$

Since the function is monotonically increasing, minimizing the above estimator of quadratic Renyi's entropy is equivalent to maximizing the following term

$$\hat{V}_2(e^{\text{tot}}) = \frac{1}{N^2}\sum_{i=1}^{N}\sum_{j=1}^{N} G_{\sigma\sqrt{2}}(e_j^{\text{tot}} - e_i^{\text{tot}}), \tag{9.56}$$

which is an estimator of the so-called quadratic information potential. Note that it can be viewed as the time average (or expectation) of $p(e^{\text{tot}})$, that is, $\hat{V}_2(e^{\text{tot}}) = (1/N)\sum_{i=1}^{N} p(e_i^{\text{tot}})$.

Given the new cost function $\overline{J}(\tilde{\mathbf{w}}) = \hat{V}_2(e^{\text{tot}})$, the steepest-ascent method is employed to reach its maximum. Take the derivative of (9.56), it holds that

$$\frac{\partial \hat{V}_2(e)}{\partial \tilde{\mathbf{w}}} = \frac{1}{\sigma^2 N^2}\cdot\sum_{i=1}^{N}\sum_{j=1}^{N} G_\sigma(\Delta e_{ij})\left(\frac{\Delta e_{ij}^2 \tilde{\mathbf{w}}}{\tilde{\mathbf{w}}^T\tilde{\mathbf{w}}+1} + \frac{\Delta e_{ij}\Delta\tilde{\mathbf{x}}_{ij}}{\sqrt{\tilde{\mathbf{w}}^T\tilde{\mathbf{w}}+1}}\right) \tag{9.57}$$

where $\Delta e_{ij}^{\text{tot}} = (e_i^{\text{tot}} - e_j^{\text{tot}})$, $\Delta\tilde{\mathbf{x}}_{ij} = (\tilde{\mathbf{x}}_i - \tilde{\mathbf{x}}_j)$.

According to the steepest-ascent method, the iterative solution for optimizing $\overline{J}(\tilde{\mathbf{w}})$ is presented as

$$\tilde{\mathbf{w}}_{\tau+1} = \tilde{\mathbf{w}}_\tau + \mu\frac{\partial \hat{V}_2(e)}{\partial \tilde{\mathbf{w}}}|_{\tilde{\mathbf{w}}_\tau}, \tag{9.58}$$

where τ denotes the iteration step, and μ denotes the learning step-size.

After the convergence of the algorithm (57) and the weight vector $\tilde{\mathbf{w}}$, the unknown parameter vector w of the system can be obtained by

$$\mathbf{w} = \frac{1}{\alpha}\tilde{\mathbf{w}}. \tag{9.59}$$

Referring to the derivation of algorithm (9.29), in an online scenario, the adaptive updating rule for the estimate of parameter vector can be obtained:

$$\tilde{\mathbf{w}}_{i+1} = \tilde{\mathbf{w}}_i + \mu \frac{\partial \hat{V}_2(e_i)}{\partial \tilde{\mathbf{w}}}|_{\tilde{\mathbf{w}}_i}, \tag{9.60}$$

where

$$\frac{\partial \hat{V}_2(e_i)}{\partial \tilde{\mathbf{w}}} = \frac{1}{\sigma^2 L^2} \cdot \sum_{j=i-L+1}^{i} G_\sigma(\Delta e_{ij}) \left(\frac{\Delta e_{ij}^2 \tilde{\mathbf{w}}}{\tilde{\mathbf{w}}_2^T \tilde{\mathbf{w}}_2 + 1} + \frac{\Delta e_{ij} \Delta \tilde{\mathbf{x}}_{ij}}{\sqrt{\tilde{\mathbf{w}}_2^T \tilde{\mathbf{w}}_2 + 1}} \right). \tag{9.61}$$

Similarly, after obtaining the weight vector $\tilde{\mathbf{w}}$, we can obtain the parameter vector w of the system from $\mathbf{w} = \tilde{\mathbf{w}}/\alpha$.

9.3.3 *Input Non-Gaussian Noises Only*

In certain filtering applications, such as during the training phase for channel equalization, the output data sequence is known exactly and is, hence, noise free. This scenario is the dual of the scenario in that the input portion of the data is noisy.

The linear system is Eq. (9.1), and the measurement model can be written as below (it is called the errors-in-input model)

$$\begin{cases} \breve{y}_i = y_i \\ \breve{\boldsymbol{x}}_i = \boldsymbol{x}_i + \mathbf{u}_1 \end{cases}, \tag{9.62}$$

where $\mathbf{u}_i \in \Re^{D\times 1}$ is the zero-mean input measurement noise of variance σ_{in}^2. The noise is supposed to be stationary and ergodic in time and independent of the input and output signal.

Assuming that the above model has N measurements $\{(\breve{\mathbf{x}}_i^T, \breve{y}_i), i = 1, 2, ..., N\}$, various algorithms are designed to estimate the parameter vector of the system $\mathbf{w}^o$ based on these data.

The TLS method considers both input and output error in the linear modeling, leading to the following optimization problem,

$$\begin{aligned} &\min_{w, \{\mathbf{e}_i^{\text{in}}\}, \{e_i^{\text{out}}\}} \sum_{i=1}^{i=N} \left(e_i^{\text{out}} \varepsilon_0\right)^2 + \left((\mathbf{e}_i^{\text{in}})^T \boldsymbol{\varepsilon}_1\right)^2 = \min_{w, \{\mathbf{e}_i^{\text{in}}\}} \sum_{i=1}^{i=N} \left((\mathbf{e}_i^{\text{in}})^T \boldsymbol{\varepsilon}_1\right)^2 \\ &s.t.\ \tilde{y}_i = (\tilde{\mathbf{x}}_i - \mathbf{e}_i^{\text{in}})^T \mathbf{w},\ \forall i \in \{1, ..., N\}, \end{aligned} \tag{9.63}$$

where $\mathbf{e}_i^{\text{in}} \in \Re^{D\times 1}$ is the error/perturbation at input and $\boldsymbol{\varepsilon}_1 = [\varepsilon_1, \varepsilon_2, ..., \varepsilon_D]^T$ is a weighting vector with positive entries. Assume that all the dimensions of the input noise intensity σ_{in}^2 are the same, this is, all $\varepsilon_i, i = 1, 2, ..., D$ are set to be equal, giving the different dimensions of input error the same weight. Thus, the above optimization problem is equivalent to the following minimization problem,

$$\min_{w} \sum_{i=1}^{i=N} \frac{\left|\tilde{y}_i - \tilde{\mathbf{x}}_i^T \mathbf{w}\right|^2}{\mathbf{w}^T \mathbf{w}}. \tag{9.64}$$

By defining

$$e_i^{\text{tot}} = \frac{(\tilde{y}_i - \tilde{\mathbf{x}}_i^T \mathbf{w})}{\sqrt{\mathbf{w}^T \mathbf{w}}}, \tag{9.65}$$

the above minimization problem (9.64) can be expressed as

$$\min_{w} \sum_{i=1}^{i=N} (e_i^{\text{tot}})^2. \tag{9.66}$$

When the error is non-Gaussian, the methods using information-theoretic measure may outperform the traditional method. Given the above discussions, under the errors-in-input model, the system parameters by minimizing the quadratic Renyi's entropy of the total error are estimated, and the MTEE algorithm was proposed. Specifically, the cost function is given by

$$J(\mathbf{w}) = H_{\alpha=2}(e^{\text{tot}}) = -\log \int p^2(e^{\text{tot}}) de^{\text{tot}}, \tag{9.67}$$

where $p(e^{\text{tot}})$ is the probability distribution function (pdf) of e^{tot}.

Using the Parzen window method, the pdf $p(e^{\text{tot}})$ can be estimated by samples of e^{tot}, $\{e_i^{\text{tot}} = (\tilde{y}_i - \tilde{\mathbf{x}}_i^T \mathbf{w})/\sqrt{\mathbf{w}^T \mathbf{w}}, i = 1, 2, ...N\}$:

$$\hat{p}(e^{\text{tot}}) = \frac{1}{N} \sum_{i=1}^{N} G_\sigma(e^{\text{tot}} - e_i^{\text{tot}}), \tag{9.68}$$

where $G_\sigma(\cdot)$ is a Gaussian kernel function with kernel size σ. Substituting (9.68) into (9.57), a convenient evaluation of the integral operator in the formulation of quadratic Renyi's entropy can be obtained as below

$$\hat{H}_2(e^{\text{tot}}) = -\log\left(\frac{1}{N^2} \sum_{i=1}^{N} \sum_{j=1}^{N} G_{\sigma\sqrt{2}}(e_j^{\text{tot}} - e_i^{\text{tot}})\right). \tag{9.69}$$

Since the function is monotonically increasing, minimizing the above estimator of quadratic Renyi's entropy is equivalent to maximizing the following term

$$\hat{V}_2(e^{\text{tot}}) = \frac{1}{N^2} \sum_{i=1}^{N} \sum_{j=1}^{N} G_{\sigma\sqrt{2}}(e_j^{\text{tot}} - e_i^{\text{tot}}), \tag{9.70}$$

which is an estimator of the quadratic information potential. Note that it can be viewed as the time average of $p(e^{\text{tot}})$, that is, $\hat{V}_2(e^{\text{tot}}) = (1/N)\sum_{i=1}^{N} p(e_i^{\text{tot}})$.

Given the new cost function $\overline{J}(\mathbf{w}) = \hat{V}_2(e^{\text{tot}})$, the steepest-ascent method is employed to reach its maximum. Take the derivative of (9.70), it holds that

$$\begin{aligned}\nabla\overline{J}(\mathbf{w}) &= \frac{\partial \hat{V}_2(e^{\text{tot}})}{\partial \mathbf{w}} = \frac{\partial\left(\frac{1}{N^2}\sum_{i=1}^{N}\sum_{j=1}^{N} G_\sigma(e_j^{\text{tot}} - e_i^{\text{tot}})\right)}{\partial \mathbf{w}} \\ &= \frac{1}{\sigma^2 N^2}\sum_{i=1}^{N}\sum_{j=1}^{N}\left[G_\sigma(e_j^{\text{tot}} - e_i^{\text{tot}})(e_i^{\text{tot}} - e_j^{\text{tot}}) \times \frac{\partial(e_j^{\text{tot}} - e_i^{\text{tot}})}{\partial \mathbf{w}}\right] \\ &= \frac{1}{\sigma^2 N^2}\cdot\sum_{i=1}^{N}\sum_{j=1}^{N} G_\sigma(\Delta e_{ij})\left(\frac{\Delta e_{ij}^2 \mathbf{w}}{\mathbf{w}^T\mathbf{w}} + \frac{\Delta e_{ij}\Delta\tilde{\mathbf{x}}_{ij}}{\sqrt{\mathbf{w}^T\mathbf{w}}}\right).\end{aligned} \tag{9.71}$$

According to the steepest-ascent method, the iterative solution for optimizing $\overline{J}(\mathbf{w})$ is presented as

$$\mathbf{w}_{\tau+1} = \mathbf{w}_\tau + \mu\frac{\partial \hat{V}_2(e)}{\partial \mathbf{w}}|_{\mathbf{w}_\tau}, \tag{9.72}$$

where τ denotes the iteration step, and μ denotes the learning step-size.

The above results are obtained in the case of batch mode, where the data points N are fixed during the estimation. In an online scenario, at time instant i, the latest L data points $\{(\tilde{\mathbf{x}}_j, \tilde{\mathbf{y}}_j)|j = i - L + 1, ..., i\}$ can be used to calculate the latest L samples of e, then estimate the QIP based on the L samples. Furthermore, to reduce calculation cost, the estimate of QIP can be approximated stochastically by dropping the time average in (9.70), leading to the expression

$$\hat{V}_2(e_i) = \frac{1}{L}\sum_{j=i-L+1}^{i} G_{\sigma\sqrt{2}}(e_j - e_i). \tag{9.73}$$

Similarly, by using the steepest-ascent method, the adaptive updating rule for the estimate of parameter vector can be obtained:

$$\mathbf{w}_{i+1} = \mathbf{w}_i + \mu\frac{\partial \hat{V}_2(e_i)}{\partial \mathbf{w}}|_{\mathbf{w}_i}, \tag{9.74}$$

where

$$\frac{\partial \hat{V}_2(e_i)}{\partial \mathbf{w}} = \frac{1}{\sigma^2 L^2}\cdot\sum_{j=i-L+1}^{i} G_\sigma(\Delta e_{ij})\left(\frac{\Delta e_{ij}^2 \mathbf{w}}{\mathbf{w}^T\mathbf{w}} + \frac{\Delta e_{ij}\Delta\tilde{\mathbf{x}}_{ij}}{\sqrt{\mathbf{w}^T\mathbf{w}}}\right). \tag{9.75}$$

9.4 Simulation Experiments and Analysis

In this simulation, the above algorithms are used to the system identification of the linear system. The linear system is given by $\boldsymbol{H} = [-0.3, -0.9, 0.8, -0.7, 0.6, 0.2, -0.5, 1.0, -0.7, 0.9, -0.4]^T$, and its normalization is $\overline{\mathbf{H}} = \mathbf{H}/\|\mathbf{H}\|$. Their convergence is compared under different noises and different learning rates. Denote the learning error of the weight vector is

$$Error(k) = 20\log(\left\|\overline{\mathbf{H}} - \hat{\mathbf{H}}(k)\right\|^2),$$

where $\hat{\mathbf{H}}(k)$ is the estimated weight vector.

9.4.1 Experiment of NG-TLS Algorithm

In this section, we will show the convergence and parameter estimation performance of the NG-TLS algorithm via the TLS filtering. In this simulation experiment, all the noises are non-Gauss noise. Here, we will compare the NG-TLS algorithm with the NG-LS algorithm and the classical TLS algorithm. All the following curves are the average effect of the 20 independent experiments, the iterative times is 1000, and the learning rate reduce from 0.05 to 0.005. The Gauss kernel function is set as 0.5, which is slightly less than the input signal strength. The signal to noise ratio (SNR) of the input signal is equal to that of the output signal. The following is the learning curves and the modulus curves of the weight vectors when the SNR is 20, 10, and 5 dB, respectively.

From Figs. 9.1, 9.2, and 9.3, we can see that the NG-TLS method show a better estimation performance, its error of the weight vector is also smaller and the estimation accuracy is higher, while the classical TLS method show the higher error of weight vector and the lower estimation accuracy under the non-Gauss noises. On the whole, the NG-TLS has the better estimation effect. When the SNR(=5 dB) is smaller, the convergence speed of the NG-TLS is faster than the of the NG-LS, and its weight vector error is smaller than that of the NG-LS method when reaching a stable state, which means that the NG-TLS method based on EIV model considering both the input non-Gauss noise and the output non-Gauss noise has the better estimation effect. When the SNR(=20 dB) is larger, we can observe that the convergence curve of the NG-TLS method intersects at a place with that of the NG-LS method, which means that on this condition the effects of two methods is similar. When the SNR is 10 dB, the convergence speed and the weight error of the stable state are the smallest, the estimation accuracy is the highest. In addition, from the modulus convergence of the weight vectors in Figs. 9.1, 9.2, and 9.3, we can see that the classical TLS method is divergent when facing non-Gauss noise, while the NG-LS method and the NG-TLS have better convergence.

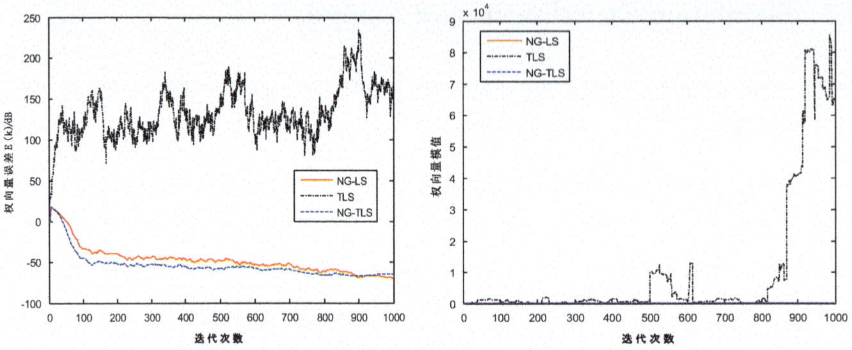

Fig. 9.1 Learning curves of different algorithms under SNR = 5 dB. **a** Error curves of the weight vectors. **b** Modulus curves of the weight vectors

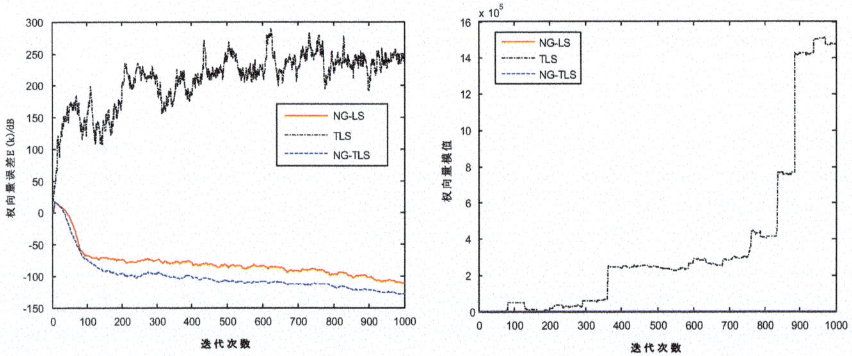

Fig. 9.2 Learning curves of different algorithms under SNR = 10 dB. **a** Error curves of the weight vectors. **b** Modulus curves of the weight vectors

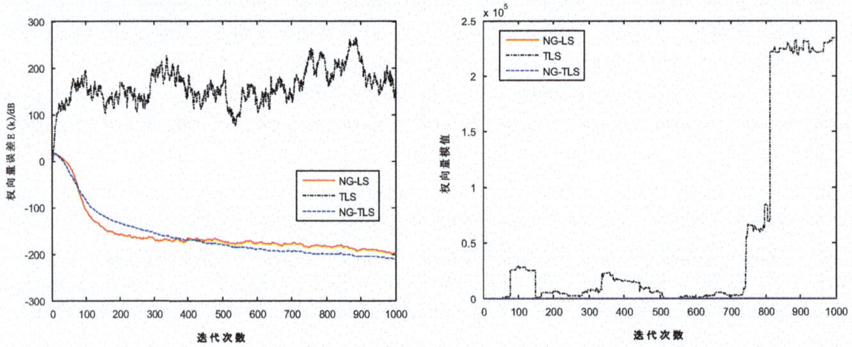

Fig. 9.3 Learning curves of different algorithms under SNR = 20 dB. **a** Error curves of the weight vectors. **b** Modulus curves of the weight vectors

9.4.2 Experiment of NG-LS-TLS Algorithm

In this section, we will test and verify the performance of the NG-LS-TLS method via several TLS filtering simulation experiments. Here, we will compare the three methods, that is NG-LS-TLS, NG-TLS and NG-LS, and contrast their actual estimated performances under non-Gauss noises with the mixed input noises.

In this simulation, Figs. 9.4, 9.5, 9.6, 9.7, 9.8, and 9.9 show that the learning error curves and the modulus curves of the weight vectors of the three methods when the learning rate $u = 0.01$, $u = 0.005$ and the SNR is 20, 10, and 5 dB, respectively.

All the following curves are the average effect of the 20 independent experiments, the iterative times is 1000, and the learning rate reduce from 0.05 to 0.005. The Gauss kernel function is set as 0.5, which is slightly less than the input signal strength. The SNR of the input signal is equal to that of the output signal, the added noises are the noises produced by the Poisson distribution, and a part of input signals are mixed

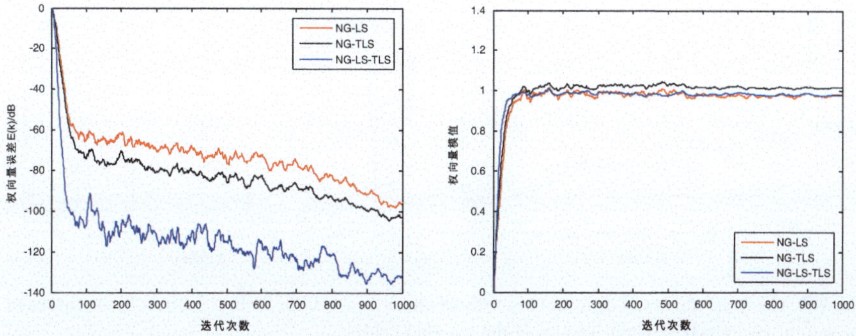

Fig. 9.4 Learning curves of different algorithms under $\mu = 0.01$, SNR = 5 dB. **a** Error curves of the weight vectors. **b** Modulus curves of the weight vectors

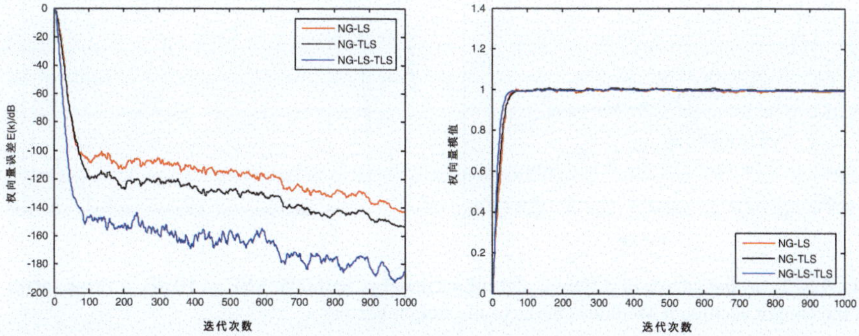

Fig. 9.5 Learning curves of different algorithms under $\mu = 0.01$, SNR = 10 dB. **a** Error curves of the weight vectors. **b** Modulus curves of the weight vectors

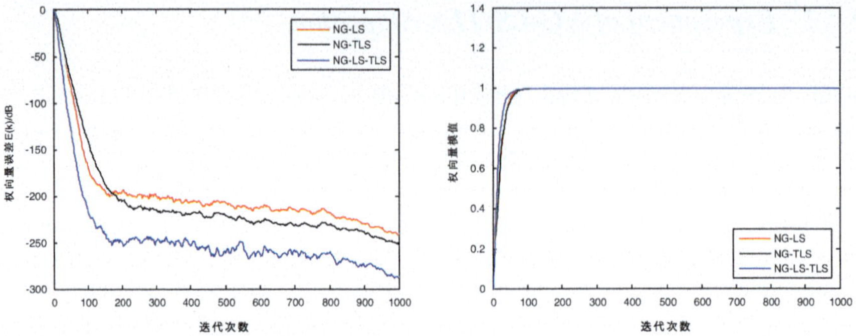

Fig. 9.6 Learning curves of different algorithms under $\mu = 0.01$, SNR = 20 dB. **a** Error curves of the weight vectors. **b** Modulus curves of the weight vectors

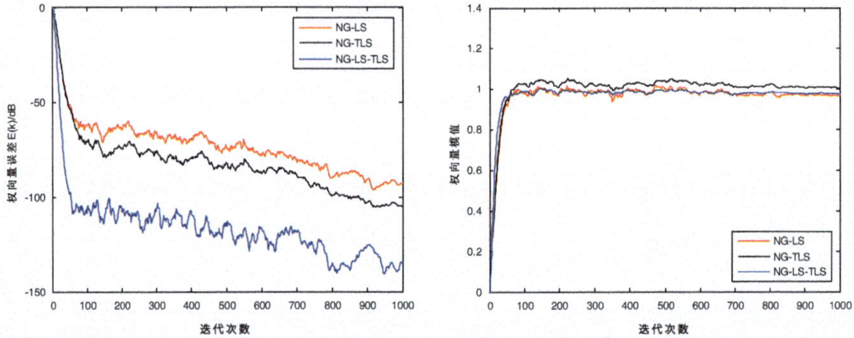

Fig. 9.7 Learning curves of different algorithms under $\mu = 0.005$, SNR = 5 dB. **a** Error curves of the weight vectors. **b** Modulus curves of the weight vectors

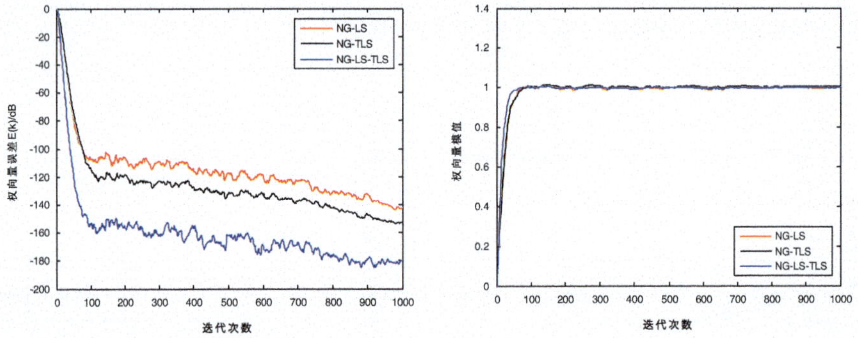

Fig. 9.8 Learning curves of different algorithms under $\mu = 0.005$, SNR = 10 dB. **a** Error curves of the weight vectors. **b** Modulus curves of the weight vectors

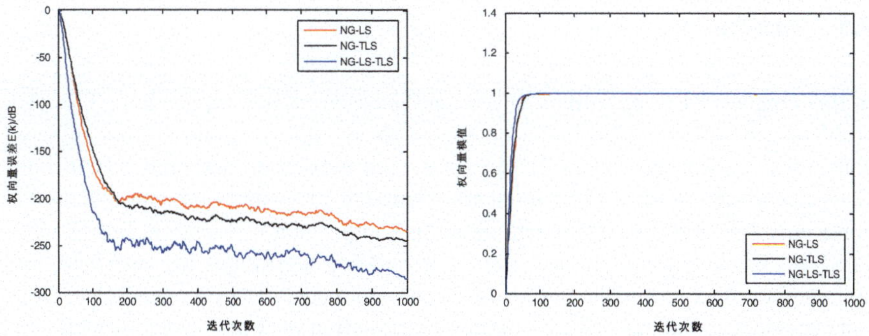

Fig. 9.9 Learning curves of different algorithms under $\mu = 0.005$, SNR $= 20$ dB. **a** Error curves of the weight vectors. **b** Modulus curves of the weight vectors

in non-Gauss noises of equivalent SNR, and a part of input signals do not contain non-Gauss noises.

Figures 9.4, 9.5, and 9.6 show that the learning error curves and the modulus convergence curves of the weight vectors of the three methods using the same learning rate $u = 0.01$, and different SNRs, respectively. Figures 9.7, 9.8, and 9.9 show that simulation effects of the three methods using the same learning rate $u = 0.005$, and different SNRs, respectively. In this simulation, the system vector is normalized by $\overline{\mathbf{H}} = \mathbf{H}/\|\mathbf{H}\|$, and the SNR of the input signal is set to be equal to that of the output signal.

From Figs. 9.4, 9.5, and 9.6, we can see that the NG-LS-TLS method shows the best estimation performance under the mixed noises, the weight vector error is also the smallest and the estimation accuracy is the highest. When the SNR is smaller, the convergence speed of the NG-TLS is better than that of the NG-LS method, and the weight vector error of the stable state is smaller than that of the NG-LS method, which means that the NG-TLS method based on EIV model considering both the input non-Gauss noise and the output non-Gauss noise has the better estimation effect.

When the SNR is 20 dB, we can see that the convergence speed of the NG-LS is slightly better than that of the NG-TLS method, but from the final weight vector error, the estimation effect of NG-TLS method is also better that that of the NG-LS method and the estimation accuracy is more higher. Further, when the SNR is 0.005 dB, from Figs. 9.7, 9.8, and 9.9, we can see that the convergence trends of the weight vector error do not much change compared to Figs. 9.4, 9.5, and 9.6, which means that the three methods have the similar estimation accuracy when the learning rate is smaller and does not much change.

9.5 Summary

In this chapter, TLS algorithms under non-Gaussian noises are studied. The TLS problems on several individual scenarios such as input and output contaminated with non-Gaussian noises (NG-TLS), partial input and output contaminated with non-Gaussian noises (NG-LS-TLS), input non-Gaussian noise only, output non-Gaussian noise only (NG-LS), and disparate input and output non-Gaussian noises (D-NG-TLS) are presented and studied, the corresponding algorithms are provided, and the simulation experiments and analyses are conducted. The proposed algorithms can provide flexible method for parameter estimation or fitting problem with non-Gaussian noise existing in both input and output data.

References

1. G.H. Golub, C.F. Van Loan, An analysis of the total least squares problem. SIAM J. Numeric. Anal. **17**(6), 883–893 (1980)
2. S. Van Huffel, J. Vandewalle, *The Total Least Squares Problem: Computational Aspects and Analysis* (SIAM, Philadelphia, PA, USA, 1991)
3. I. Markovsky, S. Van Huffel, Overview of total least-squares methods. Signal Process. **87**, 2283–2302 (2007)
4. P.C. Shen, C.G. Li, Minimum total error entropy method for parameter estimation. IEEE Trans. Signal Process. **63**(15), 4079–4090 (2025)
5. K.N. Plataniotis, D. Androutsos, A.N. Venetsanopoulos, Nonlinear filtering of non-Gaussian noise. J. Intell. Robot. Syst. **19**, 207–231 (1997)
6. B. Weng, K.E. Barner, Nonlinear system identification in impulsive environments. IEEE Trans. Signal Process. **53**(7), 2588–2594 (2005)
7. B. Chen, Y. Zhu, J. Hu, Mean-square convergence analysis of ADALINE training with minimum error entropy criterion. IEEE Trans. Neural Netw. **21**(7), 1168–1179 (2010)
8. J.C. Principe, *Information Theoretic Learning: Renyi's Entropy and Kernel Perspectives* (Springer-Verlag, Berlin, Germany, 2010)
9. D. Erdogmus, J.C. Principe, Generalized information potential criterion for adaptive system training. IEEE Trans. Neur. Netw. **13**(5), 1035–1044 (2002)
10. D. Erdogmus, J.C. Principe, An error entropy minimization algorithm for supervised training of nonlinear adaptive systems. IEEE Trans. Signal Process. **50**(7), 1780–1786 (2002)
11. D. Erdogmus, J.C. Principe, Convergence properties and data efficiency of the minimum error entropy criterion in adaline training. IEEE Trans. Signal Process. **51**(7), 1966–1978 (2003)
12. J.C. Principe, D. Xu, J.W. Fisher, Information theoretic learning, in *Unsupervised Adaptive Filtering*. ed. by S. Haykin (Wiley, New York, NY, USA, 2000), pp.265–319
13. E. Parzen, in *On estimation of a probability density function and mode, in Time Series Analysis Papers* (HoldenDay, San Francisco, CA, USA, 1967)
14. R. Arablouei, S. Werner, K. Dogancay, Analysis of the gradient descent total least- squares adaptive filtering algorithm. EEE Trans. Signal Process. **62**(5), 1256–1264 (2014)
15. D.Z. Feng, Z. Bao, L.C. Jiao, Total least mean squares algorithm. IEEE Trans. Signal Process. **46**(8), 2122–2130 (1997)

Chapter 10
Performance Analysis Methods of TLS Algorithms

10.1 Introduction

The stability and convergence are the important performances for an adaptive control system or algorithm, wherein the stability is also a prerequisite whether a system can work or not and the convergence is a metric on which the performance of adaptive algorithm is evaluated. The performance analysis such as convergence, stability of iterative or recursive TLS algorithms is an important topic. However, the convergence of neural-based TLS, MCA, or SVD learning algorithms is a difficult topic for direct study and analysis. Traditionally, the stability of these algorithms is analyzed based on the Lyapunov stability theory, and the convergence of these algorithms is indirectly analyzed based on the stochastic approximation theorem, or via corresponding deterministic continuous-time systems. In the analysis of stability of algorithms, how to choose a Lyapunov function is a difficult task, while in the convergence analysis, the stochastic approximation theorem requires that some restrictive conditions must be satisfied. One important condition is that the learning rates of the algorithms must approach zero, which is not a reasonable requirement to be imposed in many practical applications. Clearly, the restrictive condition is difficult to be satisfied in many practical applications, where a constant learning rate is usually used due to computational round-off issues and tracking requirements. All these studies of DCT formulations of learning algorithms are grounded on restrictive hypotheses so that the fundamental theorem of stochastic approximation can be applied.

However, when some of these hypotheses cannot be satisfied, how to study the convergence of the original stochastic discrete formulation? In order to analyze the convergence of neural-based TLS, MCA, or SVD learning algorithms, several methods have been proposed, i.e., the deterministic continuous-time (DCT) method, the stochastic discrete-time (SDT) method, the deterministic discrete-time (DDT) method and the fixed pointed method. The DCT method, first formalized by [1] and [2], is based on a fundamental theorem of stochastic approximation theory. Thus, it is an approximation analysis method. The SDT method is a direct analysis method, and

X. Kong and D. Feng, *Efficient Online Learning Algorithms for Total Least Square Problems*, Engineering Applications of Computational Methods 21,
https://doi.org/10.1007/978-981-97-1765-1_10

it can analyze the temporal behavior of algorithms and derive the relation between the dynamic stability and learning rate [3]. The DDT method, as a bridge between DCT and SDT methods, transforming the original SDT system into a corresponding DDT system and preserving the discrete-time nature of the original SDT systems, can shed some light on the convergence characteristics of SDT systems [4]. Zurifia proposed the DDT method [4], and Zhang et al. developed and promoted the application of the DDT method in MCA algorithms. Different from the DCT method, the DDT method allows the learning rate to be a constant and can be used to indirectly analyze the dynamic behaviors of stochastic learning algorithms. Since the DDT method is more reasonable for studying the convergence of neural-based MCA algorithms than the traditional DCT method, it has been widely used to study many neural network algorithms [5–6, 10].

However, the above algorithms all belong to traditional MCA algorithms. So far, there are few studies in the literature in which the DDT systems of generalized MCA (GMCA) or generalized minor eigenpairs extraction algorithms are analyzed. Generally, GMCA algorithms are more complex than traditional MCA algorithms, which will certainly increase the difficulty of analysis. Herein, we will derive some convergence conditions for the proposed algorithm via DDT method. The objective of this chapter is to review and study the performance analysis methods of TLS or MCA algorithms, and to place the focus on the DDT method of the convergence analysis of GMCA algorithms for adaptive TLS.

The remainder of this chapter is organized as follows. A review of performance analysis methods for neural-based MCA/TLS algorithms, including the DCT method, the SDT method, the Lyapunov function method, the fixed pointed method and the DDT method, is presented in Sect. 10.2. In Sect. 10.3, a novel self-stabilizing algorithm for generalized minor eigenpairs (GME) extraction is proposed, a DDT system of the novel GME algorithm is introduced, and numerical simulation is conducted. Section 10.4 gives the summary of this chapter.

10.2 Review of the Analysis Methods of MCA/TLS Algorithms

10.2.1 Deterministic Continuous-Time System Method

According to the stochastic approximation theory (see [1, 2]), if certain conditions are satisfied, its corresponding DCT systems can represent the SDT system effectively (i.e., their asymptotic paths are close with a large probability), and eventually, the MCA/TLS solution tends with probability 1 to the uniformly asymptotically stable solution of the ODE. From a computational point of view, the most important conditions are the following:

1. $x(t)$ is zero-mean, stationary, and bounded with probability 1.
2. $\alpha(t)$ is a decreasing sequence of *positive* scalars.

3. $\Sigma_t \alpha(t) = \infty$.
4. $\Sigma_t \alpha^P(t) < \infty$ for some p.
5. $\lim_{t\to\infty} \sup\left[\frac{1}{\alpha(t)} - \frac{1}{\alpha(t-1)}\right] < \infty$.

For example, the sequence $\alpha(t) = \text{const}\ t^{-\gamma}$ satisfies Conditions 2–5 for $0 < \gamma \leq 1$. The fourth condition is less restrictive than the Robbins–Monro condition $\Sigma_t \alpha^2(t) < \infty$, which is satisfied, for example, only by $\alpha(t) = \text{const}\ t^{-\gamma}$ with $1/2 < \gamma \leq 1$.

For example, MCA EXIN algorithm can be written as:

$$\mathbf{w}(t+1) = \mathbf{w}(t) - \frac{\alpha(t)y(t)}{\|\mathbf{w}(t)\|_2^2}\left[\mathbf{x}(t) - \frac{y(t)\mathbf{w}(t)}{\|\mathbf{w}(t)\|_2^2}\right], \tag{10.1}$$

and its corresponding deterministic continuous-time (DCT) systems is

$$\frac{d\mathbf{w}(t)}{dt} = -\frac{1}{\|\mathbf{w}(t)\|_2^2}\left[\mathbf{R} - \frac{\mathbf{w}^T(t)\mathbf{R}\mathbf{w}(t)}{\|\mathbf{w}(t)\|_2^2}\right]\mathbf{w}(t) = -\frac{1}{\|\mathbf{w}(t)\|_2^2}[\mathbf{R} - r(\mathbf{w}, \mathbf{R})\mathbf{I}]\mathbf{w}(t). \tag{10.2}$$

where $\mathbf{R} = E[\mathbf{x}(t)\mathbf{x}^T(t)]$ and $\boldsymbol{I}$ is the identity matrix. The asymptotic behavior of the MCA EXIN ODE has the following theorem.

Theorem 10.1 *Let* $\boldsymbol{R}$ *be the* $n \times n$ *autocorrelation matrix of the input data, with eigenvalues* $0 \leq \lambda_n \leq \lambda_{n-1} \leq \cdots \leq \lambda_1$ *and corresponding orthonormal eigenvectors* $\mathbf{z}_n, \mathbf{z}_{n-1}, \cdots, \mathbf{z}_1$. *If w(0) satisfies* $\mathbf{w}^T(0)\mathbf{z}_n \neq 0$ *and* λ_n *is single, then, in the limits of validity of the ODE approximation, for MCA EXIN it holds*

$$\mathbf{w}(t) \to \pm\|\mathbf{w}(0)\|_2\mathbf{z}_n \ \vee\ \mathbf{w}(t) \to \infty \tag{10.3}$$

$$\|\mathbf{w}(t)\|_2^2 \approx \|\mathbf{w}(0)\|_2^2,\ t > 0. \tag{10.4}$$

For the convergence proof using deterministic continuous-time system method, refer to the proof of Theorem 16 in [3] for details.

10.2.2 Stochastic Discrete-Time System Method

Using only the ODE approximation does not reveal some of the most important features of these algorithms [3]. For instance, it can be shown that the constancy of the weight modulus for OJAn, Luo and MCA EXIN, which is the consequence of the use of the ODE, is not valid, except, as a very first approximation, in approaching the minor component [3]. The stochastic discrete-time system method has led to the very important problem of the sudden divergence [3]. In the following, we will analyze the performance of Luo MCA algorithm using the SDT system method.

In [7], Luo proposed a MCA algorithm, which is

$$\mathbf{w}(t+1) = \mathbf{w}(t) - \alpha(t)\|\mathbf{w}(t)\|_2^2\left[y(t)\mathbf{x}(t) - \frac{y^2(t)}{\|\mathbf{w}(t)\|_2^2}\mathbf{w}(t)\right]. \tag{10.5}$$

Since (10.5) is the gradient flow of the Rayleigh quotient (RQ) and using the property of orthogonality of RQ, it holds that [3]

$$\mathbf{w}^T(t)\left\{y(t)\mathbf{x}(t) - \frac{y^2(t)}{\|\mathbf{w}(t)\|_2^2}\mathbf{w}(t)\right\} = 0, \tag{10.6}$$

i.e., *the weight increment at each iteration is orthogonal to the weight direction.* The squared modulus of the weight vector at instant $t+1$ is then given by

$$\|\mathbf{w}(t+1)\|_2^2 = \|\mathbf{w}(t)\|_2^2 + \frac{\alpha^2(t)}{4}\|\mathbf{w}(t)\|_2^6\|\mathbf{x}(t)\|_2^4\sin^2 2\vartheta_{\mathbf{xw}}, \tag{10.7}$$

where $\vartheta_{\mathbf{xw}}$ is the angle between the direction of $\boldsymbol{x}(t)$ and $\boldsymbol{w}(t)$. From (10.7), we can see that (1): Except for particular conditions, *the weight modulus always increases*, $\|\mathbf{w}(t+1)\|_2^2 > \|\mathbf{w}(t)\|_2^2$. . These particular conditions, i.e., all data in exact particular directions, are too rare to be found in a noisy environment. (2) $\sin^2 2\vartheta_{\mathbf{xw}}$ is a positive function with peaks within the interval $(-\pi, \pi]$. This is one of the possible interpretations of the oscillatory behavior of weight modulus.

10.2.3 Lyapunov Function Approach

In [8], a random-gradient-based algorithm, called OJAm, is developed for online tracking the minor component. The algorithm is as follows:

$$\mathbf{w}(t+1) = \mathbf{w}(t) - \mu(t)\{y(t)\mathbf{x}(t) - \mathbf{w}(t)y^2(t)/[\mathbf{w}^T(t)\mathbf{w}(t)]^2\}, \tag{10.8}$$

where $y(t) = \mathbf{w}^T(t)\mathbf{x}(t)$. Under certain assumptions [1, 2], the application of the stochastic approximation theories [1, 2] leads to the following averaging equation associated with (10.8):

$$\frac{d\mathbf{w}(t)}{dt} = -\mathbf{R}\mathbf{w}(t) + \mathbf{w}(t)[\mathbf{w}^T(t)\mathbf{R}\mathbf{w}(t)]/[\mathbf{w}^T(t)\mathbf{w}(t)]^2. \tag{10.9}$$

A cost or energy function of (10.9) is expressed by

$$J_{OJAm}(\mathbf{w}) = \frac{1}{2}\{\ln[\mathbf{w}^T\mathbf{R}\mathbf{w}] + [\mathbf{w}^T\mathbf{w}]^{-1}\}. \tag{10.10}$$

It is easily shown that for any constant c the set $\{\mathbf{w}|J_{OJAm}(\mathbf{w}) < c\}$ is finite. Moreover, differentiating $J_{OJAm}(\mathbf{w})$ with respect to t by the chain rule along the solution of (10.9) yields

$$\begin{aligned}\frac{dJ_{OJAm}(\mathbf{w})}{dt} &= \frac{d\mathbf{w}^T}{dt}\left[\frac{\mathbf{R}\mathbf{w}}{\mathbf{w}^T\mathbf{R}\mathbf{w}} - \frac{\mathbf{w}}{[\mathbf{w}^T\mathbf{w}]^2}\right] \\ &= \frac{1}{\mathbf{w}^T\mathbf{R}\mathbf{w}}\frac{d\mathbf{w}^T}{dt}\left[\mathbf{R}\mathbf{w} - \frac{(\mathbf{w}^T\mathbf{R}\mathbf{w})\mathbf{w}}{[\mathbf{w}^T\mathbf{w}]^2}\right] \\ &= -\frac{1}{\mathbf{w}^T\mathbf{R}\mathbf{w}}\frac{d\mathbf{w}^T}{dt}\frac{d\mathbf{w}}{dt}.\end{aligned} \tag{10.11}$$

This shows that $J_{OJAm}(\mathbf{w})$ is a Lyapunov function. According to the Lyapunov theory, (10.9) is globally asymptotically convergent. Therefore, it is concluded that the OJAm algorithm is globally asymptotically convergent.

10.2.4 Fixed Pointed Method

In [9], a novel parallel multiple MC extraction algorithm by diagonal matrix, called Diagonal-Douglas algorithm, is proposed. The algorithm is as follows:

$$\mathbf{W}(t+1) = \mathbf{W}(t) - \eta\left[\mathbf{W}(t)\mathbf{W}^T(t)\mathbf{R}\mathbf{W}(t)\mathbf{A} - \mathbf{R}\mathbf{W}(t)\left(\mathbf{W}^T(t)\mathbf{W}(t)\right)^2\right], \tag{10.12}$$

where $\boldsymbol{A} = \mathrm{diag}\{a_1, a_2,, a_r\}$ is an arbitrary diagonal matrix and its diagonal elements only need to satisfy the requirement that $a_1 > a_2 > \cdot > a_r > 0$, and r means the number of desired MCs.

Based on the optimization theory, the algorithm can only reach steady state at the fixed points, so we should firstly search all of the fixed points, which will be completed by the following theorem.

Theorem 10.2 *Suppose that* $\boldsymbol{H}_r$ *is composed by any r distinct eigenvectors of* $\boldsymbol{R}$*, then* $\mathbf{H}_r\sqrt{\mathbf{A}}$ *is the fixed point of the Diagonal-Douglas algorithm.*

The detailed proof steps of Theorem 10.2 can see the reference [9] and also see the Sect. 6.3.2 in Chap. 6.

Theorem 10.2 lists all of the fixed points, yet it does not confirm that at which fixed point the Diagonal-Douglas algorithm is stable. So we must pick out the stable one from all of the fixed points, and the main results are given by the following theorem.

Theorem 10.3 *The Diagonal-Douglas algorithm is stable if and only if* $\mathbf{W} = \mathbf{P}\sqrt{\mathbf{\Lambda}}$, *where* $\mathbf{P} \in \Re^{n \times r}$ *is composed by the first r MCs arranged by their corresponding eigenvalues in ascending order, i.e.,* $\boldsymbol{P} = [\boldsymbol{h}_1, \boldsymbol{h}_2, \cdot, \boldsymbol{h}_r]$.

Similarly, the detailed proof steps of Theorem 10.3 can see the reference [9], and also see the Sect. 6.3.2 in Chap. 6.

10.2.5 Deterministic Discrete-Time System Method

Traditionally, the convergence of neural-based MCA or TLS learning algorithms is analyzed via DCT systems based on a stochastic approximation theorem. However, there exist some restrictive conditions when using stochastic approximation theorem. One crucial condition is that the learning rate in the learning algorithm must converge to zero, which is not suitable in most practical applications because of the round-off limitation and tracking requirements [4, 5]. In order to overcome the shortcomings of the DCT method, Zurifia proposed the DDT method [4], and Zhang et al. developed and promoted the application of the DDT method in MCA algorithms.

The DDT method allows the learning rate to be a constant and can be used to indirectly analyze the dynamic behaviors of stochastic learning algorithms. The convergence of many MCA algorithms [5–6, 10] has been analyzed by the DDT method. The detailed steps of the DDT method can be found in references [5, 10]. In this chapter, we place the focus on the DDT convergence analysis of GMCA algorithms for adaptive TLS, in which the DDT method is extended to analyze the convergence of generalized MCA algorithms.

10.3 Convergence of Novel Generalized MCA Algorithm via DDT

In this section, we will analyze the convergence of a novel self-stabilizing algorithm for generalized minor eigenpairs extraction proposed by us via a DDT method.

10.3.1 Algorithm Presentation

A. ***Definitions***

Definition 10.1 Given an vector $\mathbf{w} \in \Re^{n \times 1}$ and a matrix $\mathbf{R} \in \Re^{n \times n}$, define the R-norm of vector $\mathbf{w}$ as $\|\mathbf{w}\|_{\mathbf{R}} = \sqrt{\mathbf{w}^T \mathbf{R} \mathbf{w}}$.

B. ***Proposed DDT System for MGE Extraction***

Consider the following linear neural network:

$$f(k) = \mathbf{w}^T(k)\mathbf{g}(k),$$

where $\boldsymbol{w}(k)$ denotes the weight vector, and $\boldsymbol{g}(k)$ and $f(k)$ are the input and output of this model, respectively. Our objective is to construct a weight vector updating rule so that the weight vector can reach the direction of the desired minor generalized eigenvector (MGE). Suppose that $\{\boldsymbol{y}(k)\}$ and $\{\boldsymbol{x}(k)\}$, $k = 1, 2,\ldots$ are two input signals that are sampled from two stationary stochastic processes. Let $\boldsymbol{R}_x = E[\boldsymbol{x}(k)\boldsymbol{x}^T(k)]$ and $\boldsymbol{R}_y = E[\boldsymbol{y}(k)\boldsymbol{y}^T(k)]$ denote the autocorrelation matrices of the two input signals, respectively. Clearly, $(\boldsymbol{R}_y,\boldsymbol{R}_x)$ has n non-negative generalized eigenvalues and generalized eigenvectors. For convenience, let us denote them as λ_i and v_i $(i = 1, 2,\ldots, n)$, respectively. When using the neural network method to extract the MGE, the two input signals $\{\boldsymbol{y}(k)\}$ and $\{\boldsymbol{x}(k)\}$ serve as the input of this model.

Here, by extending the Chen algorithm [11], we propose the following DDT system for estimating the MGE:

$$\begin{aligned}\mathbf{w}(k+1) = \mathbf{w}(k) + \eta\big[&\mathbf{w}^T(k)\mathbf{R}_y\mathbf{w}(k)\,\mathbf{w}(k) + \mathbf{w}(k)\\ &- \mathbf{w}^T(k)\mathbf{R}_x\mathbf{w}(k)(\mathbf{I} + (\mathbf{R}_x)^{-1}\mathbf{R}_y)\mathbf{w}(k)\big].\end{aligned} \tag{10.13}$$

Note that if the matrix $\boldsymbol{R}_x$ in (10.13) is replaced with $\boldsymbol{I}$, then the proposed algorithm reduces to the Chen algorithm. For ease of discussion, we refer to this algorithm as the generalized Chen (GChen) algorithm herein.

C. ***Adaptive Algorithm***

For some real applications, the autocorrelation matrices ($\boldsymbol{R}_y$ and $\boldsymbol{R}_x$) are not explicitly given. In this case, one commonly used method is to estimate them online from their respective input signals [$\boldsymbol{x}(x)$ and $\boldsymbol{y}(k)$]. Herein, we adopt the exponentially weighted sample method to estimate the autocorrelation matrices. The updating equations are given as follows:

$$\hat{\mathbf{R}}_y(k+1) = \beta\hat{\mathbf{R}}_y(k) + \mathbf{y}^T(k+1)\mathbf{y}(k+1), \tag{10.14}$$

$$\hat{\mathbf{R}}_x(k+1) = \alpha\hat{\mathbf{R}}_x(k) + \mathbf{x}^T(k+1)\mathbf{x}(k+1), \tag{10.15}$$

where $0 < \alpha, \beta \le 1$ are the forgetting factors. The exact values of the forgetting factors depend on the input signals. The forgetting factors can be used to implement an effective window that ensures that past samples are less significantly weighted than recent ones. The forgetting factors are set to values close to one for slow time-varying signals, and one should choose near-zero values for fast time-varying inputs.

Denote $\mathbf{Q}(k+1) = \mathbf{R}_x^{-1}(k+1)$. By applying the matrix inversion lemma to (10.15), we have

$$\mathbf{Q}(k+1) = \frac{1}{\alpha}\left[\mathbf{Q}(k) - \frac{\mathbf{Q}(k)\mathbf{x}(k+1)\mathbf{x}^T(k+1)\mathbf{Q}(k)}{\alpha + \mathbf{x}^T(k+1)\mathbf{Q}(k)\mathbf{x}(k+1)}\right]. \tag{10.16}$$

By replacing $\mathbf{R}_y$, $\mathbf{R}_x$, $\mathbf{R}_x^{-1}$ in the algorithm (10.13) by $\hat{\mathbf{R}}_y(k+1)$,$\hat{\mathbf{R}}_x(k+1)$,$\mathbf{Q}(k+1)$, respectively, an adaptive algorithm can be obtained for calculating the MGE, which is given as follows.

Algorithm 1
Adaptive version of the GChen algorithm.

Initialization: Set $\hat{\mathbf{R}}_y(0) = \hat{\mathbf{R}}_x(0) = \mathbf{0}$, $\mathbf{Q}_x(0) = \delta\mathbf{I}$, where δ is a small number and $\boldsymbol{0}$ is a null matrix whose elements are all zero. $\boldsymbol{w}(0)$ is a random $n \times 1$ vector.

Update: for $k = 1,2,3,\ldots n$, repeat the following steps.

Step 1. Calculate $\hat{\mathbf{R}}_y(k)$, $\hat{\mathbf{R}}_x(k)$, $\mathbf{Q}_x(k)$ by (10.14–10.16).

Step 2. Update the weight vector $\mathbf{w}(k)$

$$\begin{aligned}\mathbf{w}(k+1) = \mathbf{w}(k) + \eta\big[&\mathbf{w}^T(k)\hat{\mathbf{R}}_y\mathbf{w}(k)\,\mathbf{w}(k) + \mathbf{w}(k)\\ &- \mathbf{w}^T(k)\hat{\mathbf{R}}_x\mathbf{w}(k)(\mathbf{I} + \mathbf{Q}_x(k)\hat{\mathbf{R}}_y)\mathbf{w}(k)\big].\end{aligned} \tag{10.17}$$

10.3.2 Self-stabilizing Analysis

This section serves to prove the self-stabilization property of the GChen algorithm.

Theorem 10.4 If η is small enough, then the $\boldsymbol{R}_x$-norm of the weight vector in (10.13) will converge to one.

Proof From the weight vector updating Eq. (10.13), we can easily calculate the weight vector norm at time $k+1$ by the following equation:

$$\begin{aligned}\|\mathbf{w}(k+1)\|_{\mathbf{R}_x}^2 &= \mathbf{w}^T(k+1)\mathbf{R}_x\mathbf{w}(k+1)\\ &= [\mathbf{w}(k) + \eta(\mathbf{w}^T(k)\mathbf{R}_x\mathbf{w}(k)\mathbf{w}(k) + \mathbf{w}(k)\\ &\quad - \mathbf{w}^T(k)\mathbf{R}_x\mathbf{w}(k)(\mathbf{I} + \mathbf{R}_x^{-1}\mathbf{R}_y)\mathbf{w}(k))]^T\mathbf{R}_x\\ &\quad [\mathbf{w}(k) + \eta(\mathbf{w}^T(k)\mathbf{R}_x\mathbf{w}(k)\mathbf{w}(k) + \mathbf{w}(k)\\ &\quad - \mathbf{w}^T(k)\mathbf{R}_x\mathbf{w}(k)(\mathbf{I} + \mathbf{R}_x^{-1}\mathbf{R}_y)\mathbf{w}(k))]\\ &= \mathbf{w}^T(k)\mathbf{R}_x\mathbf{w}(k) + 2\eta\mathbf{w}^T(k)\mathbf{R}_x\mathbf{w}(k)\mathbf{w}(k) \times [1 - \mathbf{w}^T(k)\mathbf{R}_x\mathbf{w}(k)] + O(\eta)\\ &\approx \mathbf{w}^T(k)\mathbf{R}_x\mathbf{w}(k)\{1 + 2\eta[1 - \mathbf{w}^T(k)\mathbf{R}_x\mathbf{w}(k)]\}.\end{aligned} \tag{10.18}$$

Since η is sufficiently small, we neglect the second-order terms in the equation after the fourth equal sign. Comparing the weight vector norm at the next time with that at the current time, we obtain

$$\frac{\|\mathbf{w}(k+1)\|^2_{\mathbf{R}_x}}{\|\mathbf{w}(k)\|^2_{\mathbf{R}_x}} \approx \frac{\mathbf{w}^T(k)\mathbf{R}_x\mathbf{w}(k)\{1+2\eta[1-\mathbf{w}^T(k)\mathbf{R}_x\mathbf{w}(k)]\}}{\mathbf{w}^T(k)\mathbf{R}_x\mathbf{w}(k)} \approx 1+2\eta[1-\|\mathbf{w}(k)\|^2_{\mathbf{R}_x}]$$

$$\begin{cases} >1 \;\; if \;\; \|\mathbf{w}(k)\|^2_{\mathbf{R}_x} < 1 \\ =1 \;\; if \;\; \|\mathbf{w}(k)\|^2_{\mathbf{R}_x} = 1 \\ <1 \;\; if \;\; \|\mathbf{w}(k)\|^2_{\mathbf{R}_x} > 1 \end{cases}. \tag{10.19}$$

From (10.19), it is obvious that regardless of whether the weight vector norm is bigger or smaller than one at the current time, the weight vector norm will tend to one at the next time. Since the final convergence value has no relationship with the initial weight vector, we can conclude that the algorithm described by (10.13) is self-stabilizing.

The proof is complete.

10.3.3 Dynamic Characteristic of the GChen Algorithm

Next, the dynamic characteristic of the GChen algorithm will be studied via the DDT method. Following the classical transformation steps in [4], the DDT system of (10.17) can be easily obtained, which has the same formula as (10.13).

For convenience, let us suppose that the matrix pencil ($\boldsymbol{R}_y$, $\boldsymbol{R}_x$) has distinct generalized eigenvalues, which are arranged in a non-ascending order, i.e., $\lambda_1 > \lambda_2 > \cdot > \lambda_n \geq 0$. The corresponding generalized eigenvectors are denoted by $\boldsymbol{v}_1, \boldsymbol{v}_2, \ldots, \boldsymbol{v}_n$. Clearly, these generalized eigenvectors compose an $\boldsymbol{R}_x$-orthonormal basis of $\Re^n$. By using this basis, $\boldsymbol{w}(k)$ can be represented as

$$\mathbf{w}(k) = \sum_{i=1}^{n} z_i(k)\mathbf{v}_i, \tag{10.20}$$

where $\mathbf{z}_i(k) = \mathbf{v}_i^T\mathbf{R}_x\mathbf{w}(k)(i = 1, 2, \cdots, n)$ is the projection of $\boldsymbol{w}(k)$ onto generalized eigenvector $\boldsymbol{v}_i$. Substituting (10.20) into $f(k) = \mathbf{w}^T(k)\mathbf{g}(k)$ yields

$$z_i(k+1) = \left[1+\eta(1+\mathbf{w}^T(k)\mathbf{R}_y\mathbf{w}(k) - (\lambda_i+1)\mathbf{w}^T(k)\mathbf{R}_x\mathbf{w}(k))\right]z_i(k) \tag{10.21}$$

for all $k \geq 0$.

According to the properties of the Rayleigh Quotient [12], it holds that

$$0 \leq \lambda_n < \frac{\mathbf{w}^T(k)\mathbf{R}_y\mathbf{w}(k)}{\mathbf{w}^T(k)\mathbf{R}_x\mathbf{w}(k)} < \lambda_1, \tag{10.22}$$

for each $k \geq 0$.

To analyze the dynamic behavior of the GChen algorithm, we first prove that the weight vector norm is bounded throughout the iterations by the following two theorems.

Theorem 10.5 *Suppose that* $\eta \leq 0.4$ *and* $\eta\lambda_1 \leq 0.35$. *If the initial vector satisfies* $\mathbf{w}^T(0)\mathbf{R}_x\mathbf{v}_n \neq 0$, *and* $\|\mathbf{w}(0)\|_{\mathbf{R}_x} \leq 1$, *it holds that* $\|\mathbf{w}(k)\|_{\mathbf{R}_x} < 1 + \eta\lambda_1$ *for all* $k \geq 0$.

Proof From (10.20) and (10.21), we obtain

$$\begin{aligned}\|\mathbf{w}(k+1)\|^2_{\mathbf{R}_x} &== \sum_{i=1}^{n} z_i^2(k+1)\\ &= \sum_{i=1}^{n}\left[1+\eta(1+\mathbf{w}^T(k)\mathbf{R}_y\mathbf{w}(k)-(\lambda_i+1)\mathbf{w}^T(k)\mathbf{R}_x\mathbf{w}(k)\right]^2 z_i^2(k)\\ &< \left[1+\eta(1+\lambda_1\mathbf{w}^T(k)\mathbf{R}_x\mathbf{w}(k)-(\lambda_n+1)\mathbf{w}^T(k)\mathbf{R}_x\mathbf{w}(k)\right]^2\\ &\sum_{i=1}^{n} z_i^2(k) < \left[1+\eta(1+(\lambda_1-1))\mathbf{w}^T(k)\mathbf{R}_x\mathbf{w}(k)\right]^2\|\mathbf{w}(k)\|^2_{\mathbf{R}_x}.\end{aligned} \tag{10.23}$$

Then we have

$$\|\mathbf{w}(k+1)\|^2_{\mathbf{R}_x} < \left[1+\eta(1+(\lambda_1-1))\|\mathbf{w}(k)\|^2_{\mathbf{R}_x}\right]^2\|\mathbf{w}(k)\|^2_{\mathbf{R}_x}. \tag{10.24}$$

Over the interval [0, 1], define a differential function

$$f(s) = (1+\eta(1+(\lambda_1-1)s))^2 s, \tag{10.25}$$

where $s = \|\mathbf{w}(k)\|^2_{\mathbf{R}_x}$ and $f(s) = \|\mathbf{w}(k+1)\|^2_{\mathbf{R}_x}$.

Let $\dot{f}(s)$ be the derivative of $f(s)$. Then

$$\dot{f}(s) = (1+\eta+\eta(\lambda_1-1)s)(1+\eta+3\eta(\lambda_1-1)s). \tag{10.26}$$

From (10.26), we have that $\dot{f}(s) = 0$ if

$$s = s_1 = -\frac{1+\eta}{\eta(\lambda_1-1)} \; or \; s = s_2 = -\frac{1+\eta}{3\eta(\lambda_1-1)}. \tag{10.27}$$

Next, let us consider the following two cases.

Case 1: ($\lambda_1 > 1$): Since $\eta > 0$, $\lambda_1 > 1$, we have $s_1 < s_2 < 0$. Furthermore, we have

$$\dot{f}(s) > 0. \tag{10.28}$$

for all $0 < s < 1$.

Case 2: ($\lambda_1 < 1$): By using $\eta < 0.4$, we have $s_1 > s_2 > 1$. Then, for all $0 < s < 1$, we can obtain that

$$\dot{f}(s) > 0. \tag{10.29}$$

From (10.28) and (10.29), we can conclude that $f(s)$ is a monotonically increasing function on [0, 1]. Therefore, the maximum is attained when $s = 1$, i.e.,

$$f(s) \leq f(1) = (1 + \eta\lambda_1)^2. \tag{10.30}$$

From (10.24), we have

$$\|\mathbf{w}(k)\|_{\mathbf{R}_x} < 1 + \eta\lambda_1. \tag{10.31}$$

for all $k \geq 0$.

This completes the proof.

Theorem 10.6 *Suppose that* $\eta \leq 0.4$ *and* $\eta\lambda_1 \leq 0.35$. *If the initial vector satisfies* $\mathbf{w}^T(0)\mathbf{R}_x\mathbf{v}_n \neq 0$ *and* $\|\mathbf{w}(0)\|_{\mathbf{R}_x} \leq 1$, *then it holds that* $\|\mathbf{w}(k)\|_{\mathbf{R}_x} > c$ *for all* $k \geq 0$, *where*

$c = [1 - \eta(\eta\lambda_1)(2 + \eta\lambda_1) - \eta\lambda_1(1 + \eta\lambda_1)^2](1 + \eta\lambda_1)$.

Proof By using (10.20) and (10.21), we can obtain that

$$\begin{aligned} \|\mathbf{w}(k+1)\|^2_{\mathbf{R}_x} &== \sum_{i=1}^{n} z_i^2(k+1) \\ &= \sum_{i=1}^{n} \left[1 + \eta(1 + \mathbf{w}^T(k)\mathbf{R}_y\mathbf{w}(k) - (\lambda_i + 1)\mathbf{w}^T(k)\mathbf{R}_x\mathbf{w}(k)\right]^2 z_i^2(k). \end{aligned} \tag{10.32}$$

Next, let us consider the following two cases.

Case 1 $((1 + \mathbf{w}^T(k)\mathbf{R}_y\mathbf{w}(k) - (\lambda_i + 1)\mathbf{w}^T(k)\mathbf{R}_x\mathbf{w}(k)) > 0)$:

By using the conclusion $\|\mathbf{w}(k)\|_{\mathbf{R}_x} < 1 + \eta\lambda_1$ in Theorem 10.6, we have

$$\begin{aligned} &1 + \eta[(1 + \mathbf{w}^T(k)\mathbf{R}_y\mathbf{w}(k) - (\lambda_i + 1)\mathbf{w}^T(k)\mathbf{R}_x\mathbf{w}(k)] \\ &> 1 + \eta - \eta(1 + \lambda_1)\mathbf{w}^T(k)\mathbf{R}_y\mathbf{w}(k) \\ &\geq 1 - \eta(\eta\lambda_1)(2 + \eta\lambda_1) - \eta\lambda_1(1 + \eta\lambda_1)^2 \\ &\geq 1 - 0.4 \times 0.35 \times (2 + 0.35) - 0.35 \times (1 + 0.35)^2 \\ &= 0.0331 > 0. \end{aligned} \tag{10.33}$$

From (10.32) and (10.33), we have

$$\|\mathbf{w}(k+1)\|^2_{\mathbf{R}_x} \geq [1 - \eta\lambda_1(1 + \eta\lambda_1)^2 - \eta(\eta\lambda_1)(2 + \eta\lambda_1)]^2(1 + \eta\lambda_1)^2. \tag{10.34}$$

Case 2 ($(1 + \mathbf{w}^T(k)\mathbf{R}_y\mathbf{w}(k) - (\lambda_i + 1)\mathbf{w}^T(k)\mathbf{R}_x\mathbf{w}(k)) < 0$):
In this case, we have

$$(\lambda_i + 1)\mathbf{w}^T(k)\mathbf{R}_x\mathbf{w}(k) - 1 > \mathbf{w}^T(k)\mathbf{R}_y\mathbf{w}(k) > 0. \tag{10.35}$$

From Theorem 10.5, we have

$$\begin{aligned}
&1 - \eta[(\lambda_i + 1)\mathbf{w}^T(k)\mathbf{R}_x\mathbf{w}(k) - 1 - \mathbf{w}^T(k)\mathbf{R}_y\mathbf{w}(k)] \\
&> 1 + \eta - \eta(\lambda_1 + 1)\mathbf{w}^T(k)\mathbf{R}_x\mathbf{w}(k) \\
&\geq 1 - \eta(\eta\lambda_1)(2 + \eta\lambda_1) - \eta\lambda_1(1 + \eta\lambda_1)^2 \\
&\geq 1 - 0.4 \times 0.35 \times (2 + 0.35) - 0.35 \times (1 + 0.35)^2 \\
&= 0.0331 > 0.
\end{aligned} \tag{10.36}$$

By using (10.32) and (10.36), we obtain

$$\|\mathbf{w}(k+1)\|^2_{\mathbf{R}_x} \geq [1 - \eta\lambda_1(1 + \eta\lambda_1)^2 - \eta(\eta\lambda_1)(2 + \eta\lambda_1)]^2(1 + \eta\lambda_1)^2. \tag{10.37}$$

Denote $c = [1 - \eta(\eta\lambda_1)(2 + \eta\lambda_1) - \eta\lambda_1(1 + \eta\lambda_1)^2](1 + \eta\lambda_1)$. Then, by using $\eta \leq 0.4$ *and* $\eta\lambda_1 \leq 0.35$, we can obtain

$$\begin{aligned}
c &= [1 - \eta(\eta\lambda_1)(2 + \eta\lambda_1) - \eta\lambda_1(1 + \eta\lambda_1)^2](1 + \eta\lambda_1) \\
&> [1 - 0.4 \times 0.35 \times (2 + 0.35) - 0.35 \times (1 + 0.35)^2] \\
&\times (1 + 0.35) = 0.0447 > 0.
\end{aligned} \tag{10.38}$$

Clearly,

$$\|\mathbf{w}(k+1)\|^2_{\mathbf{R}_x} \geq c \tag{10.39}$$

for all $k \geq 0$.

This completes the proof.

Theorems 10.5 and 10.6 provide the upper and lower bounds of the weight vector norm. Next, we will prove that when the above conditions are satisfied, the weight vector tends to $a\mathbf{v}_n$ as k approaches infinity, i.e., $\lim_{k\to\infty} \mathbf{w}(k) = a\mathbf{v}_n$, where a is a constant. To complete this task, the following preliminary conclusions are needed.

Lemma 10.1 *Suppose that* $\eta \leq 0.4$ *and* $\eta\lambda_1 \leq 0.35$. *If* $\mathbf{w}^T(0)\mathbf{R}_x\mathbf{v}_n \neq 0$, *and* $\|\mathbf{w}(0)\|_{\mathbf{R}_x} \leq 1$, *for* all $k \geq 0$, *it holds that* $1 + \eta(1 + \mathbf{w}^T(k)\mathbf{R}_y\mathbf{w}(k) - (\lambda_i + 1)\mathbf{w}^T(k)\mathbf{R}_x\mathbf{w}(k)) > 0$ *for all* $k \geq 0$, *where* $i = 1, 2, \cdots, n$.

Proof By using the two conditions $\eta\lambda_1 \leq 0.35$ and $\eta \leq 0.4$, we obtain

$$\begin{aligned}
&1 + \eta[1 + \mathbf{w}^T(k)\mathbf{R}_y\mathbf{w}(k) - (\lambda_i + 1)\mathbf{w}^T(k)\mathbf{R}_x\mathbf{w}(k)] \\
&> 1 + \eta - \eta(\lambda_1 + 1)\mathbf{w}^T(k)\mathbf{R}_x\mathbf{w}(k)
\end{aligned}$$

$$\begin{aligned} &\geq 1 - \eta(\eta\lambda_1)(2 + \eta\lambda_1) - \eta\lambda_1(1 + \eta\lambda_1)^2 \\ &\geq 1 - 0.4 \times 0.35 \times (2 + 0.35) - 0.35 \times (1 + 0.35)^2 \\ &= 0.0331 > 0. \end{aligned} \tag{10.40}$$

This completes the proof.

From Lemma 10.1, we can conclude that throughout the iterations, $z_i(k)$ does not change signs in (10.21). Clearly, by using $\mathbf{w}^T(0)\mathbf{R}_x\mathbf{v}_n \neq 0$, we obtain

$$z_n(k) = \mathbf{v}_n^T\mathbf{R}_x\mathbf{w}(k) \neq 0. \tag{10.41}$$

From (10.20), for all $k \geq 0$, $\mathbf{w}(k)$ can be decomposed into

$$\mathbf{w}(k) = \sum_{i=1}^{n-1} \mathbf{z}_i(k)\mathbf{v}_i + \mathbf{z}_n(k)\mathbf{v}_n. \tag{10.42}$$

Clearly, the convergence of $\mathbf{w}(k)$ can be determined from that of $z_i(k)(i = 1, 2, \cdots, n)$. Lemmas 10.2 and 10.3 provide the convergence analysis of $z_i(k)(i = 1, 2, \cdots, n)$.

Lemma 10.2 *Suppose that* $\eta \leq 0.4$ *and* $\eta\lambda_1 \leq 0.35$. *If the initial vector satisfies* $\mathbf{w}^T(0)\mathbf{R}_x\mathbf{v}_n \neq 0$ *and* $\|\mathbf{w}(0)\|_{\mathbf{R}_x} \leq 1$, *it holds that* $\lim_{k\to\infty} z_i(k) = 0$, $(i = 2, \cdots, n-1)$.

Proof From Lemma 10.1, for all $k \geq 0$, we have

$$1 + \eta(1 + \mathbf{w}^T(k)\mathbf{R}_y\mathbf{w}(k) - (\lambda_i + 1)\mathbf{w}^T(k)\mathbf{R}_x\mathbf{w}(k)) > 0, \tag{10.43}$$

where $i = 1, 2, \cdots, n-1$.

By Theorems 10.5 and 10.6, it holds that

$$c < \|\mathbf{w}(k)\|_{\mathbf{R}_x} < 1 + \eta\lambda_1, \tag{10.44}$$

for all $k \geq 0$. Then we have

$$\begin{aligned} &\left[\frac{z_i(k+1)}{z_n(k+1)}\right]^2 \\ &= \left[1 - \frac{\eta(\lambda_i - \lambda_n)\mathbf{w}^T(k)\mathbf{R}_x\mathbf{w}(k)}{1 + \eta(1 + \mathbf{w}^T(k)\mathbf{R}_y\mathbf{w}(k) - (\lambda_n + 1)\mathbf{w}^T(k)\mathbf{R}_x\mathbf{w}(k))}\right]^2 \frac{z_i^2(k)}{z_n^2(k)} \\ &\leq \left[1 - \frac{\eta(\lambda_i - \lambda_n)\mathbf{w}^T(k)\mathbf{R}_x\mathbf{w}(k)}{1 + \eta(1 + (\lambda_1 - \lambda_n - 1)\mathbf{w}^T(k)\mathbf{R}_x\mathbf{w}(k))}\right]^2 \frac{z_i^2(k)}{z_n^2(k)}, \end{aligned} \tag{10.45}$$

for all $k \geq 0$, where $i = 2, 3, \cdots, n-1$.

Denote

$$\delta_k = \frac{\eta(\lambda_i - \lambda_n)\mathbf{w}^T(k)\mathbf{R}_x\mathbf{w}(k)}{1 + \eta(1 + (\lambda_1 - \lambda_n - 1)\mathbf{w}^T(k)\mathbf{R}_x\mathbf{w}(k))}. \tag{10.46}$$

Next, we will prove that $0 < \delta_k < 2$.
First, we will prove that $\delta_k > 0$. From $\eta \leq 0.4$ and $\eta\lambda_1 \leq 0.35$, we obtain

$$\begin{aligned}
&1 + \eta(1 + (\lambda_1 - \lambda_n - 1)\mathbf{w}^T(k)\mathbf{R}_x\mathbf{w}(k)) > 1 + \eta - \eta\mathbf{w}^T(k)\mathbf{R}_x\mathbf{w}(k) \\
&> 1 + \eta - \eta(1 + \eta\lambda_1)^2 = 1 - \eta(\eta\lambda_1)^2 - 2\eta(\eta\lambda_1) \\
&\geq 1 - 0.4 \times (0.35)^2 - 2 \times 0.4 \times 0.35 = 0.6710 > 0.
\end{aligned} \tag{10.47}$$

By using $\lambda_i > \lambda_n$ and $\mathbf{w}^T(k)\mathbf{R}_x\mathbf{w}(k) > 0$, we have $\delta_k > 0$.
Next we prove that $\delta_k < 2$.

$$\begin{aligned}
\delta_k &< \frac{\eta\lambda_1\mathbf{w}^T(k)\mathbf{R}_x\mathbf{w}(k)}{1 + \eta(1 - \mathbf{w}^T(k)\mathbf{R}_x\mathbf{w}(k))} \\
&< \frac{\eta\lambda_1(1 + \eta\lambda_1)^2}{1 + \eta - \eta(1 + \eta\lambda_1)^2} = \frac{\eta\lambda_1(1 + \eta\lambda_1)^2}{1 - 2\eta(\eta\lambda_1) - \eta(\eta\lambda_1)^2} \\
&\leq \frac{0.4 \times (1 + 0.35)^2}{1 - 2 \times 0.4 \times 0.35 - 0.4 \times (0.35)^2} \\
&= 0.9506 < 2.
\end{aligned} \tag{10.48}$$

Denote $\theta_k = (1 - \delta_k)^2$. From (10.47) and (10.48), we can obtain that $0 < \theta_k < 1$. Then, we have

$$\left[\frac{z_i(k+1)}{z_n(k+1)}\right]^2 \leq \theta_k \frac{z_i^2(k)}{z_n^2(k)} \leq \cdots \leq \prod_{j=0}^{k} \theta_j \frac{z_i^2(0)}{z_n^2(0)}, \tag{10.49}$$

for all $k \geq 0$, where $i = 1, 2, 3, \cdots, n - 1$.

Denote $\theta = \max(\theta_0, \theta_1, \cdots \theta_k, \cdots)$. It holds that $0 < \theta < 1$. Equation (10.49) can be written as

$$\left[\frac{z_i(k+1)}{z_n(k+1)}\right]^2 \leq \theta^{k+1} \frac{z_i^2(0)}{z_n^2(0)}. \tag{10.50}$$

Since $0 < \theta < 1$, then it follows that

$$\lim_{k \to \infty} \frac{z_i(k)}{z_n(k)} = 0 \quad i = 1, 2, 3 \cdots, n - 1. \tag{10.51}$$

From Theorems 10.5 and 10.6, we can conclude that $z_n(k)$ must be bounded. Then we have

$$\lim_{k\to\infty} z_i(k) = 0, \quad i = 1, 2, 3, \cdots, n-1. \tag{10.52}$$

This completes the proof.

Lemma 10.3 *Suppose that $\eta \le 0.4$ and $\eta\lambda_1 \le 0.35$. If the initial vector satisfies $\mathbf{w}^T(0)\mathbf{R}_x\mathbf{v}_n \neq 0$ and $\|\mathbf{w}(0)\|_{\mathbf{R}_x} \le 1$, it holds that $\lim_{k\to\infty} z_n(k) = \alpha$, where α is a constant.*

Proof According to Lemma 10.2, $\mathbf{w}(k)$ will converge to the MGE $\mathbf{v}_n$ as $k \to \infty$. Suppose that $\mathbf{w}(k)$ has reached to the direction of $\mathbf{v}_n$ at time k_0, i.e., $\mathbf{w}(k_0) = z_n(k_0)\mathbf{v}_n$.

By using (10.20), we have

$$\begin{aligned} z_n(k+1) &= \left[1 + \eta\left[1 + \lambda_n z_n^2(k) - (\lambda_n + 1)z_n^2(k)\right]\right]z_n(k) \\ &= \left[1 + \eta(1 - z_n^2(k))\right]z_n(k). \end{aligned} \tag{10.53}$$

By using Theorem 10.5, we have $|z_n(k)| \le 1 + \eta\lambda_1$. From Lemma 10.1, we can obtain that $z_n(k)$ does not change signs in (10.21). Next, let us consider the following two cases.

Case 1 *($0 < z_n(k) \le 1 + \eta\lambda_1$):* From (10.53), it holds that

$$\begin{aligned} &1 - z_n(k+1) \\ &= 1 - \left[1 + \eta(1 - z_n^2(k))\right]z_n(k) \\ &= 1 - [1 + \eta[1 - z_n(k)][1 + z_n(k)]]z_n(k) \\ &= [1 - z_n(k)][1 - \eta z_n(k)[z_n(k) + 1]] \end{aligned} \tag{10.54}$$

for all $k \ge k_0$. Then, by using $\eta \le 0.4$ and $\eta\lambda_1 \le 0.35$, we have

$$\begin{aligned} &\eta z_n(k)[z_n(k) + 1] \\ &\le \eta(1 + \eta\lambda_1)(1 + \eta\lambda_1 + 1) \\ &\le 0.4 \times (1 + 0.35)(1 + 0.35 + 1) \\ &< 1.269 < 2, \end{aligned} \tag{10.55}$$

for all $k \ge k_0$.

Denote $\beta = |\,1 - \eta z_n(k)[z_n(k) + 1]|$. Clearly $0 < \beta < 1$. From (10.54) and (10.55), it follows that for all $k \ge k_0$

$$|z_n(k+1) - 1| \le \beta|z_n(k) - 1| \le \cdots \le \beta^{k+1}|z_n(0) - 1|. \tag{10.56}$$

Denote $\alpha = -\ln\beta$ and $\Pi_1 = |z_n(0) - 1|$. By substituting them into (10.56), we have

$$|z_n(k+1) - 1| \le (k+1)\Pi_1 e^{-\alpha(k+1)}. \tag{10.57}$$

Given any $\varepsilon > 0$, there exists a constant $K > 1$ such that

$$\frac{\Pi_2 K e^{-\alpha K}}{(1 - e^{-\alpha})^2} \le \varepsilon. \tag{10.58}$$

for any $k_1 > k_2 > K$.

From (10.21) and (10.58), we have

$$\begin{aligned}
|z_n(k_1) - z_n(k_2)| &= \left| \sum_{r=k_2}^{k_1-1} [z_n(r+1) - z_n(r)] \right| \le \left| \sum_{r=k_2}^{k_1-1} \left[\eta z_n(r)(1 - z_n^2(r)) \right] \right| \\
&\le \sum_{r=k_2}^{k_1-1} \eta |z_n(r)||z_n(r) + 1||z_n(r) - 1| \\
&\le \eta(1 + \eta\lambda_1)(2 + \eta\lambda_1) \sum_{r=k_2}^{k_1} |z_n(r) - 1| \le \Pi_2 \sum_{r=k_2}^{k_1} r e^{-\alpha r} \\
&\le \Pi_2 \sum_{r=K}^{+\infty} r e^{-\alpha r} \le \Pi_2 K e^{-\alpha K} \sum_{r=0}^{+\infty} r \left(e^{-\alpha}\right)^{r-1} \le \frac{\Pi_2 K e^{-\alpha K}}{(1 - e^{-\alpha})^2} \le \varepsilon,
\end{aligned} \tag{10.59}$$

where $\Pi_2 = \eta(1 + \eta\lambda_1)(2 + \eta\lambda_1)|z_n(0) - 1|$.

Case 2 ($-(1+\eta\lambda_1) \le z_n(k) < 0$): In this case, we have $0 < (-z_n(k)) \le (1+\eta\lambda_1)$. From (10.53), it holds that

$$\begin{aligned}
&1 + z_n(k+1) \\
&= 1 + \left[1 + \eta(1 - z_n^2(k))\right] z_n(k) \\
&= 1 + [1 + \eta[1 - z_n(k)][1 + z_n(k)]] z_n(k) \\
&= [1 + z_n(k)][1 - \eta(-z_n(k))(1 - z_n(k))]
\end{aligned} \tag{10.60}$$

for all $k \ge k_0$. Then, by using $\eta \le 0.4$ and $\eta\lambda_1 \le 0.35$, we obtain

$$\begin{aligned}
&- \eta z_n(k)[1 - z_n(k)] \\
&\le \eta(1 + \eta\lambda_1)(1 + \eta\lambda_1 + 1) \\
&\le 0.4 \times (1 + 0.35)(1 + 0.35 + 1) \\
&< 1.269 < 2,
\end{aligned} \tag{10.61}$$

for all $k \ge k_0$.

Denote $\beta' = |\, 1 - \eta(z_n(k))[1 - z_n(k)]|$. Clearly, $0 < \beta' < 1$. By using (10.60) and (10.61), we obtain that for all $k \ge k_0$

$$|z_n(k+1) + 1| \le \beta' |z_n(k) + 1| \le \cdots \le (\beta')^{k+1} |z_n(0) + 1|. \tag{10.62}$$

Denote $\alpha' = -\ln \beta'$ and $\Pi_1' = |z_n(0) + 1|$. By substituting them into (10.62), we have

$$|z_n(k+1) + 1| \le (k+1)\Pi_1' e^{-\alpha'(k+1)}. \tag{10.63}$$

Given any $\varepsilon' > 0$, there exists a constant $K' > 1$ such that

$$\frac{\Pi_2' K' e^{-\alpha' K'}}{(1 - e^{-\alpha'})^2} \le \varepsilon'. \tag{10.64}$$

for any $k_1' > k_2' > K'$.

From (10.21) and (10.64), we have

$$\begin{aligned} |z_n(k_1) - z_n(k_2)| &\le \sum_{r=k_2}^{k_1-1} \eta |z_n(r)||z_n(r) + 1||z_n(r) - 1| \\ &\le \eta(1 + \eta\lambda_1)(2 + \eta\lambda_1) \sum_{r=k_2}^{k_1} |z_n(r) + 1| \le \frac{\Pi_2' K' e^{-\alpha' K'}}{(1 - e^{-\alpha'})^2} \le \varepsilon', \end{aligned} \tag{10.65}$$

where $\Pi_2' = \eta(1 + \eta\lambda_1)(2 + \eta\lambda_1)|z_n(0) + 1|$.

From (10.59) and (10.65), we have that $\{z_n(k)\}$ is a Cauchy sequence, i.e., there must exist a constant a such that $\lim_{k\to\infty} z_n(k) = \alpha$.

Theorem 10.7 *Suppose that* $\eta \le 0.4$ *and* $\eta\lambda_1 \le 0.35$. *If* $\mathbf{w}^T(0)\mathbf{R}_x\mathbf{v}_n \ne 0$ *and* $\|\mathbf{w}(0)\|_{\mathbf{R}_x} \le 1$, *it holds that* $\lim\limits_{k\to\infty} \mathbf{w}(k) = \alpha\mathbf{v}_n$, *where* α *is a constant.*

Proof From (10.30) and Lemmas 10.2 and 10.3, we have that

$$\begin{aligned} \lim_{k\to\infty} \mathbf{w}(k) &= \lim_{k\to\infty} \left(\sum_{i=1}^{n-1} z_i(k)\mathbf{v}_i + z_n(k)\mathbf{v}_n \right) \\ &= \lim_{k\to\infty} \sum_{i=1}^{n-1} z_i(k)\mathbf{v}_i + \lim_{k\to\infty} (z_n(k)\mathbf{v}_n) = \alpha\mathbf{v}_n. \end{aligned} \tag{10.66}$$

The proof is complete.

At this point, we have finished the proof of the convergence analysis of (10.13).

10.3.4 Numerical Simulations and Conclusions

In this part, three experiments are designed to evaluate the effectiveness of the GChen algorithm. Experiment 1 shows the dynamic behavior of the GChen algorithm. Experiment 2 serves to investigate the self-stabilization property of the GChen algorithm.

In the third experiment, the MGE is adaptively extracted from two input signals and the extraction results are compared with those obtained by some other algorithms.

To quantitatively evaluate the performances of the algorithms, two evaluation functions are adopted. The first one is the weight vector norm

$$\rho(\mathbf{w}(k)) = \|\mathbf{w}(k)\|_{\mathbf{R}_x} = \sqrt{\mathbf{w}^T(k)\mathbf{R}_x\mathbf{w}(k)}. \tag{10.67}$$

The second one is the direction cosine (DC)

$$\mathrm{DC}(k) = \frac{\left|\mathbf{w}^T(k)\mathbf{v}_n\right|}{\|\mathbf{w}(k)\|\,\|\mathbf{v}_n\|}, \tag{10.68}$$

where $\boldsymbol{v}_n$ is the desired MGE. DC is used for examining the ability of each algorithm to extract the MGE and evaluating the estimation accuracy of the algorithm. If DC converges to one, $\boldsymbol{w}(k)$ has reached the direction of the MGE, $\boldsymbol{v}_n$. When the difference between the limiting value of DC and one is small, the algorithm has good estimation accuracy.

A. ***Experiment 1***

Herein, we provide some examples to prove the correctness of the conditions obtained in Theorem 10.7. First, randomly generate two 4×4 symmetric positive-definite matrices, which are given as follows:

$$\mathbf{R}_y = \begin{bmatrix} 0.2261 & -0.2257 & -0.0045 & -0.0615 \\ -0.2257 & 0.3646 & 0.1368 & 0.0736 \\ -0.0045 & 0.1368 & 0.1763 & -0.0168 \\ -0.0615 & 0.0736 & -0.0168 & 0.4401 \end{bmatrix} \tag{10.69}$$

$$\mathbf{R}_x = \begin{bmatrix} 0.7672 & 0.0359 & -0.0878 & -0.1449 \\ 0.0359 & 0.5369 & -0.0757 & 0.0238 \\ -0.0878 & -0.0757 & 0.5129 & 0.1072 \\ -0.1449 & 0.0238 & 0.1072 & 0.6567 \end{bmatrix}. \tag{10.70}$$

The GChen algorithm is used to estimate the MGE of $(\boldsymbol{R}_y, \boldsymbol{R}_x)$, whose largest generalized eigenvalue is $\lambda_1 = 1.0818$. For the GChen algorithm, the initial weight vector [denoted by $\boldsymbol{w}(0)$] is randomly produced and its $\boldsymbol{R}_x$-norm is less than 1. Besides, we set $\eta = 0.1$, which satisfies the conditions $\eta \leq 0.4$ and $\eta\lambda_1 \leq 0.35$. The dynamic behavior of $z_i(k)$ is shown in Fig. 10.1. From Fig. 10.1, we can see that after many iterations, the convergence value of $z_i(k)$ $(i = 1,2,3)$ is zero and that of $z_4(k)$ is one. This result is consistent with Lemmas 10.2 and 10.3.

Next, we test the GChen algorithm on a 15×15 matrix pencil, whose largest generalized eigenvalue is $\lambda_1 = 9.428$. For this experiment, we set $\eta = 0.03$ and $\boldsymbol{w}(0)$ is generated in the same way as in the previous experiment. Through these settings, the requirements in Theorem 10.7 can be satisfied. The simulation results, which

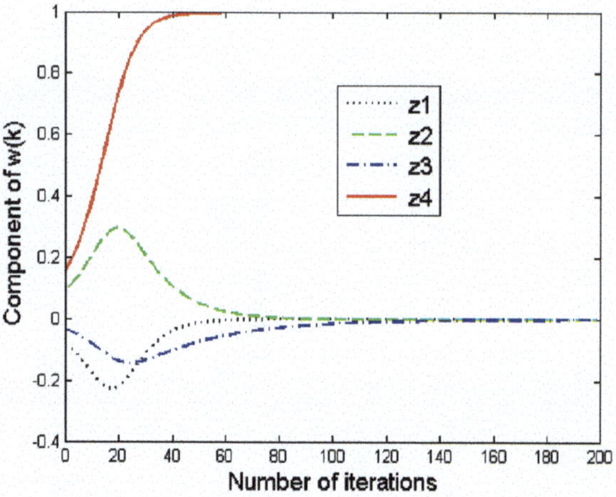

Fig. 10.1 Convergence of z_i with low-dimensional matrices

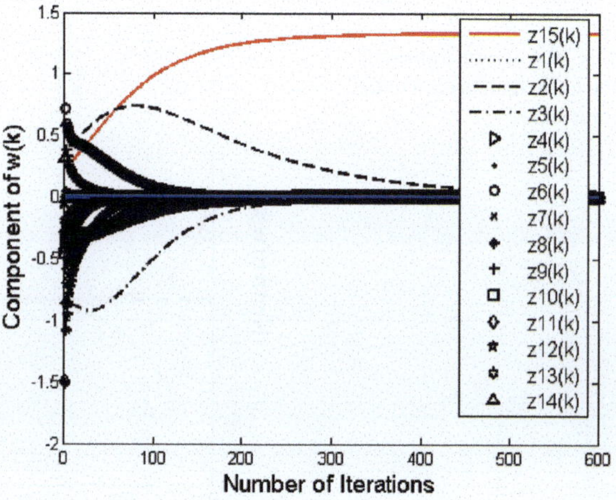

Fig. 10.2 Convergence of z_i with high-dimensional

are obtained by 100 Monte Carlo runs, are shown in Fig. 10.2. In Fig. 10.2, we can see that except the curve of z_{15}, all the curves tend to zero after approximately 500 iterations, while z_{15} converges to some constant value. This conclusion is consistent with the theoretical analyses in Lemmas 10.2 and 10.3. From this experiment, it is concluded that the GChen algorithm can address high-dimensional matrices.

The third experiment is an example in which the norm of $\boldsymbol{w}(k)$ diverges to infinity since the requirement on the learning rate in Theorem 10.7 is not satisfied. Here, we also use the two matrices (10.69) and (10.70). Different from the first experiment, this experiment adopts the learning rate $\eta = 0.5$. In this case, we have $\eta\lambda_1 = 0.5409 > 0.35$, which does not satisfy the conditions $\eta < 0.4$ and $\eta\lambda_1 \leq 0.35$. The initial vector is set as $\boldsymbol{w} = [-2.1064, -0.6199, -0.2234, 0.9457]^T$, and the simulation results are shown in Fig. 10.3. From Fig. 10.3, we see that the weight vector diverges

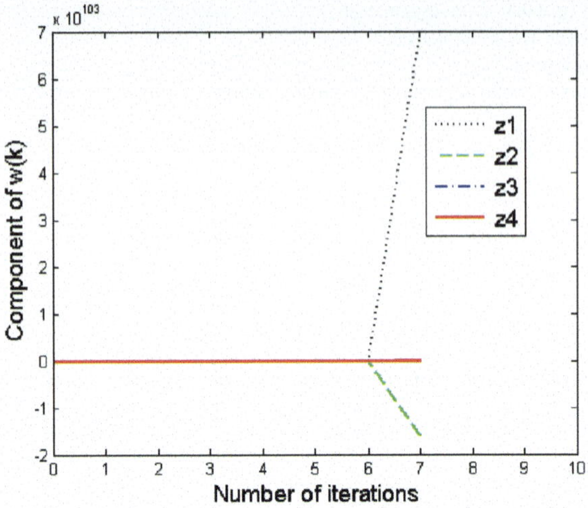

Fig. 10.3 Divergence of z_i

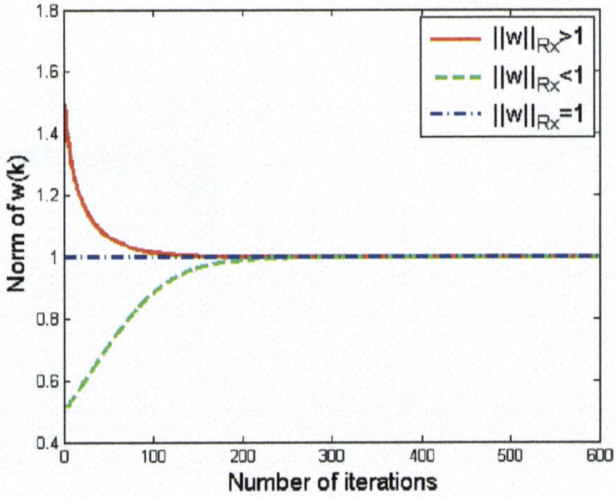

Fig. 10.4 Norm curves of the GChen algorithm under three cases

after only six iterations. Thus, this simulation verifies the correctness of the obtained conditions.

B. ***Experiment 2***

Herein, two matrices (10.69) and (10.70) are adopted, and GChen algorithm (10.13) is used. The learning rate is equal to $\eta = 0.01$. To evaluate the effect of $\boldsymbol{w}(0)$ on the final result, $\|\mathbf{w}(0)\|_{\mathbf{R}x}$ is set to three different values: one value is equal to 1, one is larger than 1, and the other is smaller than 1. The experimental results, which are evaluated from 100 runs, are shown in Figs. 10.4 and 10.5.

From Fig. 10.4, it is obvious that no matter how the initial weight vector is selected, the $\boldsymbol{R}_{\mathrm{x}}$-norm of the weight vector can converge to 1 in the GChen algorithm. From Fig. 10.5, we can see that when the GChen algorithm (10.13) is used for extracting

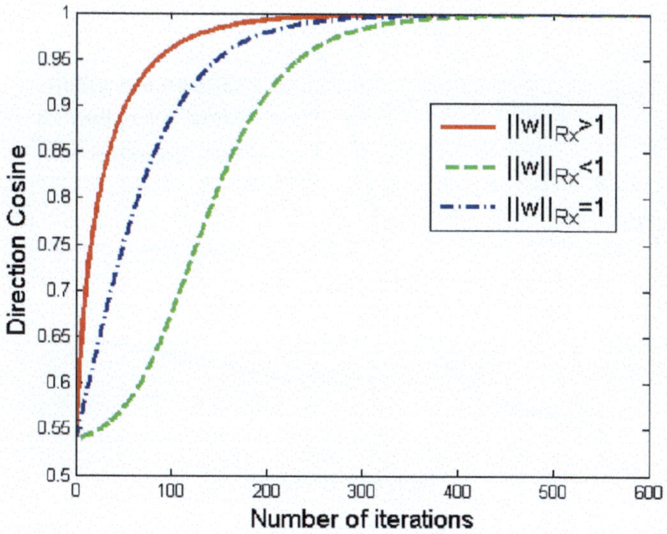

Fig. 10.5 DC curves of the GChen algorithm under three cases

the MGE of the matrix pencil, all the DC curves converge to 1. In other words, the GChen algorithm (10.13) can reach the direction of the MGE after the iterations. This experiment tests and verifies the self-stabilizing property of the GChen algorithm.

C. ***Experiment 3***

In this part, to extract the MGE online, we compare the GChen algorithm with the modified Oja-Xu algorithm (MOX) algorithm and the generalized power (GP) algorithm [13]. They are gradient algorithms. Herein, we use the following equations to generate the input signals, which are the same as those in [13]:

$$y(k) = \sqrt{2}\sin(0.62\pi k + \theta_1) + n_1(k) \tag{10.71}$$

$$x(k) = \sqrt{2}\sin(0.74\pi k + \theta_2) + \sqrt{2}\sin(0.46\pi k + \theta_3) + n_2(k), \tag{10.72}$$

where $\theta_i(i = 1,2,3)$ are the initial phases, which obey the uniform distribution on the interval $[0, 2\pi]$. $n_1(k)$ and $n_2(k)$ are two zero-mean white noises with variances $\sigma_1^2 = \sigma_2^2 = 0.1$. Let $\boldsymbol{y}(k) = [y(k),\ldots,y(k + 7)]^T$ and $\boldsymbol{x}(k) = [x(k),\ldots, x(k + 7)]^T$ be the two input vectors. Three algorithms are used to adaptively extract the MGE from the two signals. For these algorithms, the forgetting factors are set as $\alpha = \beta = 0.998$, and the initial autocorrelation matrices are set as $\hat{\mathbf{R}}_y(0) = \hat{\mathbf{R}}_x(0) = \mathbf{0}$ and $\boldsymbol{Q}_x(0) = \delta \boldsymbol{I}$, where δ is a small positive number. For GChen, MOX, GP algorithms, the learning rate is set as 0.09, 0.1, 0.05, respectively. The experimental results are shown in Figs. 10.6 and 10.7.

From Fig. 10.6, we can see that DC of the GChen algorithm converges to 1 after the iterations, which means the GChen algorithm can adaptively extract the MGE

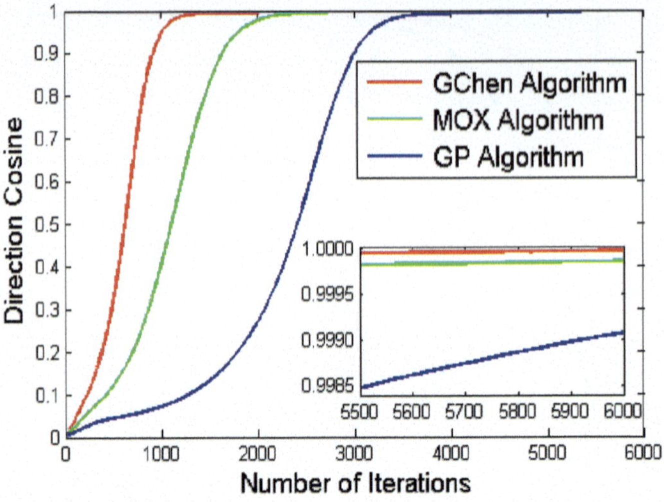

Fig. 10.6 DC curves of these algorithms

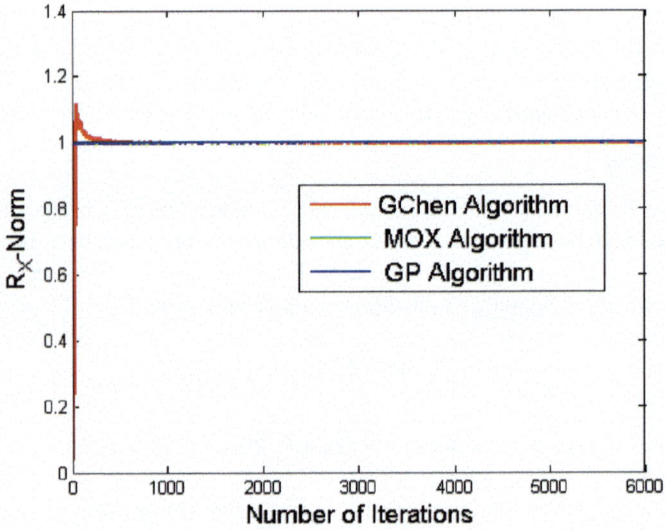

Fig. 10.7 R_X-norm curves of the algorithms

from the input signals. In addition, it is noted that among the three algorithms, the GChen algorithm has the fastest convergence speed. In the last 500 iteration steps, we observe that the convergence values of these algorithms are not equal to one and that of the GChen algorithm is the closest to one, which means that the GChen algorithm has the highest estimation accuracy for the desired MGE. In Fig. 10.7, due to the normalization operation, the weight vector norms of the GP and MOX are equal to

1 throughout the iterations. Although the normalization operation is not adopted for the GChen algorithm, the $\boldsymbol{R}_x$-norm curve also converges to one after some iterations.

10.4 Summary

In this chapter, several performance analysis methods of TLS or MCA algorithms, including the deterministic continuous-time system method, the stochastic discrete-time system method, the Lyapunov function approach, the fixed pointed method, and the deterministic discrete-time system method, are briefly reviewed and discussed. With respect to a novel self-stabilizing algorithm for generalized minor eigenpairs (GME) extraction proposed by us, a DDT system of this novel GME algorithm is presented, and the boundary condition for the convergence of the algorithm is given. In this chapter, the DDT method is expanded and applied into more complex convergence analysis of algorithms.

References

1. H. J. Kushner, D. S. Clark, in *Stochastic Approximation Methods for Constrained and Unconstrained Systems* (New York, Springer), pp. 1076
2. L. Ljung, Analysis of recursive stochastic algorithms. IEEE Trans. Autom. Control **22**(4), 551–575 (1977)
3. G. Cirrincione, M. Cirrincione, J. Herault, S. Van Huffel, The MCA EXIN neuron for the minor component analysis. IEEE Trans. Neural Netw. **13**(1), 160–187 (2002)
4. P. J. Zufiria, On the discrete-time dynamics of the basic Hebbian neural network node. IEEE Trans. Neural Netw. **13**(6), 1342–1352 (2002)
5. Y. Zhang, M. Ye, J.C. Lv, K.K. Tan, Convergence analysis of a deterministic discrete time system of Oja's PCA learning algorithm. IEEE Trans. Neural Networks **16**(6), 1318–1328 (2005)
6. X.Y. Kong, Q.S. An, H.G. Ma, C.Z. Han, Q. Zhang, Convergence analysis of deterministic discrete time system of a unified self-stabilizing algorithm for PCA and MCA. Neural Netw. **36**(8), 64–72 (2012)
7. F.L. Luo, R. Unbehauen, A minor subspace analysis algorithm. IEEE Trans. Neural Netw. **8**(5), 1149–1155 (1997)
8. D.Z. Feng, W.X. Zheng, Y. Jia, Neural network Learning algorithms for tracking minor subspace in high-dimensional data stream. IEEE Trans. Neural Netw. **16**(3), 513–521 (2005)
9. Y. B. Gao, X. Y. Kong, H. D. Dong, Novel parallel multiple minor components extraction algorithm by diagonal matrix method. Neural Process. Lett. (2023). https://doi.org/10.1007/s11063-023-11186-4
10. D. Peng, Y. Zhang, Y. Xiang, H. Zhang, A globally convergent MC algorithm with an adaptive learning rate. IEEE Trans. Neural Netw. Learn. Syst. **23**, 359–365 (2012)
11. T. Chen, S.I. Amari, Unified stabilization approach to principal and minor components extraction algorithms. Neural Netw. **14**, 1377–1387 (2001)
12. T.D. Nguyen, I. Yamada, Necessary and sufficient conditions for convergence of the DDT systems of the normalized PAST algorithms. Signal Process. **94**, 288–299 (2014)
13. T.D. Nguyen, N. Takahashi, I. Yamada, An adaptive extraction of generalized eigensubspace by using exact nested orthogonal complement structure. Multidimension. Syst. Signal Process. **24**(3), 457–483 (2013)